X-Ray Diffraction Crystallography

Yoshio Waseda • Eiichiro Matsubara
Kozo Shinoda

X-Ray Diffraction Crystallography

Introduction, Examples and Solved Problems

With 159 Figures

Springer

Professor Dr. Yoshio Waseda
Professor Kozo Shinoda
Tohoku University, Institute of Multidisciplinary Research for Advanced Materials
Katahira 2-1-1, 980-8577 Sendai, Aoba-ku, Japan
E-mail: waseda@tagen.tohoku.ac.jp; shinoda@tagen.tohoku.ac.jp

Professor Dr. Eiichiro Matsubara
Kyoto University, Graduate School of Engineering
Department of Materials Science and Engineering
Yoshida Honmachi, 606-8501 Kyoto, Sakyo-ku, Japan
E-mail: e.matsubara@materials.mbox.media.kyoto-u.ac.jp

Supplementary problems with solutions are accessible to qualified instructors at springer.com on this book's product page. Instructors may click on the link additional information and register to obtain their restricted access.

ISBN 978-3-642-16634-1 e-ISBN 978-3-642-16635-8
DOI 10.1007/978-3-642-16635-8
Springer Heidelberg Dordrecht London New York

Library of Congress Control Number: 2011923528

Cover design: eStudio Calamar Steinen

Printed on acid-free paper

Springer is part of Springer Science+Business Media (www.springer.com)

Preface

X-ray diffraction crystallography for powder samples is well-established and widely used in the field of materials characterization to obtain information on the atomic scale structure of various substances in a variety of states. Of course, there have been numerous advances in this field, since the discovery of X-ray diffraction from crystals in 1912 by Max von Laue and in 1913 by W.L. Bragg and W.H. Bragg. The origin of crystallography is traced to the study for the external appearance of natural minerals and a large amount of data have been systematized by applying geometry and group theory. Then, crystallography becomes a valuable method for the general consideration of how crystals can be built from small units, corresponding to the infinite repetition of identical structural units in space.

Many excellent and exhaustive books on X-ray diffraction and crystallography are available, but the undergraduate students and young researchers and engineers who wish to become acquainted with this subject frequently find them overwhelming. They find it difficult to identify and understand the essential points in the limited time available to them, particularly on how to estimate useful structural information from the X-ray diffraction data. Since X-ray powder diffraction is one of the most common and leading methods in materials research, mastery of the subject is essential.

In order to learn the fundamentals of X-ray diffraction crystallography well and to be able to cope with the subject appropriately, a certain number of "exercises" involving calculation of specific properties from measurements are strongly recommended. This is particularly true for beginners of X-ray diffraction crystallography. Recent general purpose X-ray diffraction equipments have a lot of inbuilt automation for structural analysis. When a sample is set in the machine and the preset button is pressed, results are automatically generated some of which are misleading. A good understanding of fundamentals helps one to recognize misleading output.

During the preparation of this book, we have tried to keep in mind the students who come across X-ray diffraction crystallography for powder samples at the first time. The primary objective is to offer a textbook to students with almost no basic knowledge of X-rays and a guidebook for young scientists and engineers engaged in full-scale materials development with emphasis on practical problem solving. For the convenience of readers, some essential points with basic equations are summarized in each chapter, together with some relevant physical constants and atomic scattering factors of elements listed in appendices.

Since practice perfects the acquisition of skills in X-ray diffraction crystallography, 100 supplementary problems are also added with simple solutions. We hope that the students will try to solve these supplementary problems by themselves to deepen their understanding and competence of X-ray crystallography without serious difficulty. Since the field of X-ray structural analysis of materials is quite wide, not all possible applications are covered. The subject matter in this book is restricted to fundamental knowledge of X-ray diffraction crystallography for powder samples only. The readers can refer to specialized books for other applications.

The production of high-quality multi-layered thin films with sufficient reliability is an essential requirement for device fabrication in micro-electronics. An iron-containing layered oxy-pnictide $LaO_{1-x}F_xFeAs$ has received much attention because it exhibits superconductivity below 43 K as reported recently by Dr. Hideo Hosono in Japan. The interesting properties of such new synthetic functional materials are linked to their periodic and interfacial structures at a microscopic level, although the origin of such peculiar features has not been fully understood yet. Nevertheless, our understanding of most of the important properties of new functional materials relies heavily upon their atomic scale structure. The beneficial utilization of all materials should be pursued very actively to contribute to the most important technological and social developments of the twenty-first century harmonized with nature. Driven by environmental concerns, the interest in the recovery or recycling of valuable metallic elements from wastes such as discarded electronic devices will grow significantly over the next decade. The atomic scale structure of various materials in a variety of states is essential from both the basic science and the applied engineering points of view. Our goal is to take the most efficient approach for describing the link between the atomic scale structure and properties of any substance of interest.

The content of this book has been developed through lectures given to undergraduate or junior-level graduate students in their first half (Master's program) of the doctoral course of the graduate school of engineering at both Tohoku and Kyoto universities. If this book is used as a reference to supplement lectures in the field of structural analysis of materials or as a guide for a researcher or engineer engaged in structural analysis to confirm his or her degree of understanding and to compensate for deficiency in self-instruction, it is an exceptional joy for us.

Many people have helped both directly or indirectly in preparing this book. The authors are deeply indebted to Professors Masahiro Kitada for his valuable advice on the original manuscript. Many thanks are due to Professor K.T. Jacob (Indian Institute of Science, Bangalore), Professor N.J. Themelis (Columbia University), Professor Osamu Terasaki (Stockholm University) and Dr.Daniel Grüner and Dr. Karin Söderberg (Stockholm University) and Dr. Sam Stevens (University of Manchester) who read the manuscript and made many helpful suggestions.

The authors would like to thank Ms. Noriko Eguchi, Ms. Miwa Sasaki and Mr. Yoshimasa Ito for their assistance in preparing figures and tables as well as the electronic TeX typeset of this book. The authors are also indebted to many sources of material in this article. The encouragement of Dr. Claus Ascheron of Springer-Verlag, Mr. Satoru Uchida and Manabu Uchida of Uchida-Rokakuho Publishing Ltd should also be acknowledged.

Sendai, Japan
January 2011

Yoshio Waseda
Eiichiro Matsubara
Kozo Shinoda

Note: A solution manual for 100 supplementary problems is available to instructors who have adopted this book for regular classroom use or tutorial seminar use. To obtain a copy of the solution manual, a request may be delivered on your departmental letterhead to the publisher (or authors), specifying the purpose of use as an organization (not personal).

Contents

Chapter 1
Fundamental Properties of X-rays

1.1 Nature of X-rays

X-rays with energies ranging from about 100 eV to 10 MeV are classified as electromagnetic waves, which are only different from the radio waves, light, and gamma rays in wavelength and energy. X-rays show wave nature with wavelength ranging from about 10 to 10^{-3} nm. According to the quantum theory, the electromagnetic wave can be treated as particles called photons or light quanta. The essential characteristics of photons such as energy, momentum, etc., are summarized as follows.

The propagation velocity c of electromagnetic wave (velocity of photon) with frequency ν and wavelength λ is given by the relation.

$$c = \nu\lambda \qquad (\mathrm{ms}^{-1}) \tag{1.1}$$

The velocity of light in the vacuum is a universal constant given as c = 299792458 m/s ($\approx 2.998 \times 10^8$ m/s). Each photon has an energy E, which is proportional to its frequency,

$$E = h\nu = \frac{hc}{\lambda} \qquad (\mathrm{J}) \tag{1.2}$$

where h is the Planck constant (6.6260×10^{-34} J · s). With E expressed in keV, and λ in nm, the following relation is obtained:

$$E(\mathrm{keV}) = \frac{1.240}{\lambda(\mathrm{nm})} \tag{1.3}$$

The momentum p is given by mv, the product of the mass m, and its velocity v. The de Broglie relation for material wave relates wavelength to momentum.

$$\lambda = \frac{h}{p} = \frac{h}{mv} \tag{1.4}$$

The velocity of light can be reduced when traveling through a material medium, but it does not become zero. Therefore, a photon is never at rest and so has no rest mass m_e. However, it can be calculated using Einstein's mass-energy equivalence relation $E = mc^2$.

$$E = \frac{m_e}{\sqrt{1-\left(\frac{v}{c}\right)^2}}c^2 \tag{1.5}$$

It is worth noting that (1.5) is a relation derived from Lorentz transformation in the case where the photon velocity can be equally set either from a stationary coordinate or from a coordinate moving at velocity of v (Lorentz transformation is given in detail in other books on electromagnetism: for example, P. Cornille, Advanced Electromagnetism and Vacuum Physics, World Scientific Publishing, Singapore, (2003)). The increase in mass of a photon with velocity may be estimated in the following equation using the rest mass m_e:

$$m = \frac{m_e}{\sqrt{1-\left(\frac{v}{c}\right)^2}} \tag{1.6}$$

For example, an electron increases its mass when the accelerating voltage exceeds 100 kV, so that the common formula of $\frac{1}{2}mv^2$ for kinetic energy cannot be used. In such case, the velocity of electron should be treated relativistically as follows:

$$E = mc^2 - m_e c^2 = \frac{m_e}{\sqrt{1-\left(\frac{v}{c}\right)^2}}c^2 - m_e c^2 \tag{1.7}$$

$$v = c \cdot \sqrt{1-\left(\frac{m_e c^2}{E + m_e c^2}\right)^2} \tag{1.8}$$

The value of m_e is obtained, in the past, by using the relationship of $m = h/(c\lambda)$ from precision scattering experiments, such as Compton scattering and $m_e = 9.109 \times 10^{-31}$ kg is usually employed as electron rest mass. This also means that an electron behaves as a particle with the mass of 9.109×10^{-31} kg, and it corresponds to the energy of $E = mc^2 = 8.187 \times 10^{-14}$ J or 0.5109×10^6 eV in eV.

There is also a relationship between mass, energy, and momentum.

$$\left(\frac{E}{c}\right)^2 - p^2 = (m_e c)^2 \tag{1.9}$$

It is useful to compare the properties of electrons and photons. On the one hand, the photon is an electromagnetic wave, which moves at the velocity of light sometimes called light quantum with momentum and energy and its energy depends upon the frequency ν. The photon can also be treated as particle. On the other hand, the electron has "mass" and "charge." It is one of the elementary particles that is a constituent of all substances. The electron has both particle and wave nature such as photon. For example, when a metallic filament is heated, the electron inside it is supplied with energy to jump out of the filament atom. Because of the negative charge of the electron, ($e = 1.602 \times 10^{-19}$ C), it moves toward the anode in an electric field and its direction of propagation can be changed by a magnetic field.

1.2 Production of X-rays

When a high voltage with several tens of kV is applied between two electrodes, the high-speed electrons with sufficient kinetic energy, drawn out from the cathode, collide with the anode (metallic target). The electrons rapidly slow down and lose kinetic energy. Since the slowing down patterns (method of loosing kinetic energy) vary with electrons, continuous X-rays with various wavelengths are generated. When an electron loses all its energy in a single collision, the generated X-ray has the maximum energy (or the shortest wavelength $= \lambda_{\mathrm{SWL}}$). The value of the shortest wavelength limit can be estimated from the accelerating voltage V between electrodes.

$$\mathrm{e}V \equiv h\nu_{\max} \tag{1.10}$$

$$\lambda_{\mathrm{SWL}} = \frac{c}{\nu_{\max}} = \frac{hc}{\mathrm{e}V} \tag{1.11}$$

The total X-ray intensity released in a fixed time interval is equivalent to the area under the curve in Fig. 1.1. It is related to the atomic number of the anode target Z and the tube current i:

$$I_{\mathrm{cont}} = AiZV^2 \tag{1.12}$$

where A is a constant. For obtaining high intensity of white X-rays, (1.12) suggests that it is better to use tungsten or gold with atomic number Z at the target, increase accelerating voltage V, and draw larger current i as it corresponds to the number of electrons that collide with the target in unit time. It may be noted that most of the kinetic energy of the electrons striking the anode (target metal) is converted into heat and less than 1% is transformed into X-rays. If the electron has sufficient kinetic energy to eject an inner-shell electron, for example, a K shell electron, the atom will become excited with a hole in the electron shell. When such hole is filled by an outer shell electron, the atom regains its stable state. This process also includes production of an X-ray photon with energy equal to the difference in the electron energy levels.

As the energy released in this process is a value specific to the target metal and related electron shell, it is called characteristic X-ray. A linear relation between the square root of frequency ν of the characteristic X-ray and the atomic number Z of the target material is given by Moseley's law.

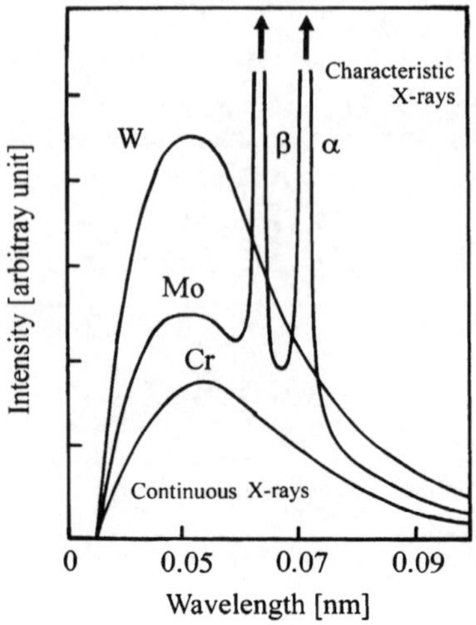

Fig. 1.1 Schematic representation of the X-ray spectrum

$$\sqrt{\nu} = B_{\mathrm{M}}(Z - \sigma_{\mathrm{M}}) \tag{1.13}$$

Here, B_{M} and σ_{M} are constants. This Moseley's law can also be given in terms of wavelength λ of emitted characteristic X-ray:

$$\frac{1}{\lambda} = R(Z - S_{\mathrm{M}})^2 \left(\frac{1}{n_1^2} - \frac{1}{n_2^2}\right) \tag{1.14}$$

Here, R is the Rydberg constant ($1.0973 \times 10^7\,\mathrm{m}^{-1}$), S_{M} is a screening constant, and usually zero for Kα line and one for Kβ line. Furthermore, n_1 and n_2 represent the principal quantum number of the inner shell and outer shell, respectively, involved in the generation of characteristic X-rays. For example, $n_1 = 1$ for K shell, $n_2 = 2$ for L shell, and $n_3 = 3$ for M shell. As characteristic X-rays are generated when the applied voltage exceeds the so-called excitation voltage, corresponding to the potential required to eject an electron from the K shell (e.g., Cu: 8.86 keV, Mo: 20.0 keV), the following approximate relation is available between the intensity of Kα radiation, I_{K}, and the tube current, i, the applied voltage V, and the excitation voltage V_{K}:

$$I_{\mathrm{K}} = B_{\mathrm{S}} i (V - V_{\mathrm{K}})^{1.67} \tag{1.15}$$

Here, B_{S} is a constant and the value of $B_{\mathrm{S}} = 4.25 \times 10^8$ is usually employed. As it is clear from (1.15), larger the intensity of characteristic X-rays, the larger the applied voltage and current.

It can be seen from (1.14), characteristic radiation is emitted as a photoelectron when the electron of a specific shell (the innermost shell of electrons, the K shell) is released from the atom, when the electrons are pictured as orbiting the nucleus in specific shells. Therefore, this phenomenon occurs with a specific energy (wavelength) and is called "photoelectric absorption." The energy, E_{ej}, of the photoelectron emitted may be described in the following form as a difference of the binding energy (E_B) for electrons of the corresponding shell with which the photoelectron belongs and the energy of incidence X-rays (hν):

$$E_{ej} = h\nu - E_B \tag{1.16}$$

The recoil of atom is necessarily produced in the photoelectric absorption process, but its energy variation is known to be negligibly small (see Question 1.6). Equation (1.16) is based on such condition. Moreover, the value of binding energy (E_B) is also called absorption edge of the related shell.

1.3 Absorption of X-rays

X-rays which enter a sample are scattered by electrons around the nucleus of atoms in the sample. The scattering usually occurs in various different directions other than the direction of the incident X-rays, even if photoelectric absorption does not occur. As a result, the reduction in intensity of X-rays which penetrate the substance is necessarily detected. When X-rays with intensity I_0 penetrate a uniform substance, the intensity I after transmission through distance x is given by.

$$I = I_0 e^{-\mu x} \tag{1.17}$$

Here, the proportional factor μ is called linear absorption coefficient, which is dependent on the wavelength of X-rays, the physical state (gas, liquid, and solid) or density of the substance, and its unit is usually inverse of distance. However, since the linear absorption coefficient μ is proportional to density ρ,(μ/ρ) becomes unique value of the substance, independent upon the state of the substance. The quantity of (μ/ρ) is called the mass absorption coefficient and the specific values for characteristic X-rays frequently-used are compiled (see Appendix A.2). Equation (1.17) can be re-written as (1.18) in terms of the mass absorption coefficient.

$$I = I_0 e^{-\left(\frac{\mu}{\rho}\right)\rho x} \tag{1.18}$$

Mass absorption coefficient of the sample of interest containing two or more elements can be estimated from (1.19) using the bulk density, ρ, and weight ratio of w_j for each element j.

$$\left(\frac{\mu}{\rho}\right) = w_1\left(\frac{\mu}{\rho}\right)_1 + w_2\left(\frac{\mu}{\rho}\right)_2 + \cdots = \sum_{j=1} w_j\left(\frac{\mu}{\rho}\right)_j \tag{1.19}$$

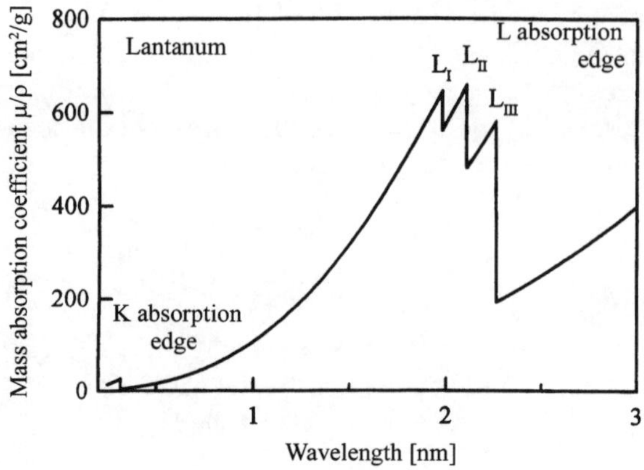

Fig. 1.2 Wavelength dependences of mass absorption coefficient of X-ray using the La as an example

Absorption of X-rays becomes small as transmittivity increases with increasing energy (wavelength becomes shorter). However, if the incident X-ray energy comes close to a specific value (or wavelength) as shown in Fig. 1.2, the photoelectric absorption takes place by ejecting an electron in K-shell and then discontinuous variation in absorption is found. Such specific energy (wavelength) is called absorption edge. It may be added that monotonic variation in energy (wavelength) dependence is again detected when the incident X-ray energy is away from the absorption edge.

1.4 Solved Problems (12 Examples)

Question 1.1 Calculate the energy released per carbon atom when 1 g of carbon is totally converted to energy.

Answer 1.1 Energy E is expressed by Einstein's relation of $E = mc^2$ where m is mass and c is the speed of light. If this relationship is utilized, considering SI unit that expresses mass in kg,

$$E = 1 \times 10^{-3} \times (2.998 \times 10^{10})^2 = 8.99 \times 10^{13} \quad \text{J}$$

The atomic weight per mole (molar mass) for carbon is 12.011 g from reference table (for example, Appendix A.2). Thus, the number of atoms included in 1 g carbon is calculated as $(1/12.011) \times 0.6022 \times 10^{24} = 5.01 \times 10^{22}$ because the numbers of atoms are included in one mole of carbon is the Avogadro's number

(0.6022×10^{24}). Therefore, the energy release per carbon atom can be estimated as:

$$\frac{(8.99 \times 10^{13})}{(5.01 \times 10^{22})} = 1.79 \times 10^{-9} \quad \text{J}$$

Question 1.2 Calculate (1) strength of the electric field E, (2), force on the electron F, (3) acceleration of electron α, when a voltage of 10 kV is applied between two electrodes separated by an interval of 10 mm.

Answer 1.2 The work, W, if electric charge Q (coulomb, C) moves under voltage V is expressed by $W = VQ$. When an electron is accelerated under 1 V of difference in potential, the energy obtained by the electron is called 1 eV. Since the elementary charge e is 1.602×10^{-19} (C),

$$\begin{aligned} 1\text{eV} &= 1.602 \times 10^{-19} \times 1 \qquad \text{(C)(V)} \\ &= 1.602 \times 10^{-19} \qquad \text{(J)} \end{aligned}$$

Electric field E can be expressed with $E = V/d$, where the distance, d, between electrodes and the applied voltage being V. The force F on the electron with elementary charge e is given by;

$$F = eE \qquad \text{(N)}$$

Here, the unit of F is Newton. Acceleration α of electrons is given by the following equation in which m is the mass of the electron:

$$\alpha = \frac{eE}{m} \qquad (\text{m/s}^2)$$

$$(1)\ \ E = \frac{10\ (\text{kV})}{10\ (\text{mm})} = \frac{10^4\ (\text{V})}{10^{-2}\ (\text{m})} = 10^6 \quad (\text{V/m})$$

$$(2)\ \ F = 1.602 \times 10^{-19} \times 10^6 = 1.602 \times 10^{-13} \quad (\text{N})$$

$$(3)\ \ \alpha = \frac{1.602 \times 10^{-13}}{9.109 \times 10^{-31}} = 1.76 \times 10^{17} \quad (\text{m/s}^2)$$

Question 1.3 X-rays are generated by making the electrically charged particles (electrons) with sufficient kinetic energy in vacuum collide with cathode, as widely used in the experiment of an X-ray tube. The resultant X-rays can be divided into two parts: continuous X-rays (also called white X-rays) and characteristic X-rays. The wavelength distribution and intensity of continuous X-rays are usually depending upon the applied voltage. A clear limit is recognized on the short wavelength side.

(1) Estimate the speed of electron before collision when applied voltage is 30,000 V and compare it with the speed of light in vacuum.
(2) In addition, obtain the relation of the shortest wavelength limit λ_{SWL} of X-rays generated with the applied voltage V, when an electron loses all energy in a single collision.

Answer 1.3 Electrons are drawn out from cathode by applying the high voltage of tens of thousands of V between two metallic electrodes installed in the X-ray tube in vacuum. The electrons collide with anode at high speed. The velocity of electrons is given by,

$$eV = \frac{mv^2}{2} \quad \rightarrow \quad v^2 = \frac{2eV}{m}$$

where e is the electric charge of the electron, V the applied voltage across the electrodes, m the mass of the electron, and v the speed of the electron before the collision. When values of rest mass $m_e = 9.110 \times 10^{-31}$ (kg) as mass of electron, elementary electron charge $e = 1.602 \times 10^{19}$ (C) and $V = 3 \times 10^4$ (V) are used for calculating the speed of electron v.

$$v^2 = \frac{2 \times 1.602 \times 10^{-19} \times 3 \times 10^4}{9.110 \times 10^{-31}} = 1.055 \times 10^{16}, \quad v = 1.002 \times 10^8 \text{ m/s}$$

Therefore, the speed of electron just before impact is about one-third of the speed of light in vacuum (2.998×10^8 m/s).

Some electrons release all their energy in a single collision. However, some other electrons behave differently. The electrons slow down gradually due to successive collisions. In this case, the energy of electron (eV) which is released partially and the corresponding X-rays (photon) generated have less energy compared with the energy ($h\nu_{max}$) of the X-rays generated when electrons are stopped with one collision. This is a factor which shows the maximum strength moves toward the shorter wavelength sides, as X-rays of various wavelengths generate, and higher the intensity of the applied voltage, higher the strength of the wavelength of X-rays (see Fig. 1). Every photon has the energy $h\nu$, where h is the Planck constant and ν the frequency.

The relationship of $eV = h\nu_{max}$ can be used, when electrons are stopped in one impact and all energy is released at once. Moreover, frequency (ν) and wavelength (λ) are described by a relation of $\lambda = c/\nu$, where c is the speed of light. Therefore, the relation between the wavelength λ_{SWL} in m and the applied voltage V may be given as follows:

$$\lambda_{SWL} = c/\nu_{max} = hc/eV = \frac{(6.626 \times 10^{-34}) \times (2.998 \times 10^8)}{(1.602 \times 10^{-19})V} = \frac{(12.40 \times 10^{-7})}{V}$$

This relation can be applied to more general cases, such as the production of electromagnetic waves by rapidly decelerating any electrically charged particle including electron of sufficient kinetic energy, and it is independent of the anode material. When wavelength is expressed in nm, voltage in kV, and the relationship becomes $\lambda V = 1.240$.

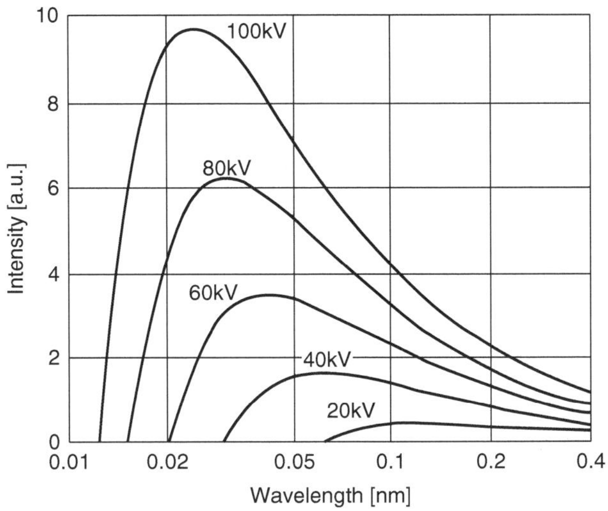

Fig. 1 Schematic diagram for X-ray spectrum as a function of applied voltage

Question 1.4 $K\alpha_1$ radiation of Fe is the characteristic X-rays emitted when one of the electrons in L shell falls into the vacancy produced by knocking an electron out of the K-shell, and its wavelength is 0.1936 nm. Obtain the energy difference related to this process for X-ray emission.

Answer 1.4 Consider the process in which an L shell electron moves to the vacancy created in the K shell of the target (Fe) by collision with highly accelerated electrons from a filament. The wavelength of the photon released in this process is given by λ, (with frequency ν). We also use Planck's constant h of (6.626×10^{-34} Js) and the velocity of light c of (2.998×10^8 ms^{-1}). Energy per photon is given by,

$$E = h\nu = \frac{hc}{\lambda}$$

Using Avogadro's number N_A, one can obtain the energy difference ΔE related to the X-ray release process per mole of Fe.

$$\Delta E = \frac{N_A hc}{\lambda} = \frac{0.6022 \times 10^{24} \times 6.626 \times 10^{-34} \times 2.998 \times 10^8}{0.1936 \times 10^{-9}}$$
$$= \frac{11.9626}{0.1936} \times 10^{-7} = 6.1979 \times 10^8 \quad \text{J/mole}$$

Reference: The electrons released from a filament have sufficient kinetic energy and collide with the Fe target. Therefore, an electron of K-shell is readily ejected. This gives the state of Fe^+ ion left in an excited state with a hole in the K-shell. When this hole is filled by an electron from an outer shell (L-shell), an X-ray photon is emitted and its energy is equal to the difference in the two electron energy levels. This variation responds to the following electron arrangement of Fe^+.

Before release	K1 L8 M14 N2
After release	K2 L7 M14 N2

Question 1.5 Explain atomic density and electron density.

Answer 1.5 The atomic density N_a of a substance for one-component system is given by the following equation, involving atomic weight M, Avogadro's number N_A, and the density ρ.

$$N_a = \frac{N_A}{M}\rho. \tag{1}$$

In the SI system, $N_a\,(m^{-3})$, $N_A = 0.6022 \times 10^{24}\,(mol^{-1})$, $\rho(kg/m^3)$, and $M\,(kg/mol)$, respectively.

The electron density N_e of a substance consisting of single element is given by,

$$N_e = \frac{N_A}{M} Z\rho \tag{2}$$

Each atom involves Z electrons (usually Z is equal to the atomic number) and the unit of N_e is also (m^{-3}).

The quantity $N_a = N_A/M$ in (1) or $N_e = (N_A Z)/M$ in (2), respectively, gives the number of atoms or that of electrons per unit mass (kg), when excluding density, ρ. They are frequently called "atomic density" or "electron density." However, it should be kept in mind that the number per m^3 (per unit volume) is completely different from the number per 1 kg (per unit mass). For example, the following values of atomic number and electron number per unit mass (=1kg) are obtained for aluminum with the molar mass of 26.98 g and the atomic number of 13:

$$N_a = \frac{0.6022 \times 10^{24}}{26.98 \times 10^{-3}} = 2.232 \times 10^{25} \qquad (kg^{-1})$$
$$N_e = \frac{0.6022 \times 10^{24}}{26.98 \times 10^{-3}} \times 13 = 2.9 \times 10^{26} \qquad (kg^{-1})$$

Since the density of aluminum is $2.70\,Mg/m^3 = 2.70 \times 10^3\,kg/m^3$ from reference table (Appendix A.2), we can estimate the corresponding values per unit volume as $N_a = 6.026 \times 10^{28}\,(m^{-3})$ and $N_e = 7.83 \times 10^{29}\,(m^{-3})$, respectively.

Reference: Avogadro's number provides the number of atom (or molecule) included in one mole of substance. Since the atomic weight is usually expressed by the number of grams per mole, the factor of 10^{-3} is required for using Avogadro's number in the SI unit system.

Question 1.6 The energy of a photoelectron, E_{ej}, emitted as the result of photoelectron absorption process may be given in the following with the binding energy E_B of the electron in the corresponding shell:

$$E_{ej} = h\nu - E_B$$

Here, $h\nu$ is the energy of incident X-rays, and this relationship has been obtained with an assumption that the energy accompanying the recoil of atom, which necessarily occurs in photoelectron absorption, is negligible.

Calculate the energy accompanying the recoil of atom in the following condition for Pb. The photoelectron absorption process of K shell for Pb was made by irradiating X-rays with the energy of 100 keV against a Pb plate and assuming that the momentum of the incident X-rays was shared equally by Pb atom and photoelectron. In addition, the molar mass (atomic weight) of Pb is 207.2 g and the atomic mass unit is $1\,\text{amu} = 1.66054 \times 10^{-27}\,\text{kg} = 931.5 \times 10^3\,\text{keV}$.

Answer 1.6 The energy of the incident X-rays is given as 100 keV, so that its momentum can be described as being $100\,\text{keV}/c$, using the speed of light c. Since the atom and photoelectron shared the momentum equally, the recoil energy of atom will be $50\,\text{keV}/c$. Schematic diagram of this process is illustrated in Fig. 1.

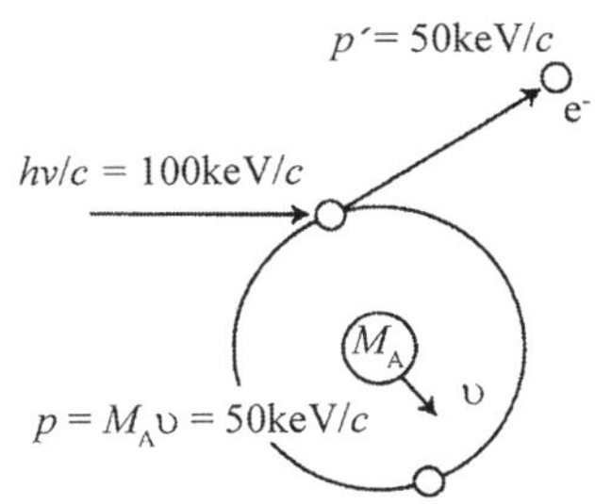

Fig. 1 Schematic diagram for the photo electron absorption process assuming that the momentum of the incident X-rays was shared equally by atom and photoelectron. Energy of X-ray radiation is 100 keV

On the other hand, one should consider for the atom that $1\,\text{amu} = 931.5\times10^3\,\text{keV}$ is used in the same way as the energy which is the equivalent energy amount of the rest mass for electron, m_e. The molar mass of 207.2 g for Pb is equivalent to

207.2 amu, so that the mass of 1 mole of Pb is equivalent to the energy of $207.2 \times 931.5 \times 10^3 = 193006.8 \times 10^3\,\text{keV}/c$.

When the speed of recoil atom is v and the molar mass of Pb is M_A, its energy can be expressed by $\frac{1}{2}M_A v^2$. According to the given assumption and the momentum described as $p = M_A v$, the energy of the recoil atom, E_r^A, may be obtained as follows:

$$E_r^A = \frac{1}{2}M_A v^2 = \frac{p^2}{2M_A} = \frac{(50)^2}{2 \times (193006.8 \times 10^3)} = 0.0065 \times 10^{-3} \quad \text{(keV)}$$

The recoil energy of atom in the photoelectron absorption process shows just a very small value as mentioned here using the result of Pb as an example, although the recoil of the atom never fails to take place.

Reference:

$$\text{Energy of 1 amu} = \frac{1.66054 \times 10^{-27} \times (2.99792 \times 10^8)^2}{1.60218 \times 10^{-19}} = 9.315 \times 10^8 \quad \text{(eV)}$$

On the other hand, the energy of an electron with rest mass $m_e = 9.109 \times 10^{-31}$ (kg) can be obtained in the following with the relationship of 1 (eV) $= 1.602 \times 10^{-19}$ (J):

$$E = m_e c^2 = \frac{9.109 \times 10^{-31} \times (2.998 \times 10^8)^2}{1.602 \times 10^{19}} = 0.5109 \times 10^6 \quad \text{(eV)}$$

Question 1.7 Explain the Rydberg constant in Moseley's law with respect to the wavelength of characteristic X-rays, and obtain its value.

Answer 1.7 Moseley's law can be written as,

$$\frac{1}{\lambda} = R(Z - S_M)^2 \left(\frac{1}{n_1^2} - \frac{1}{n_2^2}\right) \tag{1}$$

The wavelength of the X-ray photon (λ) corresponds to the shifting of an electron from the shell of the quantum number n_2 to the shell of the quantum number of n_1. Here, Z is the atomic number and S_M is a screening constant.

Using the elementary electron charge of e, the energy of electron characterized by the circular movement around the nucleus charge Ze in each shell (orbital) may be given, for example, with respect to an electron of quantum number n_1 shell in the following form:

$$E_n = -\frac{2\pi^2 m e^4}{h^2}\frac{Z^2}{n_1^2} \tag{2}$$

Here, h is a Planck constant and m represents the mass of electron. The energy of this photon is given by,

$$h\nu = E_{n_2} - E_{n_1} = \Delta E = \frac{2\pi^2 me^4}{h^2} Z^2 \left(\frac{1}{n_1^2} - \frac{1}{n_2^2} \right) \tag{3}$$

The following equation will also be obtained, if the relationship of $E = h\nu = \frac{hc}{\lambda}$ is employed while using the velocity of photon, c:

$$\frac{1}{\lambda} = \frac{2\pi^2 me^4}{ch^3} Z^2 \left(\frac{1}{n_1^2} - \frac{1}{n_2^2} \right) \tag{4}$$

If the value of electron mass is assumed to be rest mass of electron and a comparison of (1) with (4) is made, the Rydberg constant R can be estimated. It may be noted that the term of $(Z - S_M)^2$ in (1) could be empirically obtained from the measurements on various characteristic X-rays as reported by H.G.J. Moseley in 1913.

$$\begin{aligned} R = \frac{2\pi^2 me^4}{ch^3} &= \frac{2 \times (3.142)^2 \times (9.109 \times 10^{-28}) \times (4.803 \times 10^{-10})^4}{(2.998 \times 10^{10}) \times (6.626 \times 10^{-27})^3} \\ &= 109.743 \times 10^3\,(\mathrm{cm}^{-1}) = 1.097 \times 10^7\,(\mathrm{m}^{-1}) \end{aligned} \tag{5}$$

The experimental value of R can be obtained from the ionization energy (−13.6 eV) of hydrogen (H). The corresponding wave number (frequency) is 109737.31 cm^{-1}, in good agreement with the value obtained from (5). In addition, since Moseley's law and the experimental results are all described by using the cgs unit system (gauss system), 4.803×10^{-10} esu has been used for the elemental electron charge e. Conversion into the SI unit system is given by (SI unit × velocity of light × 10^{-1}) (e.g., 5th edition of the Iwanami Physics-and-Chemistry Dictionary p. 1526 (1985)). That is to say, the amount of elementary electron charge e can be expressed as:

$$1.602 \times 10^{-19}\,\mathrm{Coulomb} \times 2.998 \times 10^{10}\,\mathrm{cm/s} \times 10^{-1} = 4.803 \times 10^{-10}\,\mathrm{esu}$$

The Rydberg constant is more strictly defined by the following equation:

$$R = \frac{2\pi^2 \mu e^4}{ch^3} \tag{6}$$

$$\frac{1}{\mu} = \frac{1}{m} + \frac{1}{m_p} \tag{7}$$

Here, m is electron mass and m_P is nucleus (proton) mass.The detected difference is quite small, but the value of m_P depends on the element. Then, it can be seen from the relation of (6) and (7) that a slightly different value of R is obtained for each element. However, if a comparison is made with a hydrogen atom, there is a difference of about 1,800 times between the electron mass $m_e = 9.109 \times 10^{-31}$ kg and the proton mass which is $m_P = 1.67 \times 10^{-27}$ kg. Therefore, the relationship of (6) is usually treated as $\mu = m$, because m_P is very large in comparison with m_e.

Reference: The definition of the Rydberg constant in the SI unit is given in the form where the factor of $(1/4\pi\epsilon_0)$ is included by using the dielectric constant $\epsilon_0(8.854 \times 10^{-12}\,\text{F/m})$ in vacuum for correlating with nucleus charge Z_e.

$$R = \frac{2\pi^2\mu e^4}{ch^3} \times \left(\frac{1}{4\pi\epsilon_0}\right)^2 = \frac{me^4}{8\epsilon_0^2 ch^3}$$

$$= \frac{9.109 \times 10^{-31} \times (1.602 \times 10^{-19})^4}{8 \times (8.854 \times 10^{-12})^2 \times (2.998 \times 10^8) \times (6.626 \times 10^{-34})^3}$$

$$= \frac{9.109 \times (1.602)^4 \times 10^{-107}}{8 \times (8.854)^2 \times (2.998) \times (6.626)^3 \times 10^{-118}} = 1.097 \times 10^7\ (\text{m}^{-1})$$

Question 1.8 When the X-ray diffraction experiment is made for a plate sample in the transmission mode, it is readily expected that absorption becomes large and diffraction intensity becomes weak as the sample thickness increases. Obtain the thickness of a plate sample which makes the diffraction intensity maximum and calculate the value of aluminum for the Cu-Kα radiation.

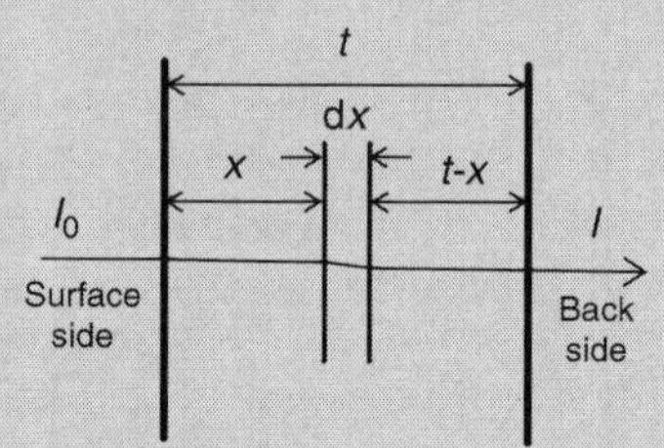

Fig. A Geometry for a case where X-ray penetrates a plate sample

Answer 1.8 X-ray diffraction experiment in the transmission mode includes both absorption and scattering of X-rays. Let us consider the case where the sample thickness is t, the linear absorption coefficient μ, the scattering coefficient S, and the intensity of incident X-rays I_0 as referred to Fig. A.

Since the intensity of the incident X-rays reaching the thin layer $\mathrm{d}x$ which is at distance of x from the sample surface is given by $I_0 \mathrm{e}^{-\mu x}$, the scattering intensity $\mathrm{d}I'_x$ from the thin layer $\mathrm{d}x$ (with scattering coefficient S) is given by the following equation:

$$\mathrm{d}I'_x = S I_0 \mathrm{e}^{-\mu x}\mathrm{d}x \qquad (1)$$

The scattering intensity $\mathrm{d}I_x$ passes through the distance of $(t - x)$ in the sample and the absorption during this passage is expressed by the form of $\mathrm{e}^{-\mu(t-x)}$. Therefore, the scattering intensity of the thin layer $\mathrm{d}x$ after passing through the sample may be given in the following form:

$$\mathrm{d}I_x = (S I_0 \mathrm{e}^{-\mu x}\mathrm{d}x)\mathrm{e}^{-\mu(t-x)} = S I_0 \mathrm{e}^{-\mu t}\mathrm{d}x \qquad (2)$$

The scattering intensity of the overall sample will be equal to the result obtained by integrating the intensity of the thin layer dx with respect to the sample thickness from zero to t.

$$I = \int_0^t SI_0 e^{-\mu t} dx = SI_0 t \cdot e^{-\mu t} \quad (3)$$

The maximum value of I is given under the condition of $dI/dt = 0$.

$$\frac{dI}{dt} = SI_0(e^{-\mu t} - t\mu e^{-\mu t}) = 0, \qquad t\mu = 1 \quad \rightarrow \quad t = \frac{1}{\mu} \quad (4)$$

We can find the values of mass absorption coefficient (μ/ρ) and density (ρ) of aluminum for Cu-Kα radiation in the reference table (e.g., Appendix A.2). The results are $(\mu/\rho) = 49.6\,\text{cm}^2/\text{g}$ and $\rho = 2.70\,\text{g/cm}^3$, respectively. The linear absorption coefficient of aluminum is calculated in the following:

$$\mu = \left(\frac{\mu}{\rho}\right)\rho = 49.6 \times 2.70 = 133.92\,(\text{cm}^{-1})$$

Therefore, the desired sample thickness t can be estimated as follows:

$$t = \frac{1}{\mu} = \left(\frac{1}{133.92}\right) = 7.47 \times 10^{-3}\,(\text{cm}) = 74.7\,(\mu\text{m})$$

Question 1.9 There is a substance of linear absorption coefficient μ.

(1) Obtain a simple relation to give the sample thickness x required to reduce the amount of transmitted X-ray intensity by half.
(2) Calculate also the corresponding thickness of Fe-17 mass % Cr alloy (density $= 7.76 \times 10^6\,\text{g/m}^3$) for Mo-K$\alpha$ radiation, using the relation obtained in (1).

Answer 1.9 Let us consider the intensity of the incident X-rays as I_0 and that of the transmitted X-rays as I. Then,

$$I = I_0 e^{-\mu x} \quad (1)$$

If the condition of $I = \frac{I_0}{2}$ is imposed, taken into account, one obtains,

$$\frac{I_0}{2} = I_e^{-\mu x} \quad (2)$$

$$\frac{1}{2} = e^{-\mu x} \quad (3)$$

When the logarithm of both sides is taken, we obtain $\log 1 - \log 2 = -\mu x \log e$. The result is $-\log 2 = -\mu x$, as they are $\log 1 = 0$, and $\log e = 1$. Here, natural logarithm is used and the required relation is given as follows:

$$x = \frac{\log 2}{\mu} \simeq \frac{0.693}{\mu} \tag{4}$$

The values of mass absorption coefficients of Fe and Cr for the Mo-Kα radiation are 37.6 and 29.9 $\mathrm{cm^2/g}$ obtained from Appendix A.2, respectively. The concentration of Cr is given by 17 mass %, so that the weight ratio of two alloy components can be set as $w_{\mathrm{Fe}} = 0.83$ and $w_{\mathrm{Cr}} = 0.17$. Then, the mass absorption coefficient of the alloy is expressed in the following:

$$\begin{aligned}\left(\frac{\mu}{\rho}\right)_{\mathrm{Alloy}} &= w_{\mathrm{Fe}}\left(\frac{\mu}{\rho}\right)_{\mathrm{Fe}} + w_{\mathrm{Cr}}\left(\frac{\mu}{\rho}\right)_{\mathrm{Cr}} \\ &= 0.83 \times (37.6) + 0.17 \times (29.9) = 36.3\,(\mathrm{cm^2/g})\end{aligned}$$

Next, note that the unit of the density of the Fe–Cr alloy is expressed in cgs, $7.76 \times 10^6\,\mathrm{g/m^3} = 7.76\,\mathrm{g/cm^3}$. We obtain,

$$\mu_{\mathrm{Alloy}} = 36.3 \times 7.76\,(\mathrm{cm^{-1}}) = 281.7\,(\mathrm{cm^{-1}})$$

$$x = \frac{0.693}{281.7} = 0.0025\,\mathrm{cm} = 0.025\,\mathrm{mm} = 25\,\mu\mathrm{m}$$

Question 1.10 Calculate the mass absorption coefficient of lithium niobate ($LiNbO_3$) for Cu-Kα radiation.

Answer 1.10 The atomic weight of Li, Nb, and oxygen (O) and their mass absorption coefficients for Cu-Kα radiation are obtained from Appendix A.2, as follows:

	Atomic weight (g)	Mass-absorption coefficient μ/ρ ($\mathrm{cm^2/g}$)
Li	6.941	0.5
Nb	92.906	145
O	15.999	11.5

The molar weight(molar mass) M per 1 mole of $LiNbO_3$ is given in the following:

$$M = 6.941 + 92.906 + (15.999 \times 3) = 147.844\,(\mathrm{g})$$

The weight ratio w_{j} of three components of Li, Nb, and O is to be obtained.

$$w_{Li} = \frac{6.941}{147.844} = 0.047, \quad w_{Nb} = \frac{92.906}{147.844} = 0.628, \quad w_{O} = \frac{47.997}{147.844} = 0.325$$

Then, the mass absorption coefficient of lithium niobate can be obtained as follows:

$$\left(\frac{\mu}{\rho}\right)_{LiNbO_3} = \sum w_j \left(\frac{\mu}{\rho}\right)_j = 0.047 \times 0.5 + 0.628 \times 145 + 0.325 \times 11.5$$
$$= 94.8\ (\text{cm}^2/\text{g})$$

Question 1.11 A thin plate of pure iron is suitable for a filter for Co-Kα radiation, but it is also known to easily oxidize in air. For excluding such difficulty,we frequently utilize crystalline hematite powder (Fe_2O_3:density 5.24×10^6 g/m^3). Obtain the thickness of a filter consisting of hematite powder which reduces the intensity of Co-Kβ radiation to 1/500 of the Kα radiation case. Given condition is as follows. The intensity ratio between Co-Kα and Co-Kβ is found to be given by 5:1 without a filter. The packing density of powder sample is known usually about 70% of the bulk crystal.

Answer 1.11 The atomic weight of Fe and oxygen (O) and their mass absorption coefficients for Co-Kα and Co-Kβ radiations are obtained from Appendix A.2, as follows:

	Atomic weight (g)	μ/ρ for Co-Kα (cm^2/g)	μ/ρ for Co-Kβ (cm^2/g)
Fe	55.845	57.2	342
O	15.999	18.0	13.3

The weight ratio of Fe and O in hematite crystal is estimated in the following:

$$M_{Fe_2O_3} = 55.845 \times 2 + 15.999 \times 3 = 159.687$$
$$w_{Fe} = \frac{55.845 \times 2}{159.687} = 0.699, \quad w_O = 0.301$$

The mass absorption coefficients of hematite crystals for Co-Kα and Co-Kβ radiations are, respectively, to be calculated.

$$\left(\frac{\mu}{\rho}\right)^{\alpha}_{Fe_2O_3} = 0.699 \times 57.2 + 0.301 \times 18.0 = 45.4\ (\text{cm}^2/\text{g})$$
$$\left(\frac{\mu}{\rho}\right)^{\beta}_{Fe_2O_3} = 0.699 \times 342 + 0.301 \times 13.3 = 243.1\ (\text{cm}^2/\text{g})$$

It is noteworthy that the density of hematite in the filter presently prepared is equivalent to 70% of the value of bulk crystal by considering the packing density, so that we have to use the density value of $\rho_f = 5.24 \times 0.70 = 3.67\,\mathrm{g/cm^3}$ Therefore, the value of the linear absorption coefficient of hematite powder packed into the filter for Co-Kα and Co-Kβ radiations will be, respectively, as follows:

$$\mu_\alpha = \left(\frac{\mu}{\rho}\right)^{\alpha}_{\mathrm{Fe_2O_3}} \times \rho_f = 45.4 \times 3.67 = 166.6\,(\mathrm{cm^{-1}})$$

$$\mu_\beta = \left(\frac{\mu}{\rho}\right)^{\beta}_{\mathrm{Fe_2O_3}} \times \rho_f = 24.1 \times 3.67 = 892.2\,(\mathrm{cm^{-1}})$$

The intensity ratio of Co-Kα and Co-Kβ radiations before and after passing through the filter consisting of hematite powder may be described in the following equation:

$$\frac{I_{\mathrm{Co-K}\beta}}{I_{\mathrm{Co-K}\alpha}} = \frac{I_0^\beta \mathrm{e}^{-\mu_\beta t}}{I_0^\alpha \mathrm{e}^{-\mu_\alpha t}}$$

From the given condition, the ratio between I_0^α and I_0^β is 5:1 without filter, and it should be 500:1 after passing through the filter. They are expressed as follows:

$$\frac{1}{500} = \frac{1}{5}\frac{\mathrm{e}^{-\mu_\beta t}}{\mathrm{e}^{-\mu_\alpha t}} \quad \rightarrow \quad \frac{1}{100} = \mathrm{e}^{(\mu_\alpha - \mu_\beta)t}$$

Take the logarithm of both sides and obtain the thickness by using the values of μ_α and μ_β.

$$(\mu_\alpha - \mu_\beta)t = -\log 100 \qquad (\because \log e = 1, \quad \log 1 = 0)$$

$$(166.6 - 892.2)t = -4.605$$

$$t = 0.0063\,(\mathrm{cm^{-1}}) = 63\,(\mu\mathrm{m})$$

Question 1.12 For discussing the influence of X-rays on the human body etc., it would be convenient if the effect of a substance consisting of multi-elements, such as water (H_2O) and air (N_2, O_2, others), can be described by information of each constituent element (H, O, N, and others) with an appropriate factor. For this purpose, the value of effective element number $\bar{Z}$ is often used and it is given by the following equation:

$$\bar{Z} = \sqrt[2.94]{a_1 Z_1^{2.94} + a_2 Z_2^{2.94} + \cdots}$$

where $a_1, a_2 \ldots$ is the electron component ratio which corresponds to the rate of the number of electrons belonging to each element with the atomic number

$Z_1, Z_2, \ldots$ to the total number of electrons of a substance. Find the effective atomic number of water and air. Here, the air composition is given by 75.5% of nitrogen, 23.2% of oxygen, and 1.3% of argon in weight ratio.

Answer 1.12 Water (H_2O) consists of two hydrogen atoms and one oxygen atom, whereas the number of electrons are one for hydrogen and eight for oxygen. The values of atomic weight per mole (molar mass) of hydrogen and oxygen (molar mass) are 1.008 and 15.999 g, respectively. Each electron density per unit mass is given as follows:

$$\text{For hydrogen} \quad N_e^H = \frac{0.6022 \times 10^{24}}{1.008} \times 1 = 0.597 \times 10^{24} \quad (g^{-1})$$

$$\text{For oxygen} \quad N_e^O = \frac{0.6022 \times 10^{24}}{15.999} \times 8 = 0.301 \times 10^{24} \quad (g^{-1})$$

In water (H_2O), the weight ratio can be approximated by $2/18$ for hydrogen and $16/18$ for oxygen, respectively. Then, the number of electrons in hydrogen and oxygen contained in 1 g water are $0.597 \times 10^{24} \times (2/18) = 0.0663 \times 10^{24}$ and $0.301 \times 10^{24} \times (16/18) = 0.2676 \times 10^{24}$,respectively, so that the number of electrons contained in 1 g water is estimated to be $(0.0663 + 0.2676) \times 10^{24} = 0.3339 \times 10^{24}$. Therefore, the electron component ratio of water is found as follows:

$$\begin{aligned} a^H &= \frac{0.0663}{0.3339} = 0.199 \\ a^O &= \frac{0.2262}{0.3339} = 0.801 \\ \bar{Z} &= \sqrt[2.94]{0.199 \times 1^{2.94} + 0.801 \times 8^{2.94}} \\ &= \sqrt[2.94]{0.199 + 362.007} = \sqrt[2.94]{362.206} = 7.42 \end{aligned}$$

Here, we use the relationship of $\bar{Z} = X^{\frac{1}{y}} \rightarrow \ln\bar{Z} = \frac{1}{y}\ln X \rightarrow \bar{Z} = e^{\frac{1}{y}\ln X}$

On the other hand, the molar masses of nitrogen, oxygen, and argon are 14.01, 15.999, and 39.948 g, respectively. Since 75.5% of nitrogen (7 electrons), 23.2% of oxygen (8 electrons), and 1.3% of argon (18 electrons) in weight ratio are contained in 1 g of air, each electron numbers is estimated in the following:

$$\text{For nitrogen} \quad N_e^N = \frac{0.6022 \times 10^{24}}{14.01} \times 0.755 \times 7 = 0.2272 \times 10^{24}$$

$$\text{For oxygen} \quad N_e^O = \frac{0.6022 \times 10^{24}}{15.999} \times 0.232 \times 8 = 0.0699 \times 10^{24}$$

$$\text{For argon} \quad N_e^{Ar} = \frac{0.6022 \times 10^{24}}{39.948} \times 0.013 \times 18 = 0.0035 \times 10^{24}$$

Therefore, the value of $(0.2272 + 0.0699 + 0.0035) \times 10^{24} = 0.3006 \times 10^{24}$ is corresponding to the number of electrons in 1 g of air. The rate to the total number of electrons of each element is as follows:

$$a^{\mathrm{N}} = \frac{0.2272}{0.3006} = 0.756$$

$$a^{\mathrm{O}} = \frac{0.0699}{0.3006} = 0.232$$

$$a^{\mathrm{Ar}} = \frac{0.0035}{0.3006} = 0.012$$

Accordingly, the effective atomic number of air is estimated in the following:

$$\begin{aligned}\bar{Z} &= \sqrt[2.94]{0.756 \times 7^{2.94} + 0.232 \times 8^{2.94} + 0.012 \times 18^{2.94}} \\ &= \sqrt[2.94]{230.73 + 104.85 + 58.84} = \sqrt[2.94]{394.42} = 7.64\end{aligned}$$

Chapter 2
Geometry of Crystals

2.1 Lattice and Crystal Systems

The origin of crystallography can be traced to the study for the external appearance of natural minerals, such as quartz, fluorite, pyrite, and corundum, which are regular in shape and clearly exhibit a good deal of symmetry. A large amount of data for such minerals have been systematized by applying geometry and group theory. "Crystallography" involves the general consideration of how crystals can be built from small units. This corresponds to the infinite repetition of identical structural units (frequently referred to as a unit cell) in space. In other words, the structure of all crystals can be described by a lattice, with a group of atoms allocated to every lattice point.

Crystals can be classified into 32 point groups on the basis of eight kinds of symmetry elements. There are seven crystal systems for classification, which consist of 14 kinds of Bravais lattices. For convenience, these relations are illustrated in Fig. 2.1. Furthermore, if it is extended to include space groups, by adding point groups, Bravais lattices, screw axis, and glide reflection axis, there will be 230 classifications in total. In other words, all crystals "belong to one of 230 space groups," the details available in other books on crystallography (see for example *International Tables for X-ray Crystallography* published by the International Union of Crystallography).

Let us consider the three-dimensional arrangement of points called a point lattice, as shown in Fig. 2.2.When the atomic position or configuration in crystal is described by a lattice point, any point indicates exactly the same environment (=identical surroundings) as any other point in the lattice. This means the lattice point can be reproduced by repeating a small unit. This small repeating unit is referred to as a unit cell (or sometime called unit lattice) where all sequences can be given by three vectors $\boldsymbol{a}$, $\boldsymbol{b}$, and $\boldsymbol{c}$ (or those lengths a, b and c) and the interaxial angles between them, α, β, and γ. The relationship between a, b, c and α, β, γ is illustrated in Fig. 2.3, and these lengths and angles are called the lattice parameters or lattice constants of the unit cell.

As shown in Fig. 2.2, there is more than one way to choose a unit cell, so that it is better to select a unit cell in the direction where three axes have the highest

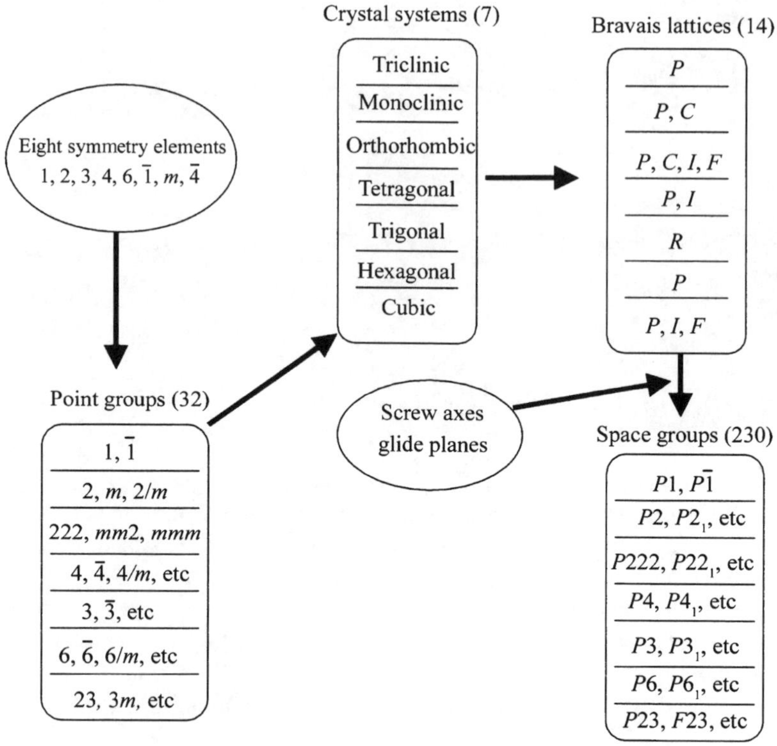

Fig. 2.1 Symmetry elements in crystals and their relationships for classification

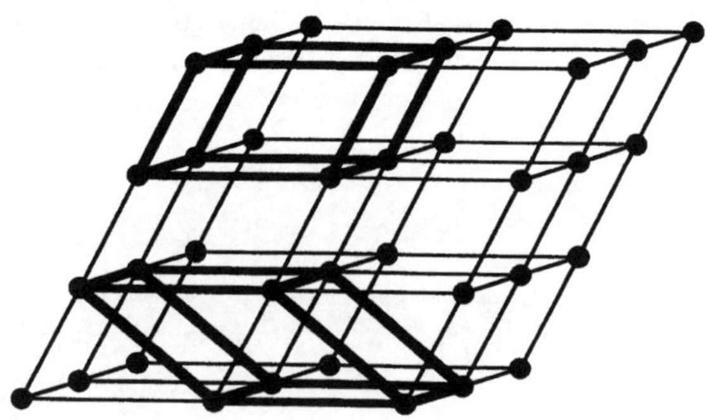

Fig. 2.2 Different ways for selecting a unit cell in a point lattice

symmetry. On the other hand, all three lengths a, b, and c in case of Fig. 2.3 have different values, and all three angles α, β, and γ are also found to be different from each other. This case called as "triclinic system" shows an axis with the altissimo symmetry of only a onefold axis (no symmetry) or $\bar{1}$ rotatory inversion (or rotoinversion) axis. For convenience, some essential points on "how atoms are arranged in a substance" are given below.

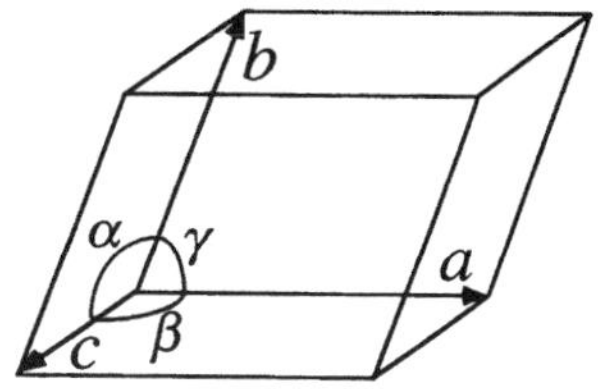

Fig. 2.3 Example of a unit cell

Four macroscopic symmetry operations or symmetry elements are well known: "reflection," "rotation," "inversion," and "rotatory inversion." For example, several planes of symmetry in a cube are readily noticed. In this symmetry of reflection, if a body shows symmetry with respect to a certain plane passing through it, reflection of either half of the body in the plane as in a mirror makes a body coincident with the other half. A body shows n-fold rotation symmetry around an axis, when a rotation by $(360/n)°$ or $(2\pi/n)$ brings it into self-coincidence. One can easily understand that a cube has a fourfold rotation axis normal to each face, a threefold axis along each body diagonal, and twofold axis joining the centers of opposite edges. In general, there are one, two, three, four and six fold axes for a rotation axis, but a onefold axis corresponds to no symmetry at all.

A body having an inversion center can bring itself into coincidence, when every point in the body is inverted or reflected at the inversion center. The corresponding points of the body are at equal distances from the center on a line drawn through the center. A cube is known to have such a center at the intersection of its body diagonals. In general, there is either one, two, three, four or sixfold axes for a rotatory-inversion axis. It is noted for an n-fold, rotatory-inversion axis exists when a body comes into coincidence with itself by coupling the rotation by $(360/n)°$ around the axis followed by inversion operation in a center lying on the axis.

By putting lattice points at the corner of these crystal systems for finding a certain minimum set of symmetry elements, seven kinds of crystal systems are obtained as shown in Table 2.1. That is, only seven different kinds of cells are necessary to cover all possible point lattices or all crystals can be classified into one of the seven crystal systems. Nevertheless, there are other ways for fulfilling the condition that each point has identical surroundings. In this regard, Auguste Bravais (physicist in France) found that there are 14 possible point lattices and no more and we use Bravais lattices as shown in Fig. 2.4. Since the unit cell including two or more lattice points is chosen in the Bravais lattice for convenience, some of the Bravais lattices can be expressed by other simple lattices. For example, the face-centered cubic lattice is also described by a trigonal (rhombohedral) lattice which contains only one lattice point (see Question 2.5).

The symbols P, F, I, etc. in Fig. 2.4 or Table 2.1 are given on the basis of the following rule. When a unit cell has only one lattice point, it is called a primitive (or simple) cell, and usually represented by P. In addition, although the trigonal (rhombohedral) crystal system can also be classified into primitive, we use R as the symbol. Other symbols are nonprimitive cells and more than one lattice point per cell is included. It may be suggested that any cell containing lattice points only at

Table 2.1 Summary of seven crystal systems and Bravais lattices

System	Axial lengths and angles	Bravais lattice	Lattice symbol
Cubic	Three equal axes at right angles $a = b = c, \alpha = \beta = \gamma = 90°$	Simple	P
		Body-centered	I
		Face-centered	F
Tetragonal	Three axes at right angles, two equals $a = b \neq c, \alpha = \beta = \gamma = 90°$	Simple	P
		Body-centered	I
Orthorhombic	Three unequal axes at right angles $a \neq b \neq c, \alpha = \beta = \gamma = 90°$	Simple	P
		Body-centered	I
		Base-centered	C
		Face-centered	F
Trigonal*	Three equal axes, equally inclined $a = b = c, \alpha = \beta = \gamma \neq 90°$	Simple	R
Hexagonal	Two equal coplanar axes at, 120° third axis at right angles $a = b \neq c, \alpha = \beta = 90°, \gamma = 120°$	Simple	P
Monoclinic	Three unequal axes, one pair not at right angles $a \neq b \neq c$, $\alpha \neq \gamma = 90° \neq \beta$	Simple	P
		Base-centered	C
Triclinic	Three unequal axes, unequally inclined and none at right angles $a \neq b \neq c, \alpha \neq \beta \neq \gamma \neq 90°$	Simple	P

*Also called rhombohedral.

the corners is primitive, whereas one containing additional points in the interior or on a cell face is nonprimitive. Symbols I and F refer to body-centered and face-centered cells, respectively. The symbols A, B, and C represent base-centered cells where the lattice point is given at the center on one pair of opposite faces A, B, or C. Here, the face of C, for example, is the face defined by b-axis and a-axis.

There are various substances and the atomic arrangements in these substances reveal a variety of crystal structures characterized by a certain periodicity. Of course, all structures cannot be covered here. However, many of elements in the periodic table are metals and about 70% of them have relatively simple crystal structure with high symmetry, such as the body-centered cubic (bcc), face-centered cubic (fcc), and hexagonal close-packed (hcp) lattices. Typical features of these three crystal structures are summarized in Fig. 2.5.

Crystals can be broadly classified into three categories from the point of view of bonding: "metallic," "ionic," and "covalent." In metallic crystals, a large number of electrons (conduction or valence electrons) are free to move the inside of the system, without belonging to specific atoms but shared by the whole system. This bonding arising from a conduction electron is not very strong. For example, the interatomic distances of alkali metals are relatively large, because the kinetic energy of conduction electrons is relatively low at the large interatomic distances. This leads to weak binding and simple structure. On the other hand, ionic crystals consist of positive and negative ions, and the ionic bond results from the electrostatic interaction

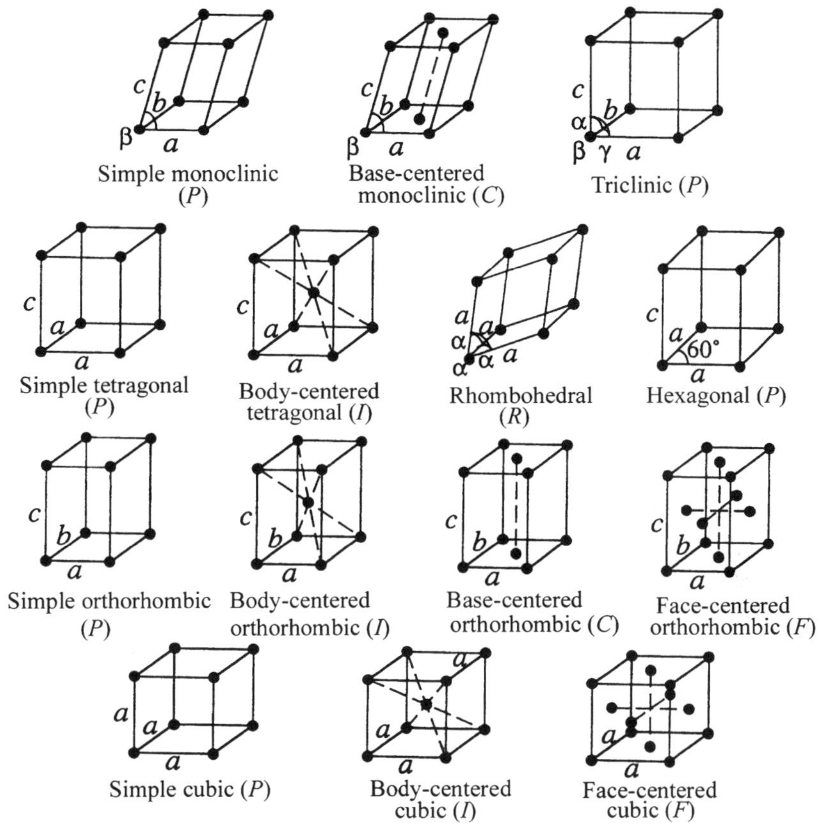

Fig. 2.4 The fourteen Bravais lattices

of oppositely charged ions in the solid state. Typical examples are metal–halogen compounds, and two typical structures found for ionic crystals are sodium chloride and cesium chloride structures. In sodium chloride, Na^+ and Cl^- ions are arranged to form the structure as shown in Fig. 2.6.

In ionic crystals, ionic arrangements that minimize electrostatic repulsion and maximize electrostatic attraction are preferred. In many cases, the negative ions (anions) of large size are densely arranged so as to avoid their direct contact, and the positive ions (cations) of small size occupy the positions equivalent to the vacant space produced by anions. Therefore, the correlation is recognized between crystal structure and the size ratio, for example, the ratio of ionic radii $= r_c/r_a$, where r_c and r_a are the radii of cation and anion, respectively. When the value of r_c/r_a is 0.225, one can find the tetrahedral arrangement with the coordination number of 4, and the octahedral arrangement with the coordination number of 6 in the $r_c/r_a =$ 0.414 case. Thus, the value of r_c/r_a has a critical value for ionic configurations. For actual ionic crystals, the arrangement is quite likely to avoid direct contact of the same electric-charged ions mainly arising from energetic constraints. Therefore,

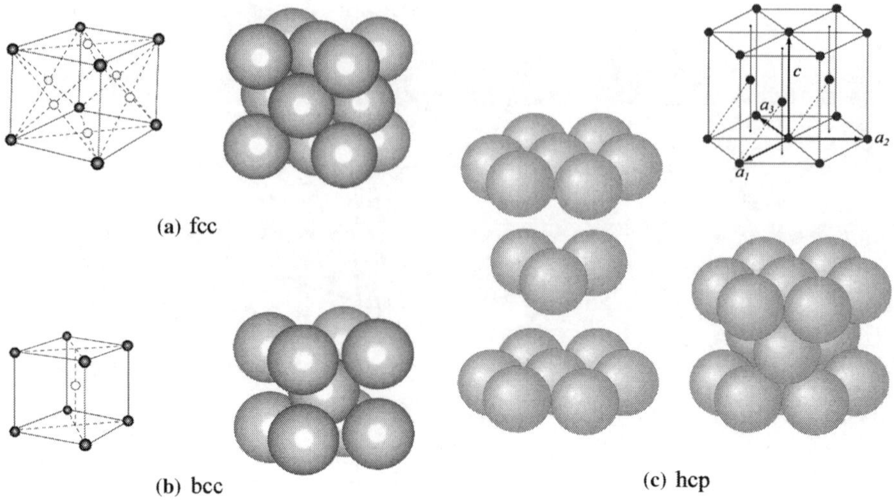

Fig. 2.5 Typical crystal structures of metallic elements

Fig. 2.6 Crystal structure of sodium chloride

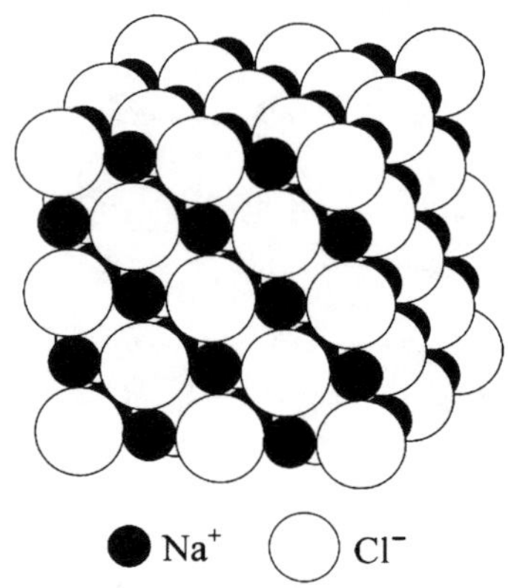

many combinations of r_c/r_a found in an actual ionic crystal show a little bit larger values than the critical values based on spherical models.

The covalent bond, having strong directional properties, is the classical electron pair, and silicon and germanium are included in this category. These crystals have the diamond structure with atoms bonded to four nearest neighbors at tetrahedral angles.

2.2 Lattice Planes and Directions

The key points for describing crystal planes and directions are discussed below. In order to show a lattice plane, Miller indices are usually employed. Miller indices are defined as the reciprocals of the fractional intercepts which the plane makes with the crystallographic axes. For example, if a plane is described by the Miller indices of $(h\ k\ l)$, the plane makes fractional intercepts of $1/h$, $1/k$, and $1/l$ with the axes a, b, and c, respectively. This reciprocal symbolism enables us to give the Miller indices being zero, when a plane is parallel to an axis. For example, the center position of a body-centered cubic lattice is expressed by $\frac{1}{2}\frac{1}{2}\frac{1}{2}$ and the position of surface-centered lattice as $\frac{1}{2}\frac{1}{2}0$; $\frac{1}{2}0\frac{1}{2}$; $0\frac{1}{2}\frac{1}{2}$. Some generalized rules for presentation are as follows:

(1) The distance from the origin to the intersection of the desired plane with each crystal axis is determined from the basis of unit length, such as a lattice parameter. As shown in Fig. 2.7, the a-axis intersects at the unit length of $1/h$.
(2) The reciprocals of three numbers are taken and let the minimum integer ratio $(h\ k\ l)$ be the index of the corresponding plane.
(3) If the desired plane is parallel to a certain axis, the distance from the origin in the axis to the intersection becomes infinite. In that case, the index is expressed by zero. For example, $(h\ 0\ 0)$ represents a plane parallel to b-axis and c-axis.
(4) Although a set of planes parallel to it can be found for every plane, Miller indices usually refer to that plane in the set which is nearest to the origin.
(5) When a plane intercepts at the negative side in any axis, such negative value is represented by writing a bar over the Miller indices, for example, $(\bar{h}\ \bar{k}\ \bar{l})$.
(6) There are sets of equivalent lattice planes related by symmetry, for example, the planes of a cube, (100), (010), $(\bar{1}00)$, $(0\bar{1}0)$, (001), and $(00\bar{1})$.They are called "planes of a form" and the expression of {001} is used. The number of the equivalent lattice planes in one plane of a form corresponds to the multiplicity factor and they are given for seven crystal systems in Table 2.2.

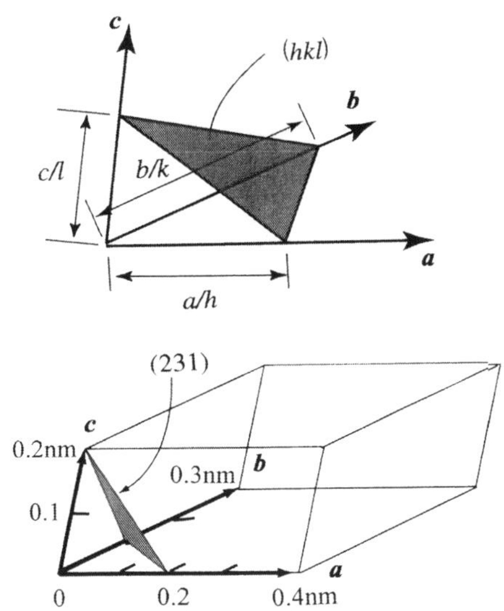

Fig. 2.7 Example of Miller indices for plane

Table 2.2 Multiplicity factors for crystalline powder samples

Cubic		*hkl*	*hkk*	*hk*0	*hh*0	*hhh*	*h*00		
		48*	24	24*	12	8	6		
Hexagonal		$hk \cdot l$	$hh \cdot l$	$h0 \cdot l$	$hk \cdot 0$	$hh \cdot 0$	$h0 \cdot 0$	$00 \cdot l$	
		24*	12	12	12*	6	6	2	
Trigonal	Referred to rhombohedral axes	*hkl*	$\bar{k}kh$	*hkk*	*hk*0	$\bar{k}hh$	*hhh*	*hh*0	*h*00
		12*	12*	6	12*	6	2	6	6
	Referred to hexagonal axes	$hk \cdot l$	$hh \cdot l$	$h0 \cdot l$	$hk \cdot 0$	$hh \cdot 0$	$0h \cdot 0$	$00 \cdot l$	
		12*	12*	6	12*	6	6	2	
Tetragonal		*hkl*	*hhl*	*hh*0	*hk*0	*h*0*l*	*h*00	00*l*	
		16*	8	4	8*	7	4	2	
Orthorhombic		*hkl*	*hk*0	*h*00	0*k*0	00*l*	*h*0*l*	0*kl*	
		8	4	2	2	2	4	4	
Monoclinic		*hkl*	*hk*0	0*kl*	*h*0*l*	*h*00	0*k*0	00*l*	
(Orthogonal axis: *b*)		4	4	4	2	2	2	2	
Triclinic		*hkl*	*hk*0	0*kl*	*h*0*l*	*h*00	0*k*0	00*l*	
		2	2	2	2	2	2	2	

*In some crystals, planes having these indices comprise of two forms with the same spacing but different structure factor. In such case, the multiplicity factor for each form is half the value given here.

On the other hand, the direction of crystal lattice is given by any coordinates $u\ v\ w$ on a line passing through the origin. Note that the indices are not necessarily integer, because this line will also pass through the point of $2u\ 2v\ 2w$, etc. Nevertheless, the direction is described using the method based on the Miller indices. For example, translation of the origin is carried out to a certain point, and if the set of $u\ v\ w$ is found the minimum integer, when the shift is made by moving the point by ua in the direction of a-axis, vb in the direction of b-axis, and wc in the direction of c-axis, the indices of the direction of the line is expressed as $[u\ v\ w]$ in a square bracket. Negative indices are written with a bar over the number, for example, $[\bar{u}\,\bar{v}\,\bar{w}]$. The equivalent direction related by symmetry is called "directions of a form" and described by $\langle u\,v\,w\,\rangle$, similar to the plane case. As already described, the direction indices are not necessarily integer. Nevertheless, since all of $[\frac{1}{2}\frac{1}{2}1]$, [112], [224], etc. indicate the same direction, they are usually described by [112]. For convenience, some examples of the plane indices and direction indices are shown in Fig. 2.8.

With respect to the hexagonal system, a slightly different method for plane indexing is employed: the so-called Miller–Bravais indices refer to plane indices with four axes such as $(h\ k\ i\ l)$, instead of Miller indices. The unit cell of a hexagonal lattice is given by two equal and coplanar vectors of a_1 and a_2 with 120° to one another, and a third axis c at right angle, as shown in Fig. 2.9. In addition, the third axis of a_3, lying on the basal plane of the hexagonal prism, is symmetrically related to a_1 and a_2 and then it is often used with the other two. The complete hexagonal lattice is obtained by repeated translations of the points at the unit cell corners by

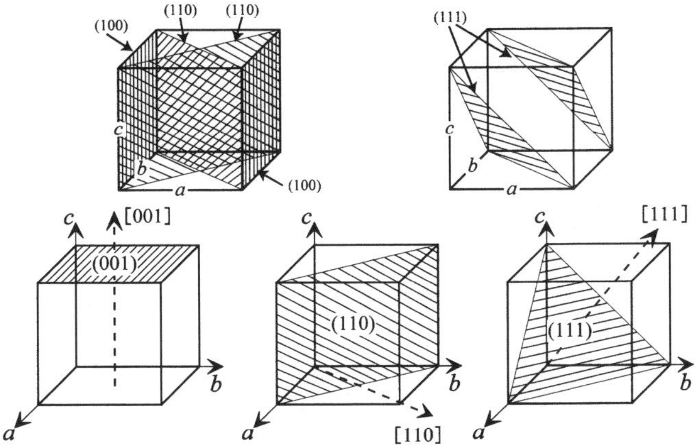

Fig. 2.8 Example of some indices for planes and directions in cubic system

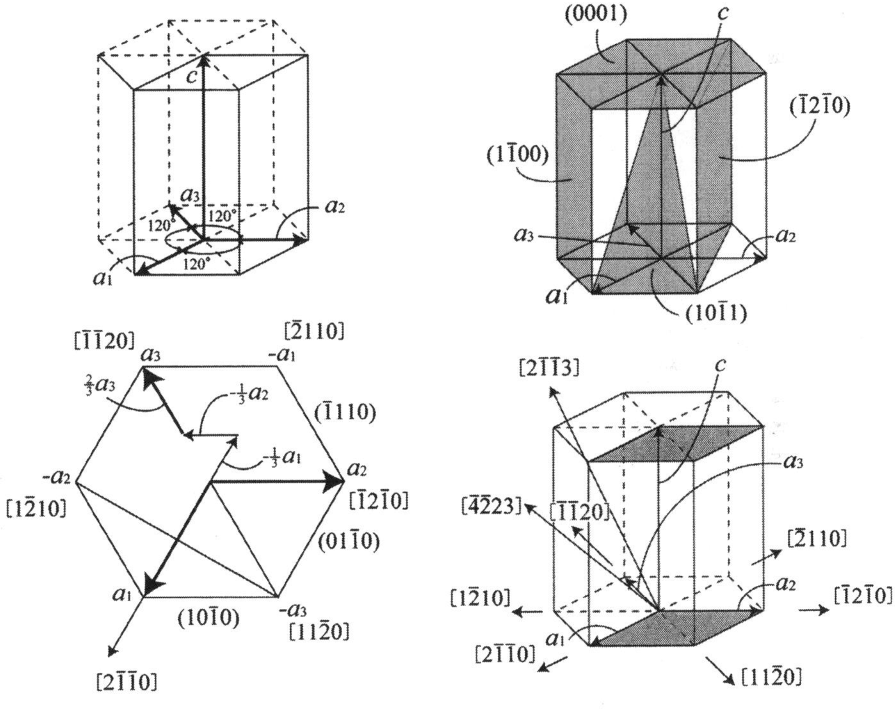

Fig. 2.9 Example of the indices for unit cell, planes, and directions in hexagonal system

the vectors a_1, a_2, and c. It is noted for the Miller–Bravais indices that the relation of $i = -(h + k)$ between h and k is always satisfied. This is because the value of i depends on the h and k values, since the intercepts of a plane on a_1 and a_2 determine its intercept on a_3 (see Fig. 2.9).

The indices of directions in the hexagonal system, the notation using four indices $[u\ v\ t\ w]$, is used, and in this case there is a relation $t = -(u + v)$. The indices of some planes and directions in the hexagonal system are illustrated in Fig. 2.9. More details such as interaction of three indices and four indices in the hexagonal crystal lattice are given in Question 2.11.

2.3 Planes of a Zone and Interplanar Spacing

As shown earlier, there are sets of equivalent planes by symmetry in any crystal lattice and they are called planes of a form. Atoms in crystals can be arranged not only on a lattice plane but also on a group of straight lines which are mutually parallel. This straight line is called "zone axis" and all the planes parallel to the direction of this line are called "planes of a zone." Such planes have quite different indices and spacings, but their parallelism to a line is satisfied. For example, the plane of a zone which belongs to the zone axis [001] in a cubic system is shown in Fig. 2.10.

If a plane belongs to a zone whose indices are $(h\ k\ l)$ and the indices of zone axis are $[u\ v\ w]$, the following relation is satisfied:

$$hu + kv + lw = 0. \tag{2.1}$$

Let us consider any two planes to be planes of a zone, when they are both parallel to their line of intersection. If these two planes are denoted by $(h_1\ k_1\ l_1)$ and $(h_2\ k_2\ l_2)$, the indices of their zone axis $[u\ v\ w]$ are given by the following relations:

$$u = k_1 l_2 - k_2 l_1, \qquad v = l_1 h_2 - l_2 h_1, \qquad w = h_1 k_2 - h_2 k_1. \tag{2.2}$$

The value of the interplanar spacing d is a function of both the plane indices $(h\ k\ l)$ and the lattice parameters $(a,\ b,\ c,\ \alpha,\ \beta,\ \text{and}\ \gamma)$. The relationship between the plane indices $(h\ k\ l)$ and the interplanar spacing d depends on crystal systems.

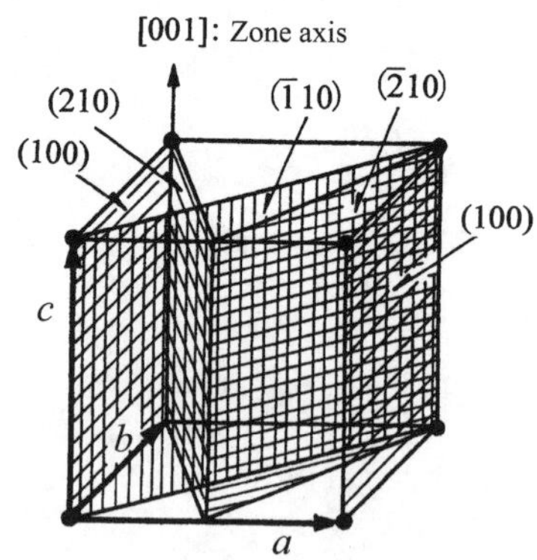

Fig. 2.10 Example of the planes belonging to the zone axis [001] in cubic system

For example, the interplanar spacing d of the plane of $(h\ k\ l)$ for the cubic and tetragonal systems is given in the following equations:

$$d = \frac{a}{\sqrt{h^2 + k^2 + l^2}} \qquad \text{(cubic)} \tag{2.3}$$

$$d = \frac{a}{h^2 + k^2 + l^2(a^2/c^2)} \qquad \text{(tetragonal)} \tag{2.4}$$

It may be worth mentioning that lower indices the plane has, larger the value of interplanar spacing becomes, and the density of the lattice points in the corresponding plane also becomes large.

2.4 Stereographic Projection

To display the angular relationships between planes and directions in a crystal distributed over three dimensions, various methods are employed. For this purpose, the stereographic projection based on spherical projection is very common in crystallography, because this projection method enables us to permit graphical solution of angular problems between planes.

In spherical projection, the direction of a plane when placing the crystal at the center of the sphere is represented by a point that the straight line drawn in the direction that passes through the center of the sphere intersects the surface of the sphere. The sphere is called a reference sphere or a projection sphere. The direction of any plane can be represented by the inclination of the normal to that plane. Then, all the planes in a crystal can be described by a set of plane normals radiating from one point within the crystal. If a reference sphere is placed about this point, the plane normals intersect the surface of the sphere in a set of points called poles. The pole position on the sphere represents the direction of the corresponding plane. The plane can also be represented by the trace (line) the extended plane makes in the sphere surface.

The spherical projection can accurately represent the symmetry of the angular relationships between planes and directions as well as zone, but the use of sphere is not always convenient, because the measurement of angles on a flat sheet is more convenient in comparison with measurements on the surface of a sphere. For this purpose, the stereographic projection is widely used. The method is similar to that used by geographers who want to transfer a world map from a terrestrial globe to a sheet of an atlas. Particularly, the equiangular stereographic projection is preferred in crystallography, because it preserves angular relationships faithfully, although area is distorted.

Stereographic projection is one of the perspective projection methods. As shown in Fig. 2.11, the projection plane is normal to the line NS that connects two poles, N (north pole) and S (south pole), and at the midpoint of the diameter AB of projection sphere a crystal C is placed. A light source is set at S (south pole), whereas the observer views the projection from N (north pole) just opposite the light source.

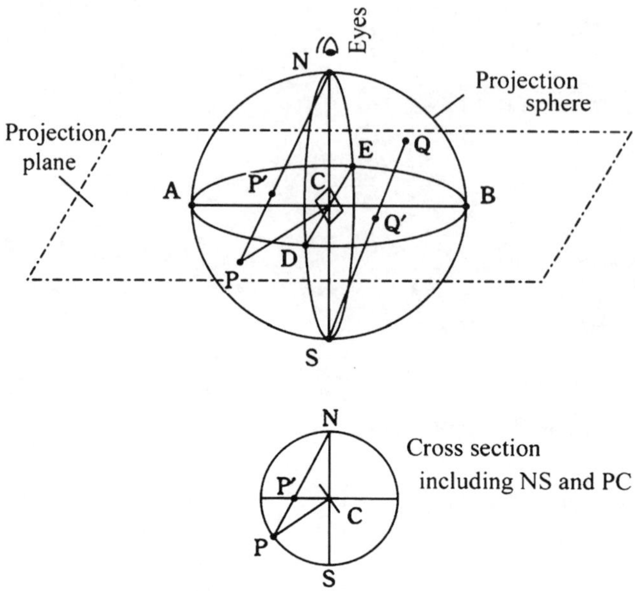

Fig. 2.11 Fundamentals of stereographic projection

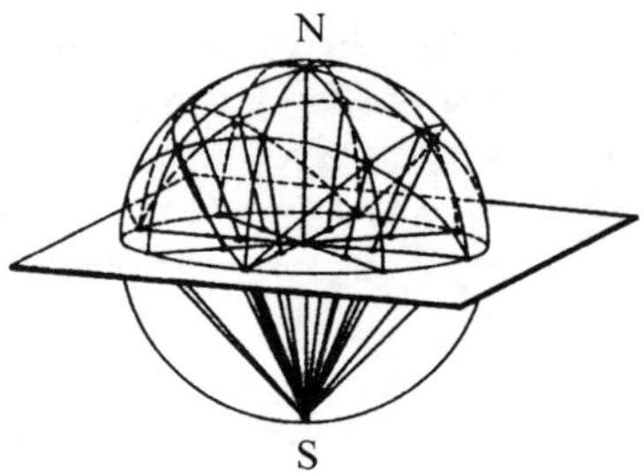

Fig. 2.12 The intersection of lines drawn between the poles of planes in the northern hemisphere and the south poles is recorded on the equatorial plane in the stereographic projection

If a certain plane of the crystal has its pole at P, the stereographic projection of P can be obtained as P′, by drawing the line NP, and it will intersect the projection plane. Alternatively stated, if a pole P is located in the southern hemisphere, its stereographic projection P′ is made from the arctic (north pole) N being the point of perspective and P′ corresponds to the intersection of a straight line NP with a projection plane. If a pole Q is in the northern hemisphere, consider the intersection Q′ with the straight line SQ which makes the Antarctic (south pole) S the point of perspective. It may be added that the stereographic projection of the pole Q is the shadow cast denoted by Q′ on the projection plane when a light source is placed at S. As shown in Fig. 2.12, a line for each of the poles in the northern hemisphere is projected to the south pole and its intersection with equatorial plane of the equator can be marked with a point.

By this method, all poles can be depicted inside an equatorial circle (basic circle). In this case, it is required to distinguish the projecting point with N being the point of perspective and the projecting point with S being the point of perspective. Such issue is easily resolved by using different symbols, for example, • for the former and ∘ for the latter. Great circles on the reference sphere project as circular arcs, whereas small circles project as circles, but their projected center does not coincide with their projection center. One can also select any arbitrary plane perpendicular to NS besides the equatorial plane as a projection plane. In this case, only the diameter of the basic circle changes but the relative positions of projections are unchanged.

The net graphics obtained by projecting meridian circles and latitude circles at every 1° or 2° on the equatorial plane is called polar net. Polar net is used for obtaining the projecting point on the equatorial plane with respect to a point on a projection sphere. When considering a terrestrial globe, the longitudinal lines correspond to great circles, whereas the latitude lines are small circles, except the equator. The net graphics obtained by projecting the meridian circles and latitude circles on one meridian circle is called Wulff net. In this case, the longitude lines are drawn by the stereographic projection of the great circles connecting the north and south poles of the net at interval of 2°, and they are displayed with thick arc (line) at every 10°. The latitude lines on the Wulff net are obtained by the stereographic projection of the small circle extending from side to side at intervals of 2°, corresponding to the intersection of a projection sphere with a plane perpendicular to NS axis. The Wulff net is quite convenient for estimating the angle between two planes.

In the analysis using the Wulff net, a tracing paper containing the stereographic projection is usually placed on a copy of the Wulff net and the centers are made coincident and fixed by a tack. The stereographic projection is made on a tracing paper with the basic circle of the same diameter as that of the Wulff net. The center of the stereographic projection always coincide with the Wulff net center. Although the uncertainty of the angle determined from the Wulff net analysis is about 1°, it is sufficient in most cases. It should be remembered that the angle between two poles is taken to average the angle between two normals, n and it is not the dihedral angle (ϕ) between the corresponding two planes P_1 and P_2. Since the poles of the planes lie on great circle (the zone circle), these two angles are simply related as $n = 180° - \phi$, as shown in Fig. 2.13. Some essential points of the stereographic projection are given below.

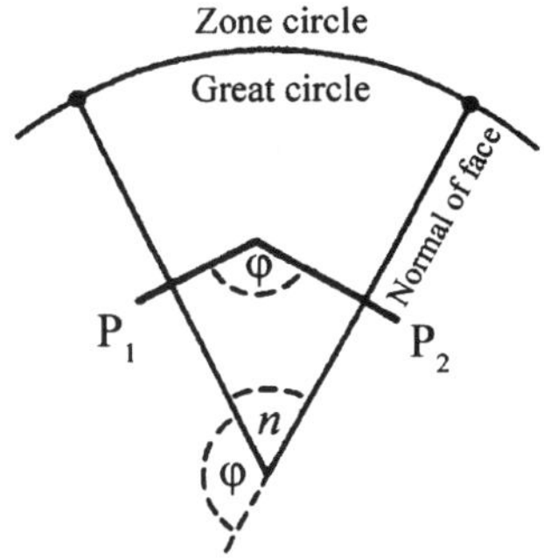

Fig. 2.13 Relationship between the angle of intersection of normal and the dihedral angle formed by two planes P_1 and P_2

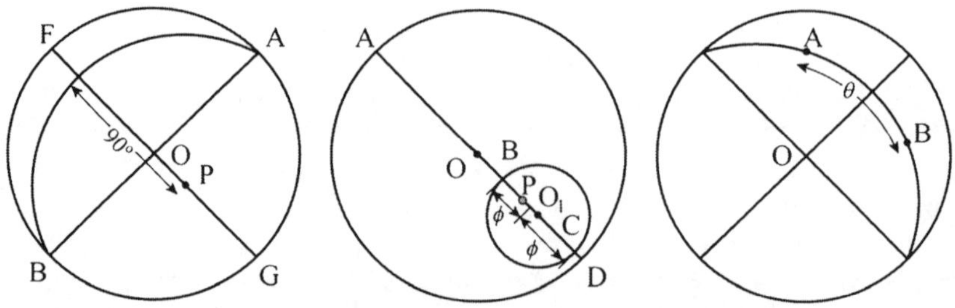

Fig. 2.14 Methods for obtaining (**a**) the pole of great circles, (**b**) the pole of small circles, and (**c**) the angle between the two lattice planes A and B using the stereographic projection

In the stereographic projection, circles on the reference (projection) sphere project as circles or two circular arcs, if they do not pass through the points, N and S, being the point of perspective. Whereas, circles project as straight lines through the center of the projection, if they pass through the points, being the perspective point. In other words, projected great circles always pass the intersection between the basic circle and the straight line passing through a center of the basic circle. This is called "theorem of corresponding circle to circle." On the other hand, the angle given by two stereographically projected great circles (it also includes the straight line case) is equal to the spherical angle of two great circles on the reference sphere. This is called "theorem of conformal mapping."

To obtain the pole of a great circle on the projection plane, with respect to a great circle ACB in Fig. 2.11 (see also Fig. 2.14a), the diameter FOG which is perpendicular to the diameter AOB is put on the equator of the Wulff net, and P is taken from C at 90°. On the other hand, to obtain the pole P of a small circle on the projection plane (see Fig. 2.14b), the diameter $AOBO_1CD$ is taken with the center of a small circle as O_1; then, this is put on the equator of the Wulff net and set point P dividing equally the angle between BC into two parts. To know the angle between the lattice planes A and B, put on the projection on the Wulff net at first, and as shown in Fig. 2.14c, it is rotated and it is made for A and B so as to get on one meridian circle. Next, if we measure the value of angle on the meridian, it is equivalent to the angle of interest.

Rotation of the stereographic projection is readily made by using the polar net and the Wulff net. In addition, the so-called standard projection is very useful for discussing problems of crystal orientation, because it gives the relative orientation of all the important planes in the crystal at a single look. Such projections are obtained by choosing some important lattice planes of low indices as the projection plane, such as the lattice planes of (110), (100), (111), or (0001), which are frequently encountered. In this process, the projection of main poles is made by placing a crystal so as to coincide the directions of [100], [110], [111], or [0001] with the north–south NS axis given in Fig. 2.11. Some standard projections, such as cubic crystals on (001) and on (011) are available in textbooks or handbooks for X-ray analysis. When this is utilized, one can obtain information about the relative relationships of main lattice planes in a crystal. It is also extremely convenient for dealing with problems of crystal orientation. Some selected examples are given in Questions 2.19–2.22.

2.5 Solved Problems (21 Examples)

Question 2.1 Illustrate (100), (110), (111), and (112) planes in cubic lattice and direction indices of [010], [111], $[\bar{1}00]$, and [120].

Answer 2.1 About given Miller indices, planes are shown in Fig. 1 and directions are illustrated in Fig. 2.

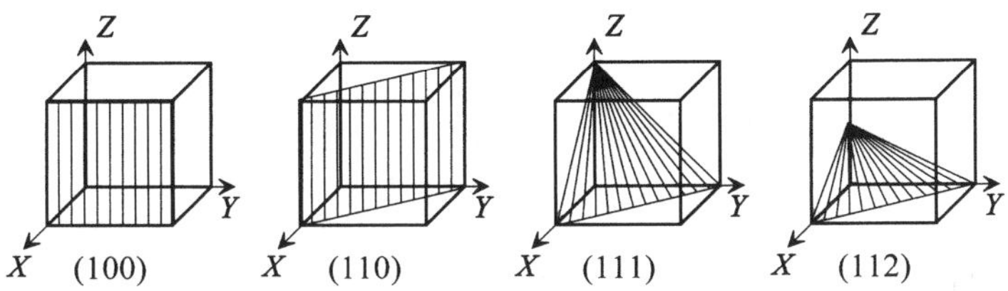

Fig. 1 Examples of the plane indices in cubic lattice

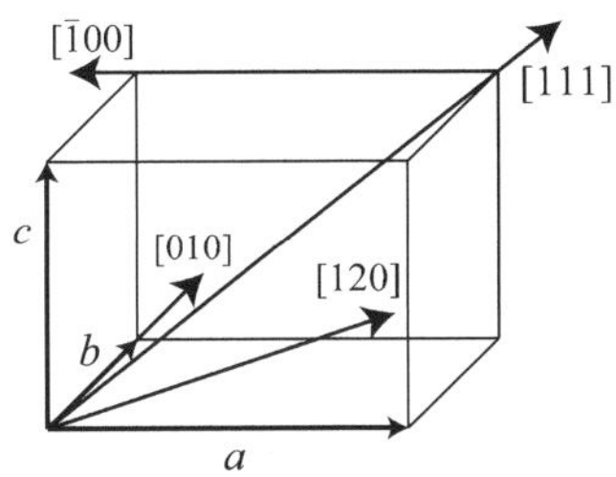

Fig. 2 Examples of the direction indices in cubic lattice

In addition, there are sets of equivalent lattice planes related by symmetry, for example, the faces of a cube, (100), (010), $(\bar{1}00)$, $(0\bar{1}0)$, (001), and $(00\bar{1})$. They are called planes of a form and the expression of {001} is used. Similarly, with respect to the direction, it is shown for [110], [101], [011], etc. as $\langle 110 \rangle$.

Question 2.2 Answer the following questions about body-centered cubic (bcc) structure with the lattice parameter "a."

(1) Obtain the volume of void, supposing the case where the spherical atoms of radius r_A are arranged in each lattice point. Calculate also the porosity and packing fraction.

(2) The position of the maximum void in this body-centered cubic lattice is known to be corresponding to the tetrahedral site (1/2, 1/4, 0), and to equivalent position. Obtain the radius of maximum sphere that fits to this space.

Answer 2.2

(1) In body-centered cubic (bcc) structure, atoms contacting each other are seen on diagonals and then we obtain the following relationship:

$$4 \times r_{\mathrm{A}} = a\sqrt{3}$$
$$r_{\mathrm{A}} = \frac{\sqrt{3}}{4}a$$

In a unit cell of bcc structure, there are two atoms: one atom at eight corners $(8 \times 1/8 = 1)$ and one atom at the center. Therefore, the volume V_{A} occupied by atoms is described by

$$V_{\mathrm{A}} = 2 \times \frac{4}{3}\pi\left(\frac{\sqrt{3}}{4}a\right)^3 = \frac{\sqrt{3}\pi}{8}a^3.$$

The unit cell volume is expressed by a^3, so that the void volume V_{H} is as follows:

$$V_{\mathrm{H}} = a^3 - V_{\mathrm{A}} = \left(1 - \frac{\sqrt{3}}{8}\pi\right)a^3.$$

The porosity in the bcc lattice is given in the following:

$$\left(\frac{V_{\mathrm{H}}}{a^3}\right) = \left(1 - \frac{\sqrt{3}}{8}\pi\right)^3 = 0.32.$$

Therefore, the packing fraction of bcc lattice is 0.68.

(2) If the radius of the sphere which fits to void is r_{X}, the following relationship is obtained by geometric conditions (see Fig. 1):

$$(r_{\mathrm{A}} + r_{\mathrm{X}})^2 = \left(\frac{a}{4}\right)^2 + \left(\frac{a}{2}\right)^2.$$

Next, using the relation of $r_{\mathrm{A}} = \frac{\sqrt{3}}{4}a$ in the bcc lattice, the radius of the maximum sphere which fits to void is given by following equations:

$$\left(\frac{\sqrt{3}}{4}a + r_{\mathrm{X}}\right)^2 = \left(\frac{a}{4}\right)^2 + \left(\frac{a}{2}\right)^2 = \left(\frac{5}{16}a\right)^2, \qquad r_{\mathrm{X}} = \left(\frac{\sqrt{5} - \sqrt{3}}{4}\right)a$$

Reference: You will understand that the maximum radius which fits to void in the bcc lattice is about 30% of the radius of the constituent atom in the following result:

$$\frac{r_X}{r_A} = \frac{\left(\frac{\sqrt{5}-\sqrt{3}}{4}\right)a}{\frac{\sqrt{3}}{4}a} = \frac{\sqrt{5}}{\sqrt{3}} - 1 = 0.29.$$

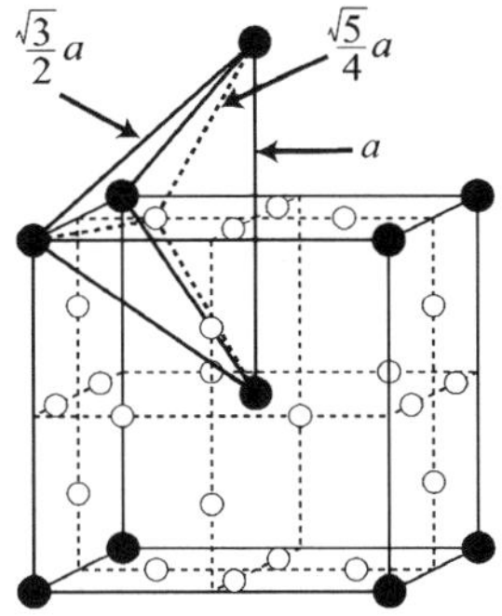

Fig. 1 Tetrahedral voids in the bcc lattice. (*Filled circle*) Metal atoms, (*open circle*) Tetrahedral void

Question 2.3 At 278 K, iron (Fe) is found to show bcc structure with a lattice parameter of 0.2866 nm. Obtain the density of iron from this information.

Answer 2.3 The bcc structure includes two atoms per unit cell. If Avogadro's number is N_A, one mole of iron includes $N_A/2$ unit cells. Therefore, the volume V per mole of Fe (atomic volume) is given by

$$V = \frac{(0.2866 \times 10^{-9})^3}{0.6022 \times 10^{24}/2}.$$

The atomic weight M (molar mass) per 1 mol of Fe is 55.845 g is obtained from Appendix A.2. Therefore, from the relationship of $V = M/\rho$, we can estimate the density value of ρ as follows:

$$\rho = \frac{55.845 \times 2}{(0.2866 \times 10^{-9})^3 \times 0.6022 \times 10^{24}} = 7.88 \times 10^6\ \text{g/m}^3.$$

Reference: The experimental value of density for Fe is 7.87×10^6 g/m^3. Since some defects such as vacancy and dislocation are usually included in an actual crystal, there are some differences between the density estimated from the X-ray structure data and the experimental value.

Question 2.4 Beryllium (Be) mineral is expressed by a chemical formula ($3BeO \cdot Al_2O_3 \cdot 6SiO_2$), and it is revealed that the structure is hexagonal with the lattice parameters $a = 0.9215\,\text{nm}$ and $c = 0.9169\,\text{nm}$, and density $2.68 \times 10^6\,\text{g/m}^3$. Obtain the numbers of molecules contained in a unit cell.

Answer 2.4 At first, we obtain the molecular weight of beryllium mineral using a chemical formula from the molecular weight per 1 mol of the individual oxide component:

$$BeO = 25.01\,\text{g}, \qquad Al_2O_3 = 101.96\,\text{g}, \qquad SiO_2 = 60.08\,\text{g}$$
$$3BeO + Al_2O_3 + 6SiO_2 = 537.47\,\text{g/mol}$$

If this molecular weight is divided by Avogadro's number, one obtain the value equivalent to the weight of one beryllium mineral molecule.

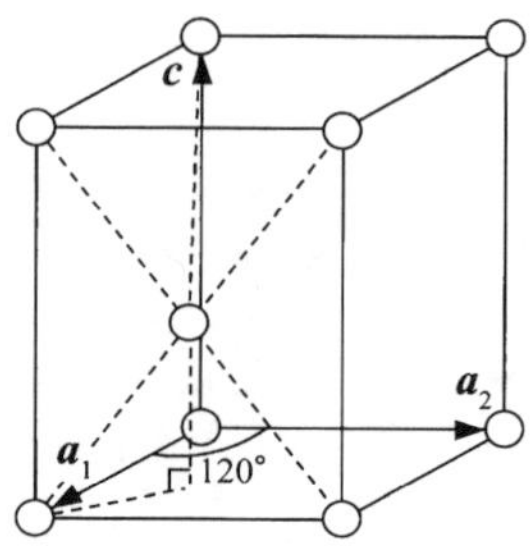

Fig. 1 Geometric feature found in hexagonal system

Next, we estimate the volume of a unit cell for beryllium mineral from the given values of lattice parameters. As readily seen in Fig. 1, in a unit cell of hexagonal system, the value of c is given by twice the height of $\left(\sqrt{\frac{2}{3}}a\right)$ for the regular tetrahedron of length a of one side, and the area of the parallelogram which corresponds to the base is given by $\left(\sqrt{\frac{3}{4}}a^2\right)$. Therefore, the volume V of a unit cell of hexagonal system is given in the following equation (see also Appendix A.6):

$$\begin{aligned} V &= \frac{\sqrt{3}}{2}a^2c = 0.866a^2c \\ &= 0.866 \times (0.9215 \times 10^{-9})^2 \times (0.9169 \times 10^{-9}) = 0.6743 \times 10^{-27}\,[\text{m}^3]. \end{aligned}$$

The product of the volume of a unit cell and the density corresponds to the weight of one beryllium mineral molecule, so that if this value is compared with the value calculated from molecular weight and Avogadro's number, the number of molecules in a unit cell will be obtained:

$$\frac{0.6743 \times 10^{-27} \times 2.68 \times 10^{6}}{\left(\frac{537.47}{0.6022\times 10^{24}}\right)} = 2.02.$$

Thus, the number of molecules in a unit cell is estimated to be two.

Question 2.5 (1) Illustrate that a trigonal cell (it is also called rhombohedral cell) is recognized in the face-centered cubic (fcc) cell. (2) The fcc structure is known in the close-packed arrangement of spheres with the identical size and its layer stacking sequence of ABCABC type. Illustrate that the layer stacking sequence of ABAB type found in the hexagonal close-packed (hcp) structure is also detected.

Answer 2.5 If six atoms in the six cell faces of the fcc lattice and two atoms of both sides of a diagonal line in a cubic are tied, a trigonal (rhombohedral) cell is built up as shown in Fig. 1. When the lattice parameter of fcc lattice is set as "a'," the lattice parameter of trigonal lattice is given by $a'/\sqrt{2}$. The lattice parameters of the trigonal system are

$$a = b = c = a'/\sqrt{2}, \quad \alpha = \beta = \gamma = 60^{\circ}.$$

Next, if you look at the (111) plane centering on the [111] direction of the fcc lattice, a part of the (0002) plane of hexagonal close-packed (hcp)-type stacking will be recognized, as seen in Fig. 2. In this case, when two layers corresponding to the (111) plane are considered as A and B layers, the atomic position of the corner of a cubic lattice in the diagonal line corresponds to the C layer.

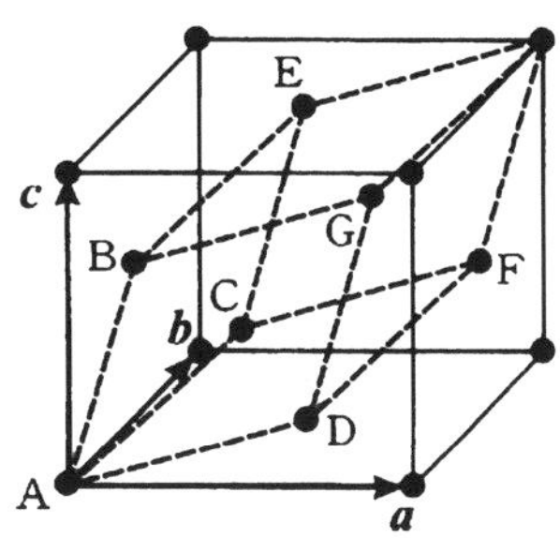

Fig. 1 The trigonal (rhombohedral) lattice recognized in fcc lattice

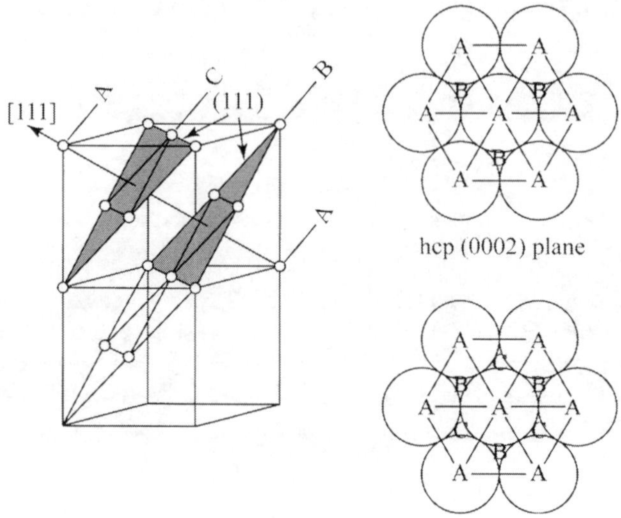

Fig. 2 Stacking of atoms in fcc lattice and hcp lattice

Question 2.6 The atomic weight per 1 mol of copper (Cu) with face-centered cubic (fcc) structure and the density at 298 K are 63.54 g and 8.89×10^6 g/m^3, respectively. Estimate the nearest-neighbor distance of Cu atoms.

Answer 2.6 In the fcc lattice, four atoms are known to be included in a unit cell. When Avogadro's number is denoted by N_A, 1 mol Cu (63.54 g) includes $N_A/4$ unit cells. If the lattice parameter is set as "a," the volume of 1 mol Cu (=the atomic volume) V can be expressed as $V = a^3 N_A/4$. On the other hand, we obtain the relationship of $a^3 N_A/4 = M/\rho$; using the atomic weight M and density ρ, the lattice parameter can be estimated as follows:

$$a^3 = \frac{4 \times 63.54}{0.6022 \times 10^{24} \times 8.89 \times 10^6} \qquad a = 3.621 \times 10^{-10}\,\text{m}.$$

The nearest-neighbor distance r of Cu atoms can be calculated since Cu atoms are in contact along the diagonal line of a cell face in the fcc structure:

$$r = a/\sqrt{2} = 2.560 \times 10^{-10}\,\text{m} = 0.2560\,\text{nm}.$$

Question 2.7 Answer the following questions about gold (Au), which has fcc structure with the lattice parameter $a = 0.4070$ nm.

(1) Obtain the nearest-neighbor distance, the second nearest-neighbor distance, and their coordination numbers.
(2) Obtain the values of density and packing fraction when the density of gold atoms are considered as hard spheres.
(3) Obtain the maximum radii of the spheres which just fit the octahedral and tetrahedral voids produced in the fcc structure consisting of hard spheres.

Answer 2.7

(1) When you look at the characteristic feature of the fcc lattice, the distance between the atoms which occupy the corners of a unit cell is equal to the lattice parameter "a," and the distance between the atom located at a center position of the cell face and the atom which occupies a corner position is $a\sqrt{2}/2 = a/\sqrt{2}$. Thus, the nearest-neighbor distance is estimated to be $r_1 = a/\sqrt{2}$ and the second nearest-neighbor distance is $r_2 = a$. It is seen that $a/\sqrt{2} < a$.

$$r_1 = \frac{a}{\sqrt{2}} = \frac{0.4070}{1.4142} = 0.2878\,\text{nm} \qquad r_2 = a = 0.4070\,\text{nm}.$$

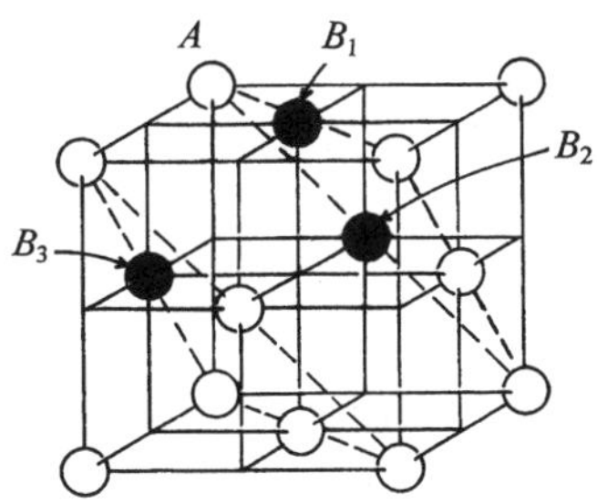

Fig. 1 Geometric feature found in fcc lattice

With respect to the coordination number in the nearest-neighbor distance, let us consider the atom A occupying the corner position (see Fig. 1). At the distance of r_1 from atom A, there are four atoms marked by B_1 which occupy the center position of the cell face, considering the four equivalent planes for a unit cell around A. One can also find two planes cross at right angles to this face and consider the situation in both upper and lower sides. For example, there are four atoms marked by B_2 in both sides and similarly we find four atoms corresponding to B_3. Thus, the total nearest neighbors are $3 \times 4 = 12$.

You may also estimate the coordination numbers in nearest-neighbor region from the characteristic features of the fcc lattice, which has close-packed

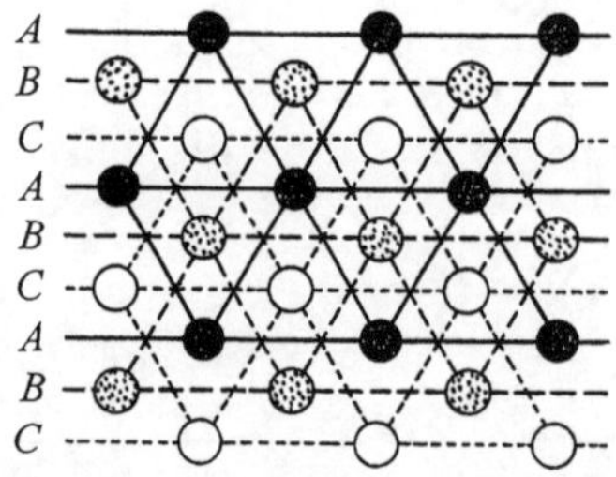

Fig. 2 Stacking of atoms in fcc lattice

arrangement of spheres with the identical sizes and layer stacking sequence of ABCABC type. When considering the environments with respect to the sphere which is located in the layer B with the help of Fig. 2, there are three atoms in both layers of A and C and six atoms in layer of B, so that we can find of total of 12 atoms in the nearest-neighbor distance. For the second nearest-neighbor case, let us consider the environments around the A atom occupying the corner position in Fig. 1. There are four atoms in the plane and two atoms at both sides. Therefore, the coordination number of the second nearest-neighbor atoms is 6.

(2) The volume V of a unit cell for fcc structure is given by a^3. On the other hand, one atom located at a corner is shared by eight cells and atom in the face center is shared by two cells. Then the number of atoms "n" which belongs to a unit cell is $n = \frac{1}{8} \times 8 + \frac{1}{2} \times 6 = 4$. Since the atomic weight of 1 mol. Au is 196.97 g from Appendix A.2, the mass m of one atom of Au can be computed using Avogadro's number as follows:

$$m = \frac{196.97}{0.6022 \times 10^{24}} = 327.0 \times 10^{-24}\ [\text{g}].$$

Then, we obtain the density value of ρ:

$$\rho = \frac{4m}{a^3} = \frac{4 \times (327.0 \times 10^{-24})}{(0.4070 \times 10^{-9})^3} = \frac{1.308 \times 10^{-21}}{0.0674 \times 10^{-27}} = 19.41 \times 10^6\ \text{g/m}^3.$$

(**Reference:** Measured density value $= 19.28 \times 10^6\ \text{g/m}^3$.)
In fcc structure, the atoms are in contact along the diagonal line of a cube face, so that the lattice parameter a is related to the atomic radius r by the relation $4r = a\sqrt{2}$:

$$r = \frac{\sqrt{2}}{4}a \quad \rightarrow \quad r = \frac{a}{2^{3/2}}.$$

The volume V' of four atomic spheres contained in a unit cell is obtained:

$$V' = 4 \times \left(\frac{4}{3}\pi r^3\right) = 4 \times \left(\frac{4}{3}\pi\right) \times \left(\frac{a}{2^{3/2}}\right)^3 = \frac{16}{3}\pi \times \frac{a^3}{16\sqrt{2}} = \frac{\pi a^3}{3\sqrt{2}}.$$

Since the volume V of a unit cell is given by a^3, the packing fraction η is obtained from the relation of V'/V:

$$\eta = \frac{\pi a^3}{3\sqrt{2}} \Big/ a^3 = \frac{\pi}{3\sqrt{2}} = 0.741.$$

(3) Let us consider the case where the hard sphere with the atomic radius $r = a/2^{3/2}$ is arranged in the fcc lattice. As shown in Fig. 3a, the void is found at the center of a unit cell as well as the midpoint of each edge-line. The void at the center of unit cell is surrounded by six spheres and constitutes the octahedral interstitial site. Since such void position forms the fcc lattice with the lattice parameter "a," the octahedral void has four equivalent positions per unit cell.

The tetrahedral void of the regular tetrahedron surrounded by four spheres is shown in Fig. 3b. Such tetrahedral voids form a simple cubic lattice with the lattice parameter $a/2$. This tetrahedral void has eight equivalent positions per unit cell.

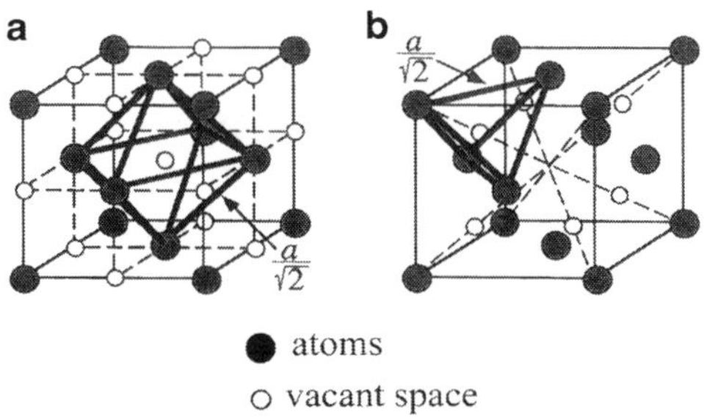

Fig. 3 (**a**) Octahedral void and (**b**) tetrahedral void in fcc lattice

Next, we will estimate the maximum radius of the sphere which is just fit to the octahedral void, r_o, and tetrahedral void, r_t, in fact structure. With the help of the relationships found in Fig. 3a, geometry of the octahedral void is illustrated in Fig. 4 and we obtain the value of r_o in the following way:

$$\mathrm{AC} = 2(r + r_o) = (\mathrm{AB}^2 + \mathrm{BC}^2)^{1/2} = 2\sqrt{2}r$$
$$r_o = r(\sqrt{2} - 1) = 0.414r$$

With respect to the tetrahedral voids, a relationship found in Fig. 5 is helpful. For triangle ABC, we find $\mathrm{AC} = a\sqrt{2}$ and $\mathrm{AB} = \sqrt{3}r$. The relationship of AB = BC is also recognized.

Considering triangle ABM (or CBM), the distance BM is given as $\mathrm{BM} = \sqrt{2}r/2$. When the center of the tetrahedral void is denoted by O, the position of O corresponds to the center of the distance BM, so that it is given by $\mathrm{OM} = \sqrt{2}r/2$.

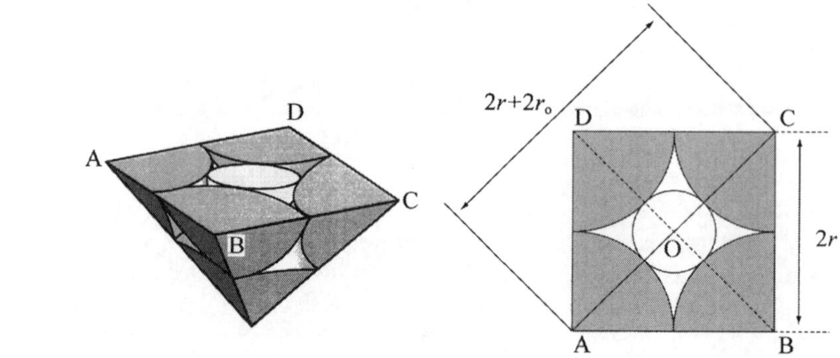

Fig. 4 Geometry of the octahedral void in fcc lattice

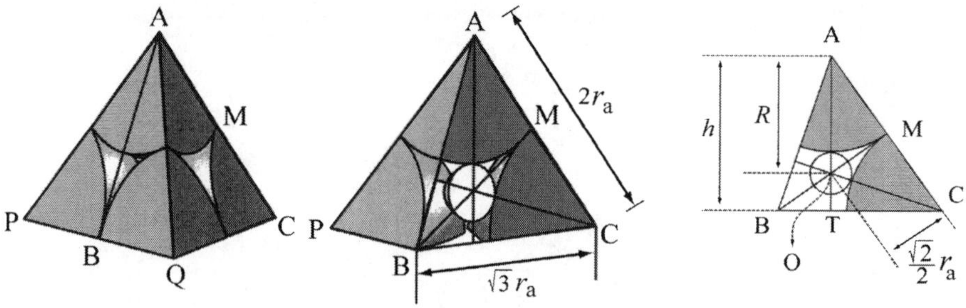

Fig. 5 Geometry of the tetrahedral void in fcc lattice

On the other hand, we take triangle AOM that AM is a half of AC and $r = a/2^{3/2}$ and the angle AMO is 90 degrees, so that from the relationship of $\mathrm{AO}^2 = \mathrm{AM}^2 + \mathrm{OM}^2$, we obtain

$$\begin{aligned}\mathrm{AO}^2 &= \left(\frac{a}{2^{3/2}}\right)^2 + \left(\frac{\sqrt{2}r}{2}\right)^2 = \left(\frac{2\sqrt{2}r}{2\sqrt{2}}\right)^2 + \left(\frac{r}{\sqrt{2}}\right)^2 \quad (\because a = 2\sqrt{2}r)\\ &= \left(1 + \frac{1}{2}\right) r^2 = \frac{3}{2} r^2\\ \mathrm{AO} &= \sqrt{\frac{3}{2}} r.\end{aligned}$$

Using the relationship of $\mathrm{AO} = r + r_t$, the desired value is obtained as follow:

$$r + r_t = \sqrt{\frac{3}{2}} r, \quad r_t = \left(\sqrt{\frac{3}{2}} + 1\right) r = 0.225r.$$

From these results, we obtain that a sphere of radius about 41% of main constituent atoms can fit into the octahedral void in fcc structure and a sphere with radius about 23% of main constituent atoms can fit into the tetrahedral void.

Question 2.8 If spheres of equal size are used to fill space, there are two ways for arranging spheres; in square form and in hexagonal form.

(1) Compute the percentage of void of these two cases for two-dimensional array of spheres.
(2) Explain the packing fraction of three-dimensional array of spheres.

Answer 2.8 In two-dimensional array of spheres, each sphere is found to be in contact with four spheres in the square form. On the other hand, each sphere contacts with six spheres in the hexagonal form, as illustrated in Fig. 1. This implies that the coordination numbers of the nearest neighbors are 4 in the square form and 6 in the hexagonal form, respectively.

(1) When the radius of a sphere is given by r, the area produced by one set of sphere array in the square form is expressed as $2r \times 2r = 4r^2$. Since the area occupied by a sphere is πr^2, the percentage of void area A_V in the square form is as follows:

$$A_V = \frac{4r^2 - \pi r^2}{4r^2} = 1 - \frac{\pi}{4} = 0.215.$$

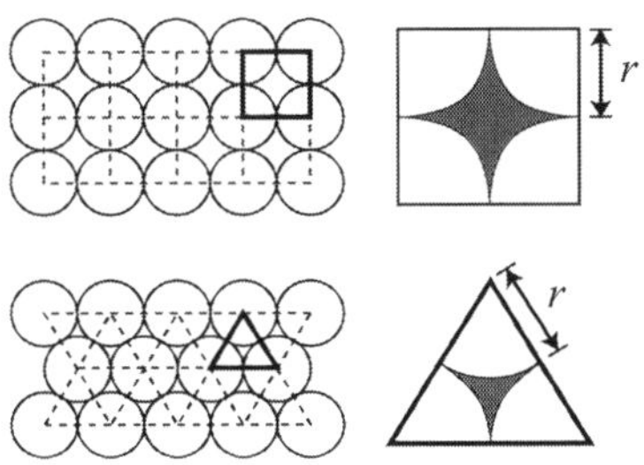

Fig. 1 Two-dimensional array of spheres in the square and hexagonal forms

In the hexagonal form, our notice focuses the area of an equilateral triangle with one edge-line of $2r$ (see Fig. 1). Since the height of the equilateral triangle is $\sqrt{3}r$, the area of this triangle is given by $\sqrt{3}r^2$. The sum of an (adjacent) interior angle of the equilateral triangle is 180°. This suggests that the area occupied by a sphere is expressed with $\pi r^2/2$, because it is equivalent to half of a sphere. Therefore, the percentage of void area A_V in the hexagonal form is

$$A_V = \frac{\sqrt{3}r^2 - \dfrac{\pi r^2}{2}}{\sqrt{3}r^2} = 1 - \frac{\pi}{2\sqrt{3}} = 0.093.$$

(2) With respect to the packing fraction of three-dimensional array of the equal size spheres, the value in the hexagonal form is higher than the square case. This is readily found from the results of Question (1).

Question 2.9 Iron (Fe) is present as γ-phase characterized by face-centered cubic (fcc) structure with the lattice parameter of 0.3647 nm at temperatures near 1,273 K. The Fe–C alloy containing 2.0 mass% of carbon (C) can form either the interstitial and substitutional solid solution. Calculate the value of density in these two cases and compare with the experimental value of 7.65×10^6 g/m^3.

Answer 2.9 From Appendix A.2, the atomic weight of Fe and C are 55.845 g and 12.011 g, respectively. For the Fe–C alloy containing 2.0 mass% of carbon (C), the atomic percent of the constituent elements can be calculated.

At first, we compute the molar values of each component such as $98.0/55.845 = 1.7549$ for Fe and $2.0/12.011 = 0.1665$ for C, and $1.7549 + 0.1665 = 1.9214$ for alloy. Then, the desired values are given by the following:

$$\text{Atomic \% of Fe} \quad (1.7549/1.9214) \times 100 = 91.33 \text{ at \%}.$$

$$\text{Atomic \% of C} \quad (0.1665/1.9214) \times 100 = 8.67 \text{ at \%}.$$

Density is mass per unit volume. Taking into consideration that four atoms per unit cell are included in fcc structure, the density can be calculated if Fe forms a solid solution with C of either interstitial or substitutional type.

(1) Density in interstitial-type solid solution is given by (mass of Fe + mass of C)/ unit volume.

$$\begin{array}{ll} \text{Mass} & 4 \times \{55.845 + (8.67/91.33) \times 12.011\} = 227.94\,\text{g} \\ \text{Volume} & (0.3647)^3 \times 10^{-27} \times 0.6022 \times 10^{24} = 29.21 \times 10^{-6}\,\text{m}^3 \\ \text{Density} & 227.94/(29.21 \times 10^{-6}) = 7.80 \times 10^6\,\text{g/m}^3 \end{array}$$

(2) Density in substitutional-type solid solution may be calculated as (mass/unit volume)

$$\begin{array}{ll} \text{Mass} & 4 \times (0.9133 \times 55.845 + 0.0867 \times 12.011) = 208.18\,\text{g} \\ \text{Volumes} & (0.3647)^3 \times 10^{-27} \times 0.6022 \times 10^{24} = 29.21 \times 10^{-6}\,\text{m}^3 \\ \text{Density} & 208.18/(29.21 \times 10^{-6}) = 7.13 \times 10^6\,\text{g/m}^3 \end{array}$$

By comparing the two calculated density values with measured density of 7.65×10^6 g/m^3, it can be concluded that the Fe–C alloy forms an interstitial solid solution.

Question 2.10 Copper (Cu) is known to form substitutional solid solution with nickel (Ni) and show face-centered cubic (fcc) structure. For Cu (lattice parameter: 0.3625 nm) which contains 0.001 mass% of Ni, calculate the distance between Ni atoms in this solid solution.

Answer 2.10 From Appendix A.2, the atomic weights per mole for Cu and Ni are 63.55 and 53.69 g, respectively. First calculate the atomic percent of Cu and Ni in the alloy. The moles of each component are obtained as $0.001/58.69 = 0.000017$ for Cu and $99.999/63.55 = 1.57355$ for Ni, and $1.57355 + 0.000017 = 1.573567$ for the alloy.

$$\text{Atomic \% of Ni} \quad (0.000017/1.573567) \times 100 = 0.0011$$
$$\text{Atomic \% of Cu} \quad 100 - 0.0011 = 99.9989$$

Considering the concentrations of the two components in atomic percent, if there are one million atoms of the alloy in total, 999,989 are Cu atoms and 11 are nickel atoms. This is approximately equivalent to there being one nickel atom as an impurity in 100,000 Cu atoms.

Four atoms are contained in a unit cell of fcc structure, 100,000 Cu atoms form 25,000 unit cells. If Ni atoms are thought to replace the position of Cu atoms at random, one Ni atom will be homogeneously distributed in these 25,000 unit cells. In other words, Ni atoms are separated, at least, by multiples of the unit cell; in the present case, $(25{,}000)^{1/3} = 29.24$ unit cells. Therefore, the distance between the impurity Ni atoms contained in Cu is estimated to be 29.24× the lattice parameter (0.3615 nm) = 10.57 nm. This result suggests that the impurity atoms are located in a relatively close region even for the cases where the impurity content is very small at the level of 0.001%. The impurity effect, for example at the level of 0.001%, cannot necessarily be ignored.

Reference: If there are 100 atoms denoted by A, 25 unit cells are known to be formed in fcc structure. Let us consider the A–B binary alloy containing 25 at.% of B atoms with fcc structure. When B atoms replace the position of A atom at random, one atom per unit cell will be replaced with B atom in this alloy. On the other hand, if B atoms are not distributed randomly, but occupy some designated positions, an ordered structure (frequently called superlattice) will be formed.

Question 2.11 A slightly different method than the usual Miller indices is employed for indexing planes in the hexagonal system. It is called the Miller–Bravais indices. Explain the essential points of the Miller–Bravais indices including some of its merits for indexing both planes and directions.

Answer 2.11 As shown in Fig. 1, a unit cell of a hexagonal lattice is provided by two equal and coplanar vectors a_1 and a_2 which are 120° to one another and the third axis c perpendicular to vectors a_1 and a_2. The complete hexagonal lattice is given by repeated translations of the points at the unit cell corners by the vectors a_1, a_2, and c. The plane indexed by the usual Miller indices is allowed for hexagonal system. However, it can sometimes cause confusion. For example, "Is the plane of (100) completely equivalent to the plane of $(\bar{1}10)$?" Miller–Bravais indices refers to plane indices with four axes such as $(h\ k\ i\ l)$, instead of Miller indices. Then, the use of Miller–Bravais indices enables us to provide one way by showing (100) $\rightarrow$ $(1\bar{1}00)$ and $(\bar{1}10) \rightarrow (10\bar{1}0)$, respectively.

The key points of Miller–Bravais indices for hexagonal system are summarized below:

(1) As shown in Fig. 1, we employ four axes, by adding the vector a_3 to vectors a_1 and a_2. Vector c remains perpendicular to vectors a_1 and a_2.
(2) Direction of the planes is decided in the same manner as for Miller indices, except for the use of four coordinate axes a_1, a_2, a_3, and c.
(3) As a result, the plane indices refer to plane indices with four digit number such as $(h\ k\ i\ l)$; they always have a relationship $i = (h + k)$.

Third point suggests that the value of i depends on h and k values, so that by replacing the index i by a dot, a different plane symbol written as $(hk \cdot l)$ is sometimes employed. However, this method is not strongly recommended because the advantage of Miller–Bravais indices (for example, similar indices to similar planes) is not always retained. For example, all six side planes of a hexagonal prism of Fig. 2 are considered crytallographically equivalent and such mutual relationship can be readily recognized from $(10\bar{1}0)$, $(01\bar{1}0)$, $(\bar{1}100)$, $(\bar{1}010)$, $(0\bar{1}10)$, and $(1\bar{1}00)$ in the Miller–Bravais notation. Such mutual relationship is not obtained directly from the abbreviated symbols ; (10·0), (01·0), $(\bar{1}1{\cdot}0)$, $(\bar{1}0{\cdot}0)$, $(0\bar{1}{\cdot}0)$ and $(1\bar{1}{\cdot}0)$.

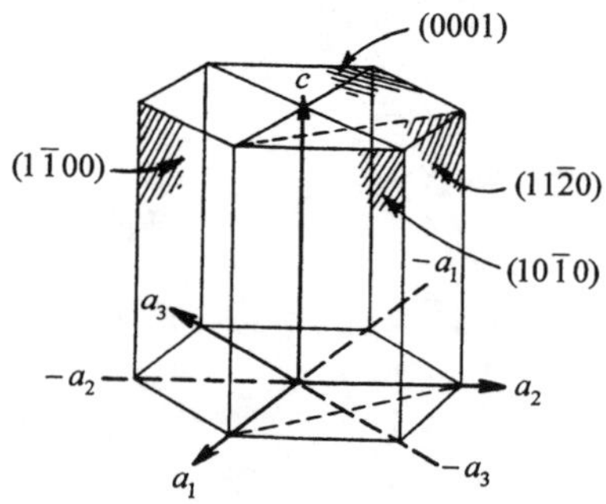

Fig. 1 Method for describing the planes in hexagonal system

However, the Miller–Bravais indices for directions in hexagonal system is slightly complicated. The directions in hexagonal system are simply described by three basic vectors a_1, a_2, and c, then we use only three digits such as $[UVW]$ to represent a direction referred to the three axes. Some examples for direction indices are illustrated in Fig. 2 together with some planes. Just as the case for plane,

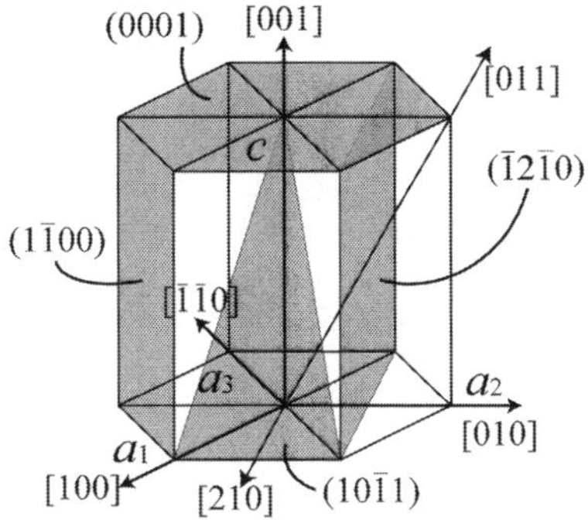

Fig. 2 Typical planes in hexagonal system

we can use four indices $[u\ v\ t\ w]$. For convenience, some additional notes are given below.

The use of four indices for directions in hexagonal system is based on four component vectors, parallel to a_1, a_2, a_3, and c. The third index is known to be equivalent to the sum of the first and second indices with change in sign, $t = -(u + v)$. When $[UVW\ ]$ are the direction indices referred to three axes, $[u\ v\ t\ w]$ corresponds to the four axes case. The relationships between these two indices are:

$$
\begin{array}{ll}
U = u - i & u = (2U - V)/3 \\
V = v - t & v = (2V - U)/3 \\
W = w & t = -(u + v) = -(U + V)/3 \\
 & w = W
\end{array}
$$

Therefore, we find that $[100] \rightarrow [2\bar{1}\bar{1}0]$ and $[210] \rightarrow [10\bar{1}0]$. To facilitate understanding, Fig. 3 shows a simple example of the straight line which passes along the origin on the bottom plane.

The coordinates of the desired point in crystal lattice are generally given by distances of parallel translations with respect to each axes required to reach the points from the origin. For describing the directions in hexagonal system, we frequently make a detour so as to obtain the distance equal to the negative value of the sum of the distances x and y, $(-(x + y))$ moved by parallel translations with respect to a_1 and a_2, respectively. It is noteworthy that this is equivalent to the distance moved by parallel translation with respect to a_3 (see Fig. 3). The direction is given by the ratio of the distances moved in parallel to each axes. The slightly complicated procedure for the hexagonal system is based on the objective of providing similar indices to similar directions.

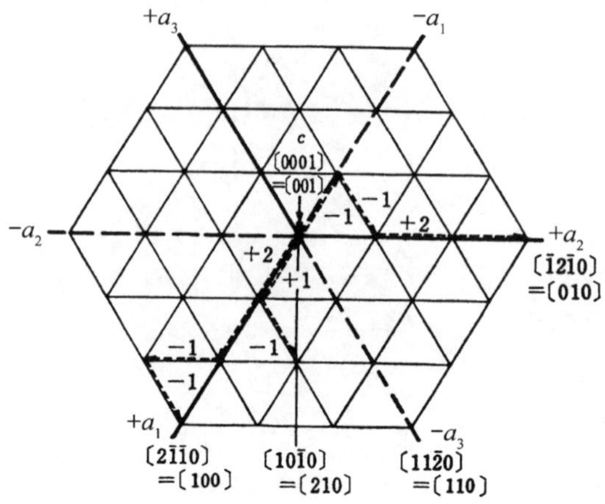

Fig. 3 Examples of the directions in hexagonal system

Question 2.12 Calculate the bond angle, θ, of O–Si–O in silica (SiO_2) assuming that Si and O atoms form regular tetrahedron with Si at the center.

Answer 2.12 Let us consider the case where the peak positions of a cube characterized by the length of a are connected to form regular tetrahedron as shown in Fig. 1.

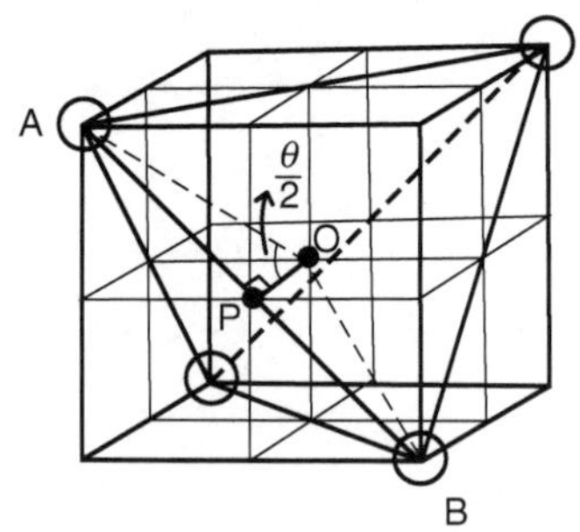

Fig. 1 Geometry in regular tetrahedron

The position of O in this figure corresponds to a center of both cube and regular tetrahedron. The positions denoted by A and B represent corners of regular tetrahedron and P gives the midpoint of regular tetrahedron which is characterized by the length of AB. The angle AOB is equivalent to the desired angle, θ , of the O–Si–O bond. Therefore, the following relationship is readily found in triangle AOP:

$$\mathrm{AP} = \frac{\mathrm{AB}}{2} = \frac{a\sqrt{2}}{2} \qquad \mathrm{OP} = \frac{a}{2} \qquad \tan\frac{\theta}{2} = \frac{a\sqrt{2}/2}{a/2} = \sqrt{2}$$

$$\frac{\theta}{2} = 54.74^\circ \qquad \theta = 109.48^\circ$$

Question 2.13 Silica (SiO_2) is known to form chain-like, sheet-like or network structures on the basis of tetrahedral unit of SiO_4^{4-}. Figure A shows the atomic arrangement of β-cristobalite at 583 K, which is the high temperature phase where Si atoms form diamond structure and each Si is surrounded by four oxygen atoms to form a tetrahedral unit.

(1) Estimate the radius of SiO_4^{4-} ion assuming the condition that the lattice parameter of β-cristobalite is given by 0.716 nm and the radius of oxygen ion is 0.140 nm.
(2) Estimate the radius ratio assuming the condition that Si^{4+} ions make contact with O^{2-} ions.

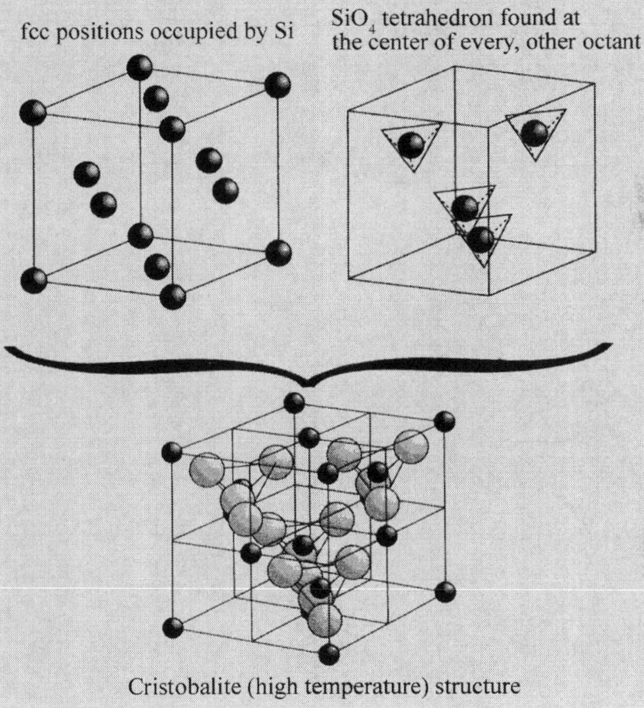

Fig. A Structure of β-cristobalite

Answer 2.13

(1) If the distance between Si–Si is given by R, it is found equivalent to the distance between (000) and $(1/4, 1/4, 1/4)$, from geometric condition:

$$(R)^2 = \left(\frac{1}{4}a\right)^2 + \left(\frac{1}{4}a\right)^2 + \left(\frac{1}{4}a\right)^2 = 3\left(\frac{a}{4}\right)^2.$$

O^{2-} ion combines with two Si^{4+} ions and the condition of keeping the distance of R for two silicon ions, thus we obtain the following relationship:

$$R = 2r_{Si} + 2r_O = \left[3\left(\frac{0.716}{4}\right)^2\right]^{1/2} = 0.310\,\text{nm}.$$

$$\text{The radius of } SiO_4^{4-} = 0.155\,\text{nm}.$$

$$2r_{Si} = 0.310 - 2 \times (0.140)$$

$$2r_{Si} = 0.03 \qquad r_{Si} = 0.015\,\text{nm} \qquad (\text{Reference} \quad Si^{4+} = 0.041\,\text{nm})$$

(2) Since the length of edge of SiO_4^{4-} tetrahedron is $2r_O$ (see Fig. 1) and the bond angles of a tetrahedron is given by 109.48° (see Question 2.12), the following relationship is readily found:

$$\sin\left(\frac{109.48}{2}\right) = \frac{r_O}{r_O + r_{Si}} = 0.816$$

$$r_O = 0.816(r_O + r_{Si}) = \frac{0.816}{(1 - 0.816)} r_{Si}$$

$$\frac{r_{Si}}{r_O} = \frac{1 - 0.816}{0.816} = 0.225$$

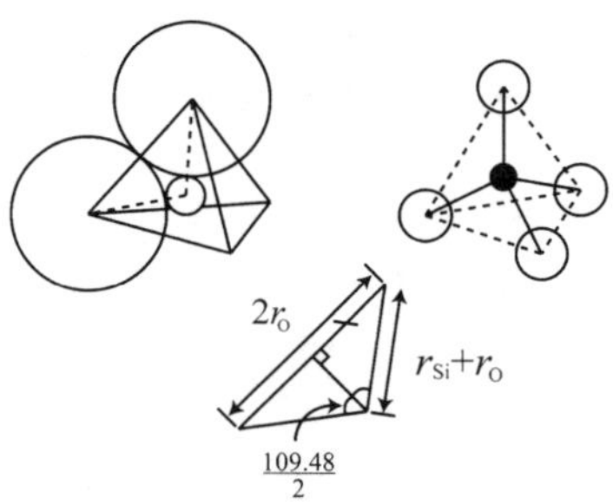

Fig. 1 Geometry of SiO_4^{4-} tetrahedron

Question 2.14 In ionic crystals, anions of the relatively larger size are densely arranged so as to avoid their direct contact, whereas cations of relatively smaller size occupy the positions equivalent to the vacant space produced by anions. For this reason, if the radii of cation and anion are described by r_c and r_a, respectively, some correlations are recognized between the coordination numbers and the size ratio of r_c/r_a. Estimate the specific values of r_c/r_a for cases that cations are surround by anions with the coordination numbers of 3, 4, 6, and 8.

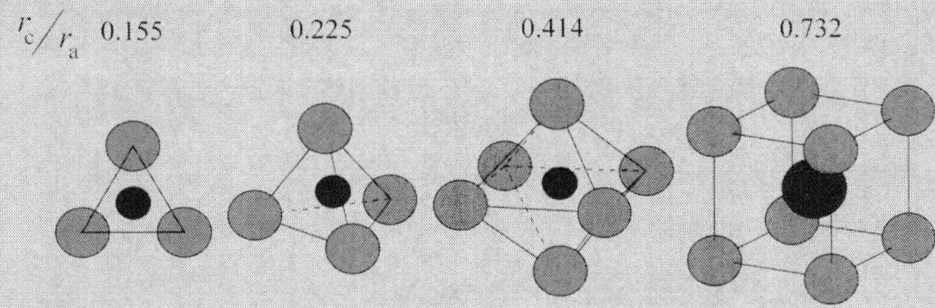

Fig. A Four typical geometries for arraying equal spheres

Answer 2.14

(1) When the coordination number is 3, a cation is likely to occupy the position equivalent to the center of an equilateral triangle of ABC formed by three anions as shown in Fig. 1. The height and the center of gravity for the equilateral triangle with its one edge being $2r_a$ are given by $\sqrt{3}r_a$ and the height $\times$ $\frac{2}{3}$, respectively. Then, the value of $\frac{2}{3} \times \sqrt{3}r_a$ is exactly equivalent to $r_a + r_c$

$$r_a + r_c = \frac{2\sqrt{3}}{3} \times r_a \quad \Rightarrow \quad r_c = \left(\frac{2\sqrt{3}}{3} - 1\right) r_a \qquad \frac{r_c}{r_a} = \frac{2\sqrt{3}}{3} - 1 = 0.155$$

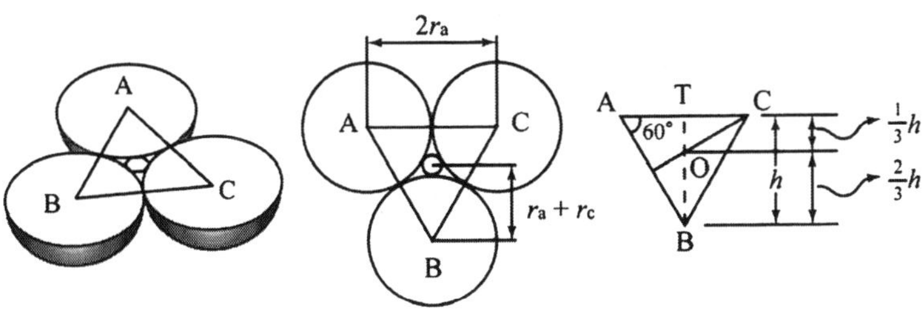

Fig. 1 Geometric relation for the 3-coordination case

(2) When the coordination number is 4, a cation is at the center of a regular tetrahedron, formed by four anions as shown in Fig. 2. Consider the triangle ABC corresponding to the plane where the vacant space existed in the regular tetrahedron with its one edge being $2r_a$. We find the relationship, AB = BC = $\sqrt{3}r_a$, so that ΔABC is an isosceles triangle. With respect to the triangle ΔABM, AM = r_a and AB = $\sqrt{3}r_a$. Hence BM = $\sqrt{2}r_a$. If O is the center of the vacant space the distance of AO is exactly equivalent to $r_a + r_c$.

Since the center O of the vacant space is also the center of regular tetrahedron, O is equivalent to the midpoint of BM. Therefore, $\mathrm{OM} = \frac{1}{2}\mathrm{BM} = \frac{\sqrt{2}}{2} r_\mathrm{a}$. Next, if triangle AOM is considered, we obtain the following relationships:

$$(\mathrm{AO})^2 = (r_\mathrm{a}+r_\mathrm{c})^2 = (\mathrm{OM})^2 + \left(\frac{1}{2}\mathrm{AM}\right)^2 = \left(\frac{\sqrt{2}}{2}r_\mathrm{a}\right)^2 + (r_\mathrm{a})^2 = \left(\sqrt{\frac{3}{2}}r_\mathrm{a}\right)^2,$$

$$r_\mathrm{a}+r_\mathrm{c} = \sqrt{\frac{3}{2}}r_\mathrm{a} \quad\Rightarrow\quad r_\mathrm{c} = \left(\sqrt{\frac{3}{2}}-1\right)r_\mathrm{a} \qquad \frac{r_\mathrm{c}}{r_\mathrm{a}} = \sqrt{\frac{3}{2}}-1 = 0.2247.$$

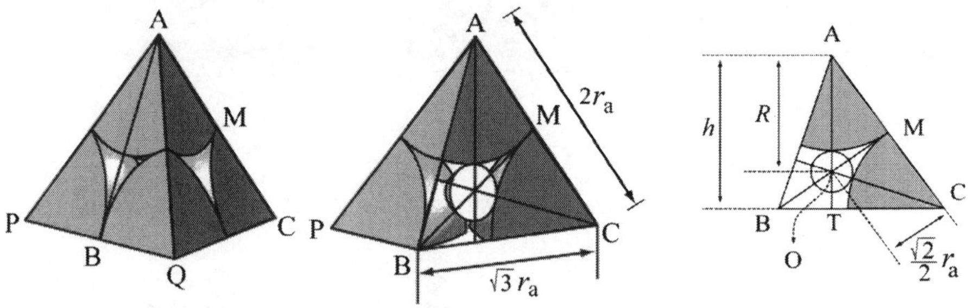

Fig. 2 Geometric relation for the 4-coordination case

(3) When the coordination number is 6, the cation is located at the center of an octahedron formed by six anions. As shown in Fig. 3, the distance of AC or BD, corresponding to the diagonal line of the cross-sectional view of ABCD portion of octahedron, is exactly equal to twice the value of $(r_\mathrm{a}+r_\mathrm{c})$. When considering that one edge of the cross-sectional view of the square ABCD is $2r_\mathrm{a}$, we obtain the following result:

$$2(r_\mathrm{a}+r_\mathrm{c}) = 2\sqrt{2}r_\mathrm{a} \quad\Rightarrow\quad r_\mathrm{c} = (\sqrt{2}-1)r_\mathrm{a} \qquad \frac{r_\mathrm{c}}{r_\mathrm{a}} = (\sqrt{2}-1) = 0.414.$$

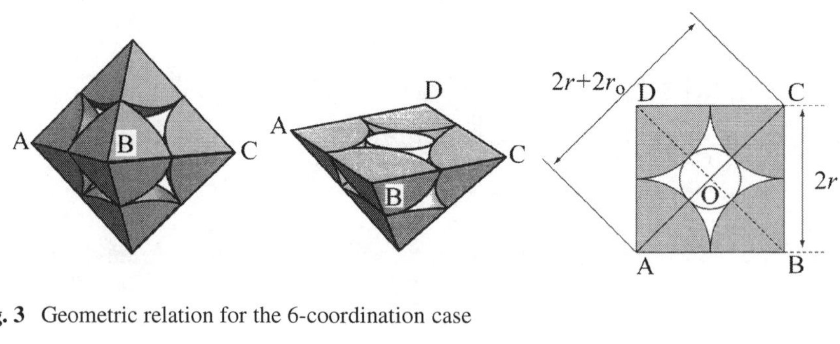

Fig. 3 Geometric relation for the 6-coordination case

(4) For 8-coordination, it is difficult to visualize the cation in the vacant space of a polyhedron formed by anions. Nevertheless, if the appropriate value of r_c/r_a is given, one can obtain the ionic arrangements in which a cation is located at a center of body-centered cubic (bcc) lattice formed by eight anions. In this case, the diagonal line of the cross-sectional view of ABGF portion of bcc lattice with one edge of $2r_a$ just corresponds to twice the value of $(r_a + r_c)$ as readily seen in Fig. 3. Here, the lengths of each edge of square ABGF are $2r_a$ and $2\sqrt{2}r_a$, respectively. Therefore,

$$2(r_a + r_c) = \sqrt{3} \times (2r_a) \qquad r_a + r_c = \sqrt{3}r_a \quad \Rightarrow \quad r_c = (\sqrt{3} - 1)r_a$$

$$\frac{r_c}{r_a} = \sqrt{3} - 1 = 0.732$$

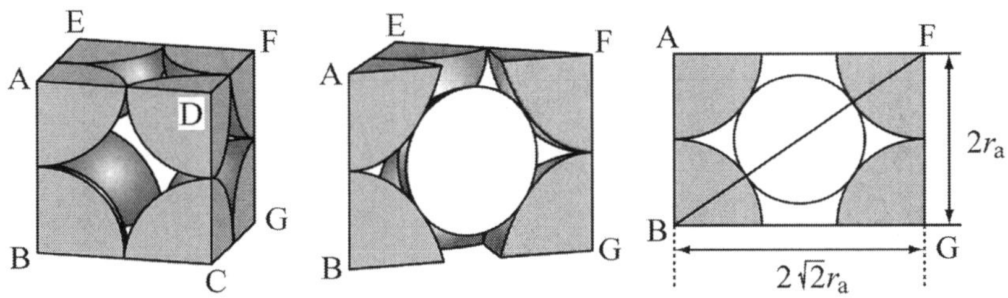

Fig. 4 Geometric relation for the 8-coordination case

Reference: $(BF)^2 = (BG)^2 + (FG)^2$

$$[2(r_c + r_a)]^2 = (2\sqrt{2}r_a)^2 + (2r_a)^2 = (12r_a)^2 = (2\sqrt{3}r_a)^2.$$

Question 2.15 Caesium chloride (CsCl) crystal has cubic structure in which Cs^+ ions occupy the center position of a unit cell and its corner positions are occupied by Cl^- ions. The density of caesium chloride is 3.97×10^6 g/m^3.

(1) If the ionic radii of Cs^+ ion and Cl^- ion are 0.169 and 0.181 nm, respectively, compute the lattice parameter and compare with the value estimated from density.

(2) There is a threshold value of the ratio of the radii of positive/negative ions for alkali halides having CsCl-type structure. The radius r^+ of a positive ion needs such as to just fit the space formed by eight negative ions of the radius of r^- without their direct contact. Calculate the minimum value of the ratio of r^+/r^-.

Answer 2.15

(1) If Cs^+ and Cl^- ions are represented by black and white circles, respectively, geometry of a unit cell of the CsCl is shown in Fig. 1. Let a be the lattice parameter. Thus, AB $= a$, AC $= \sqrt{3}a$. The Cs^+ and Cl^- ions are arranged along the diagonal AC such that

$$\mathrm{AC} = 2(r^+ + 2r^-) = 2(0.169 + 0.181) = \sqrt{3}a,$$

$$a = \frac{2(0.169 + 0.181)}{\sqrt{3}} = 0.404\,\mathrm{nm}.$$

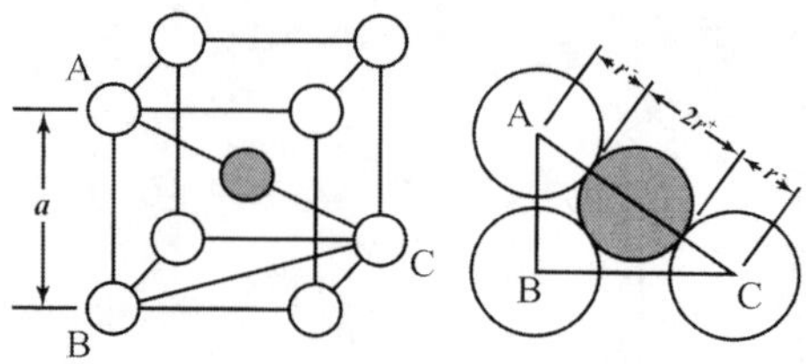

Fig. 1 Geometry of CsCl-type structure

The unit cell accommodates one Cs^+ ion and one Cl^- ion, because the ion at corners is shared by eight unit cells ($8 \times \frac{1}{8} = 1$). The molecular weight of CsCl is $132.90 + 35.45 = 168.35$ g (sum of atomic weights). Therefore, the mass m of the unit cell is

$$m = \frac{168.35}{0.6022 \times 10^{24}} = 2.796 \times 10^{-22}\,\mathrm{g}$$

Using the relation between the density (ρ) and the unit cell volume (a^3)

$$a^3 = \frac{m}{\rho} = \frac{2.796 \times 10^{-22}}{3.97 \times 10^6} = 0.0704 \times 10^{-27}\,\mathrm{m}^3,$$

$$a = \sqrt[3]{0.0704 \times 10^{-27}} = 0.413 \times 10^{-9}\,\mathrm{m} = 0.413\,\mathrm{nm}.$$

This value of $a = 0.413$ nm is larger by 2% than the lattice parameter $a = 0.404$ nm calculated from their ionic radii. Thus, the lattice parameter of ionic crystals computed from measured density is realistic. The radii of ions such as Cs^+ and Cl^- complied in handbooks are usually derived from mean values of interionic distance obtained from various ionic crystals. The lattice parameter of CsCl crystal determined by X-ray diffraction is $a = 0.4123$ nm (see Appendix A.9).

(2) The following relationship may be obtained by referring to the condition of Fig. 1:

$$AC = 2(r^+ + r^-) = a\sqrt{3}, \quad AB = 2r^- = a.$$

When two relationships are coupled, the radius ratio obtained is equal to that estimated in Question 2.14 for the coordination number 8.

$$\frac{AC}{AB} = \frac{r^+ + r^-}{r^-} = \sqrt{3}, \qquad \frac{r^+}{r^-} = \sqrt{3} - 1 = 0.732.$$

From this result, if the ratio of the ionic radii between cations and anions is smaller than 0.732, direct contact of anions will be allowed, and then this makes an ionic crystal unstable due to the strong repulsion between ions with the same electric charge.

Question 2.16 Water (H_2O) crystallizes to form ice, when cooled below 273 K (0°C) under 1 atmospheric pressure. Ice has hexagonal crystal structure with the lattice parameters of $a = 0.453$ nm and $c = 0.741$ nm. The density of ice at 1 atmospheric pressure at 273 K is 0.917×10^6 g/m^3. Estimate the number of water molecules (H_2O) contained in a unit cell of ice crystal.

Answer 2.16 In hexagonal crystal system, a unit cell is given by a prism with a vertical axis perpendicular to a rhombus-shaped base and the equal edges of which are at 60° and 120° with respect to each other (see Fig. 2.9). The length of the rhombus edge is designated by two equal vectors, a_1 and a_2. Considering that a_1 and a_2 axes make an angle of 60°, the volume V of a unit cell can be obtained as follows (refer to Appendix A.5):

$$V = a^2 \sin 60° \times c = (0.453)^2 \times (0.866) \times (0.741) = 0.132 \times 10^{-27}\ \text{nm}^3$$

The molecular weight of water (m) obtained from Avogadro's number (N_A) and density (ρ) assuming one molecule per unit cell is;

$$m = \rho N_A V = 0.917 \times 10^6 \times 0.6022 \times 10^{24} \times 0.132 \times 10^{-27} = 72.9\,\text{g}$$

The molecular weight of water is $1.008 \times 2 + 15.999 = 18.015$ g, using values of the atomic weights compiled in Appendix A.2. Then, we obtain $72.9/18.015 = 4.05$. This suggests that four water molecules are included in a unit cell of ice crystal. As a result, each oxygen is quite likely to be surrounded by four oxygen in ice crystal (see Fig. 1).

Note: The error in the measurement of lattice parameter and density is responsible for non-integer value 4.05. Water can crystallize in other structures such as trigonal lattices, tetragonal lattices etc., depending on temperature and pressure.

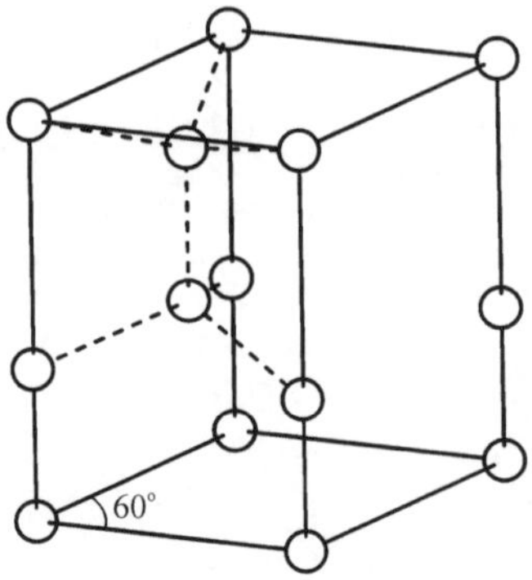

Fig. 1 Arrangement of water molecules in ice crystal

Question 2.17 Many compounds with the formula ABX_3 have the perovskite structure and a typical example is the natural mineral $CaTiO_3$. In the Perovskite structure, A atom occupy the center of a cubic unit cell and B and X atoms occupy the cell corners and face centers, respectively. There are two types of perovskite structure. Explain the essential points of two types including correlations of the unit cell. Also answer the following questions.

(1) In barium titanate ($BaTiO_3$), titanium occupy the center of a cubic unit cell, whereas barium and oxygen share the corner and the center of the cell faces, respectively. Assuming that titanium atoms share the vacant space formed by Ba-O lattice, what is the nature of the coordination and the resulting radius ratio constraint?
(2) Why does titanium occupy the vacant space located in this position.

Answer 2.17 Two types of the perovskite structures are characterized by two different unit cells. For example, MoF_3 and ReO_3 belong to the so-called ReO_3 structure, in which Mo atoms share all corner positions of a cubic unit cell, whereas F atoms occupy the midpoint of all edge-lines (see Fig. 1). Perovskite A-type structure can be obtained by adding A of ABX_3 to the center of a cubic unit cell of the ReO_3 base consisting of B and X. The structure can also be generated by adding X to the midpoint of all edge-lines of a cubic unit cell in which A and B form the CsCl type atomic arrangement. Such correlations are readily seen in Fig. 1.

In the structure of intermetallic compounds such as Cu_3Au and Cu_3Pt, Au atoms occupy all corner positions of the fcc lattice, whereas Cu atoms share all center positions of cell faces of fcc lattice. Perovskite B-type structure is obtained by adding B to the center of a cubic unit cell of the Cu_3Au base consisting of A and X. Such correlations are illustrated in Fig. 2. It is also noted that the description of layer, as shown in Fig. 3, facilitates our understanding of the characteristic of the two perovskite structures.

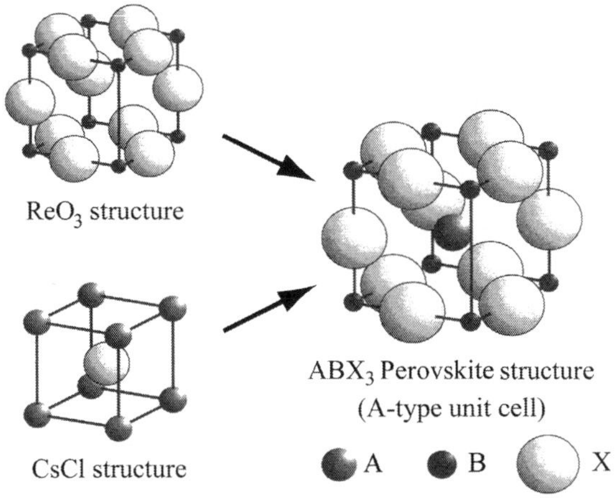

Fig. 1 Perovskite A-type structure

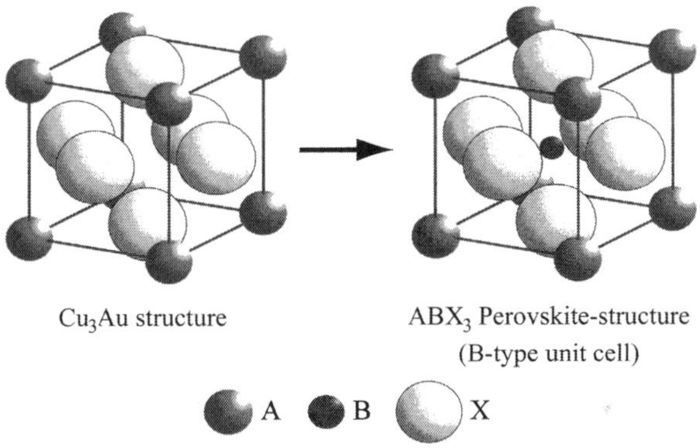

Fig. 2 Perovskite B-type structure

Since the atomic arrangement in perovskite structure is characterized by a formation where A and X are closely packed and B fits into its vacant space as shown in Figs. 1 and 2, it is desirable that the magnitude of A and X are equal and B is small.

(1) The structure of barium titanate ($BaTiO_3$) can be readily understood by referring to the B-type unit cell in Fig. 2. Ba^{2+} ions occupy the position of (000) corresponding to A-site, whereas Ti^{4+} ions corresponding to the B site occupy the position $\left(\frac{1}{2}\frac{1}{2}\frac{1}{2}\right)$. Three oxygens (O) keep the positions $\left(0\frac{1}{2}\frac{1}{2}\right)$, $\left(\frac{1}{2}0\frac{1}{2}\right)$, and $\left(\frac{1}{2}\frac{1}{2}0\right)$ of X. Here, B is found to be surrounded by six X. Incidentally, there are 12X distributed around A. Therefore, Ti^{4+} ions share the center position of octahedron consisting of six oxygen atoms. There are four positions corresponding to such vacant space given by octahedron in an fcc type unit cell formed by A and X (refer to Question 2.7) and one of them (25%) is occupied by Ti^{4+} ion.

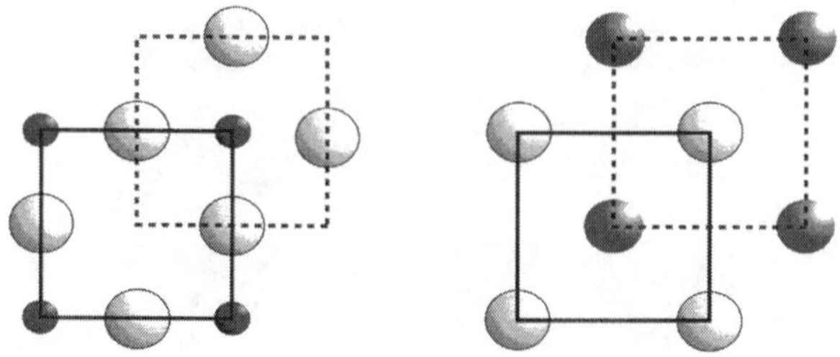

Fig. 3 Layer structure found in perovskite

(2) The octahedral position except for the octahedral void surrounded by six oxygen atoms in a unit cell is found at the midpoint of edge-lines in a unit cell and it will be surrounded by two Ba^{2+} ions and four oxygen atoms. Since the structure becomes unstable when two kinds of positive ions such as Ba^{2+} and Ti^{4+} are approaching, it is thought that Ti^{4+} ions preferentially fit into the octahedral void surrounded by six oxygens. The radii of Ba^{2+} ion and O^{2-} ion are 0.135 and 0.140 nm, respectively, so that barium titanate shows a slight deviation from an ideal perovskite structure. The dipole resulting from this lattice distortion is responsible for the ferroelectric properties.

Question 2.18 Explain the procedure of stereographic projection for an octahedron by setting a plane of projection to the equator.

Answer 2.18 By setting the plane of projection to the equatorial plane, the equatorial plane becomes the basic circle of stereographic projection. Figure 1 shows the relation when a light source is placed at south pole and the projection is made to the equatorial plane are shown in Fig. 2.

In the present case, the stereographic projection of a pole located on the upper half of the northern hemisphere is obtained by finding the intersection on the projection plane of the line connecting the target pole and south pole S. Conversely, one can get the stereographic projection of the pole located on the lower half of the southern hemisphere by finding the intersection on the projection plane of the line connecting the target pole and north pole N. To show the hemisphere in which the point is located, signs such as ⊕ ⊖ and • ∘, are used for differentiation.

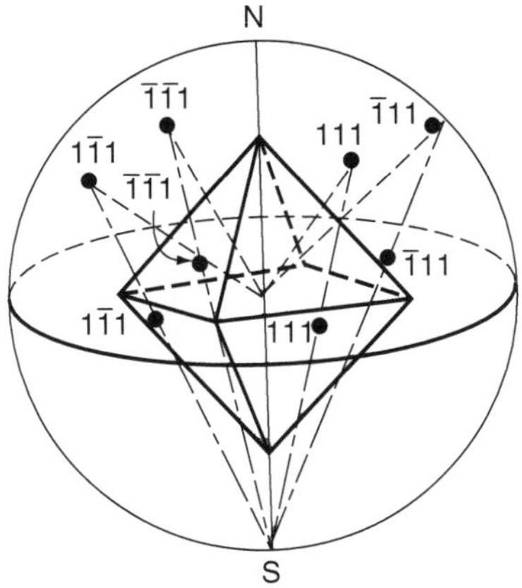

Fig. 1 Procedure of Stereographic projection of octahedron

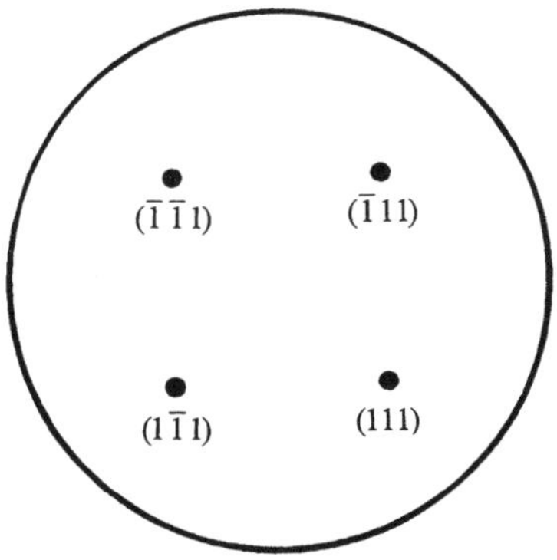

Fig. 2 The results of projection to equatorial plane

Question 2.19 Explain two types of nets, such as the polar net and the Wulff net. In addition, explain the procedure that 40° rotation of a pole P_1 about any Q_1 axis on a plane of projection.

Answer 2.19

(1) The polar net is a figure (refer to Fig. 1) of the meshes at every 2° (or 1°) obtained by projecting all meridian circles and latitude circles on an equatorial plane. The pole net is useful for finding the projecting point to the equatorial plane about the point on a projection sphere with the given coordinates (γ, ϕ) as shown in Fig. 2.

(2) The Wulff net is a figure of the meshes at every 2° (or 1°) obtained by projecting a sphere drawn with parallels of latitude and longitude on a plane parallel to the north–south axis of the sphere. For example, the longitude lines correspond

Fig. 1 Polar net

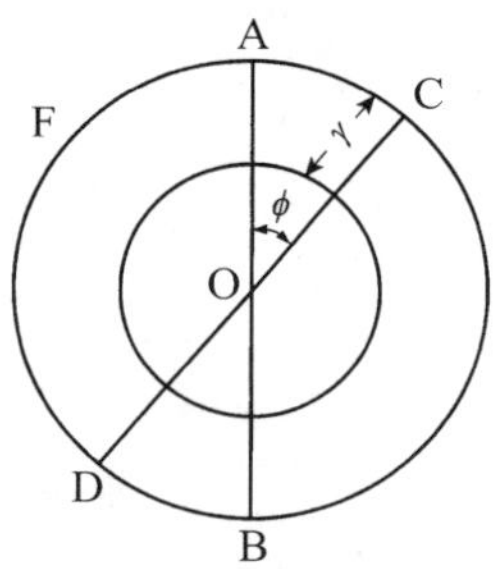

Fig. 2 Example of the use of the polar net

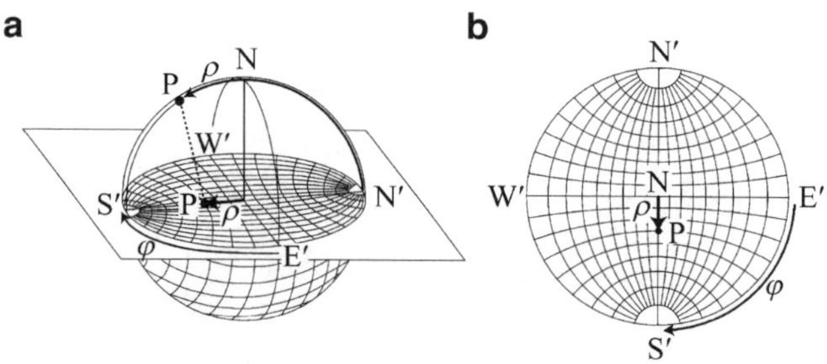

Fig. 3 Projection of the grid net of globe for producing the Wulff net

to great circles connecting the north and south poles of the net and the latitude lines are small circles extending from side to side of the net. Figure 3 shows the stereographic projection of the grid net of sphere (the NS direction perpendicular to the N′S′ one) for drawing the Wulff net. This drawing includes the positions of the angular coordinates ϕ and the corresponding pole distance ρ. Wulff net (refer to Fig. 4) is used, for example, to estimate the angle between the lattice planes A and B, as shown in Fig. 5.

Fig. 4 Wulff net

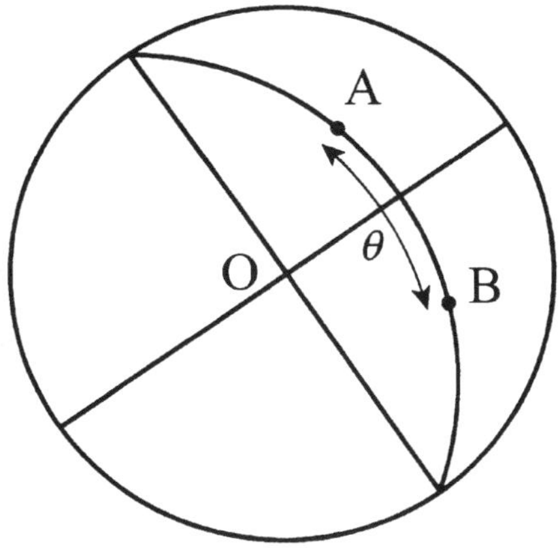

Fig. 5 Example of the use of the Wulff net

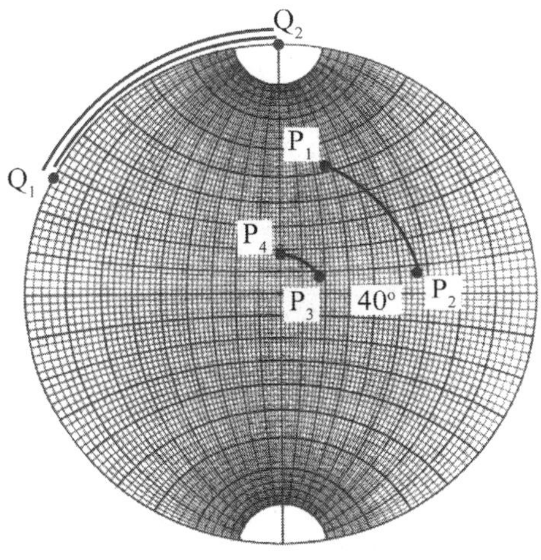

Fig. 6 Example of the rotating operation of a pole P_1 by 40° about an arbitrary axis Q_1 on the projection plane

To rotate by 40° around an axis perpendicular to a projection plane (corresponding to the case of rotating projection about the NS axis in Fig. 2.11), put the stereographic projection on the pole net and rotate projection at required angle (40°) about a central axis. On the other hand, to rotate the pole P_1 by 40° around the arbitrary axis Q_1 on projection plane (40 degrees rotation around the arbitrary axis on the circle ADBE in Fig. 2.11), the following procedures are made using the Wulff net. Rotate so as to fit the north–south axis of the Wulff net with the axis Q_1. As a result, Q_1 moves to Q_2, and P_1 moves to P_2, respectively, as shown in Fig. 6. Next, P_2 moves to P_3, by moving the pole P_2 only by 40° along the latitude line on the Wulff net. Furthermore, rotate reversely the projection only by the same angle as the first operation, and return the axis Q_3 to Q_1. As a result, P_3 is set to P_4.

Question 2.20 Find the geocentric angle between Sendai, Japan (38° of north latitude and 141° of east longitude) and Los Angeles, USA (33° of north latitude and 120° of west longitude) using the Wulff net.

Answer 2.20 Put the tracing paper on a copy of the Wulff net. Fix the centers by a tack as shown in Fig. 1. This makes the centers always coincident. Subsequently, the positions of Sendai (•) and Los Angeles (X) are marked in the tracing paper.

Next, rotate the tracing paper so as to get the marked two points on the longitude line of great circle of Wulff net. In Fig. 1, two points marked by • and X will be on the same longitude line of the great circle when rotating the tracing paper from position ① to position ②. We obtain the value of 76° for the angle from the scale of Wulff net.

Note: An uncertainty of 1° (or 2°) is generally encountered when using the Wulff net of approximately 90 mm in diameter.

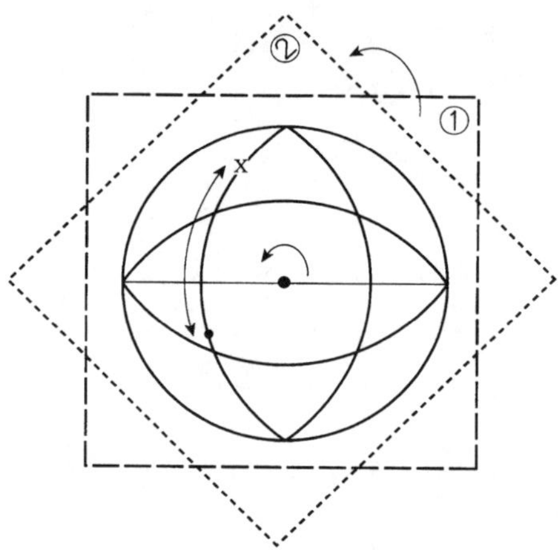

Fig. 1 Method for obtaining the angle of two positions with the Wulff net

Question 2.21 In the standard stereographic projection of cubic crystals, when setting a pole of (111) on the center of basic circle, find other poles of {111}, {100}, and {110}.

Answer 2.21 The results are summarized in Fig. 1. Since the orientation of any plane in crystal can be represented by the inclination of the normal to that plane itself, the indices of a pole located at the position of 90° relative to each zone (i.e., the set of planes) present the zone axis.

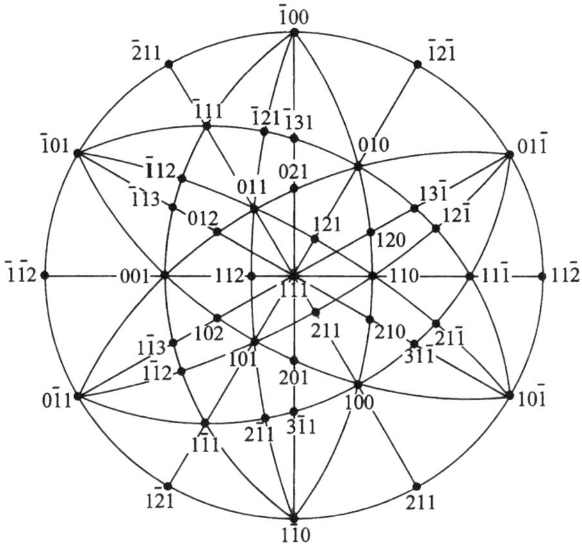

Fig. 1 Standard stereographic projection of cubic crystals on a (111) pole

Putting a crystal on the center of a reference sphere, the plane normal intersects the surface of the sphere in a set of points called poles. Each plane is represented by the intersection between normal and the reference sphere using the Miller indices. If the axis of a zone is given by the indices $[u\ v\ w]$, and any plane belongs to that zone denoted by the indices $(h\ k\ l)$, the well-known relation $(hu + kv + lw = 0)$ called Weiss rule for the zone is satisfied. It may be noted that the Weiss rule is independent of the crystal system. If two planes of $(h_1k_1l_1)$ and $(h_2k_2l_2)$ belong to one zone axis of $[u\ v\ w]$, the following relationships are obtained:

$$h_1u + k_1v + l_1w = 0 \quad \text{and} \quad h_2u + k_2v + l_2w = 0,$$

$$(ph_1 + qh_2)u + (pk_1 + qk_2)v + (pl_1 + ql_2)w = 0,$$

where p and q are arbitrary integers. In other words, if a zone axis $[u\ v\ w]$ contains two planes $(h_1k_1l_1)$ and $(h_2k_2l_2)$, planes represented by $p(h_1k_1l_1) + q(h_2k_2l_2)$ also belong the same zone. For example, the planes $(\bar{1}20)$ and (520) set as $p = 1$ and $q = 2$ belong to a zone [001] = [002] including the planes of (100) and $(\bar{1}10)$.

Reference: If any two nonparallel planes of $(h\ k\ l)$ and $(h'\ k'\ l')$ cross on one straight line and its direction is given by $[u\ v\ w]$ as shown in Fig. 2, this direction is called zone axis [u v w]. The indices of the zone axis [u v w] are related to the indices of the two planes and the vectors of a, b, and c which define a unit cell.

$$\begin{pmatrix} u \\ v \\ w \end{pmatrix} = \begin{pmatrix} kl' - lk' \\ lh' - hl' \\ hk' - kh' \end{pmatrix} \bullet \left(ua + vb + wc = \begin{vmatrix} a & b & c \\ h & k & l \\ h' & k' & l' \end{vmatrix} \right)$$

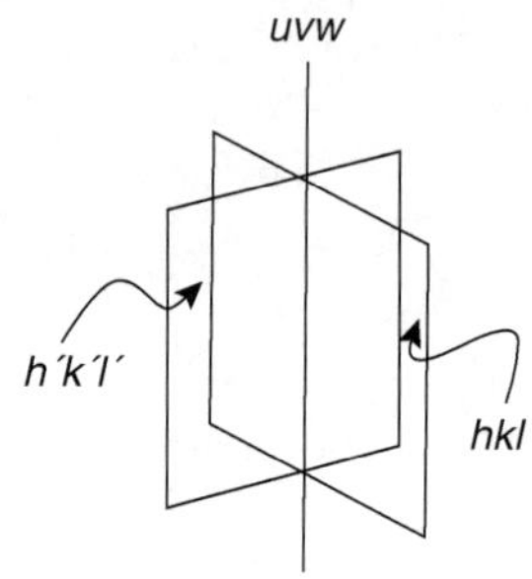

Fig. 2 The line of intersection of two planes relevant to zone axis

Chapter 3
Scattering and Diffraction

An X-ray beam is an electromagnetic wave characterized by an electric field vibrating at constant frequency, perpendicular to the direction of movement. This variation of the electric field gives electrons (charged particles) a sinusoidal change with time at the same frequency. As a result of periodic acceleration and deceleration of the electron, a new electromagnetic wave, i.e., X-rays are generated. In this sense, X-rays are scattered by electrons. This phenomenon is called Thomson scattering. On the other hand, the physical phenomenon called "diffraction as a function of atomic position" is also found when an X-ray beam encounters a crystal whose atomic arrangement shows the long range periodicity. The intensity of diffracted X-rays depends on not only the atomic arrangement but also the atomic species. When considering diffraction of X-rays from a crystal, one needs information about "atomic scattering factors" which provide a measure of the scattering ability of X-rays per atom. Since the nucleus of an atom is relatively heavy compared with an X-ray photon, it does not scatter X-rays. The scattering ability of an atom depends only on electrons, their number, and distribution.

3.1 Scattering by a Single Electron

In X-ray scattering by electrons, the scattered X-rays have the same frequency (wavelength) as the incident beam and is "coherent" with the incident X-rays. As shown in Fig. 3.1, if the incident X-ray beam traveling along the X-axis meets a single electron with mass m (kg) and charge e coulombs (C) located at the origin O, the intensity I of the scattered X-rays at position P in the X-Z plane at a distance r (m) from the origin may be expressed by the "Thomson equation":

$$I = I_0 \left(\frac{\mu_0}{4\pi}\right)^2 \left(\frac{e^4}{m^2 r^2}\right) \sin^2 \alpha = I_0 \frac{K}{r^2} \left(\frac{1 + \cos^2 2\theta}{2}\right) \tag{3.1}$$

where I_0 is the intensity of incident X-rays, $\mu_0 = 4\pi \times 10^{-7}$ (m kgC^{-2}), and α is the angle between the scattering direction and that of the acceleration of electron. On the other hand, 2θ in (3.1) is the angle between the line of OP connecting from

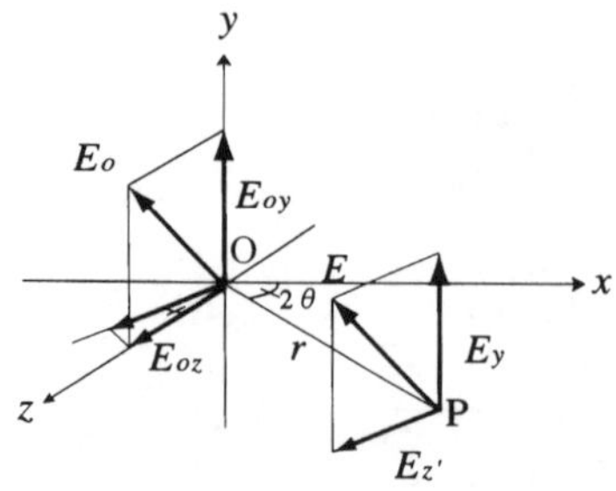

Fig. 3.1 The relationship of the components of the electric vector of the incident radiation at a point O to the components of that of the scattered radiation at a point of observation P

the origin O to the point of measurement P, and the X-axis, the direction of the incident X-ray beam. The constant $K = (2.8179 \times 10^{-15})^2\,(\mathrm{m}^2)$ is equivalent to the square of the electron radius r_e in classical electromagnetic theory. X-rays are scattered by an electron in all directions, but (3.1) clearly shows that the intensity of scattered X-rays decreases as the inverse square of the distance from the electron at origin, as well as a function of the scattering angle. In addition, the intensity of scattered X-rays is larger both in forward and backward directions compared to the direction at right angle to the incident X-ray beam. The term in the parenthesis in the last expression of (3.1) is called the "polarization factor."

For example, the ratio of (I/I_0) at the position of 0.01 m away from an electron at origin is extremely small about 7.94×10^{-26}. Nevertheless, the intensity of scattered X-rays can be amplified for detection without any difficulty, because the number of electrons contained in 1 mg of substance is of the order of $10^{20} \sim 10^{21}$ and such a large number of electrons interfere with each other.

Finding the absolute value of the scattering intensity from one electron using (3.1) is a very difficult task either by measurement or by calculation. However, for most applications only the relative values of intensities of X-ray scattering and X-ray diffraction are required. It may be assumed that all terms in (3.1) except for the polarization factor may be considered as a constant.

There is another way for X-ray scattering by an electron and it is characterized by a quite different mechanism from Thomson scattering. Compton scattering occurs when X-rays encounter a loosely bound or free electrons. It is relatively easy to understand Compton scattering by considering X-rays as particles (photons) which have energy $\mathrm{h}\nu_0$ rather than as waves.

As shown in Fig. 3.2, if a photon collides with a loosely bound electron and the collision between photon and electron is considered an elastic one, in a way similar to two billiard balls, the electron is knocked aside at an angle ϕ and the direction of the incident X-ray photon is altered by an angle 2θ (it is called recoil phenomenon). In this collision process, a part of the energy $\mathrm{h}\nu_0$ of the incident X-ray photon is converted to kinetic energy of the electron. As a result, the energy of the incident X-ray photon after collision becomes $h\nu$ which is smaller than $\mathrm{h}\nu_0$. For this reason, the wavelength λ after collision becomes slightly longer than the wavelength of the

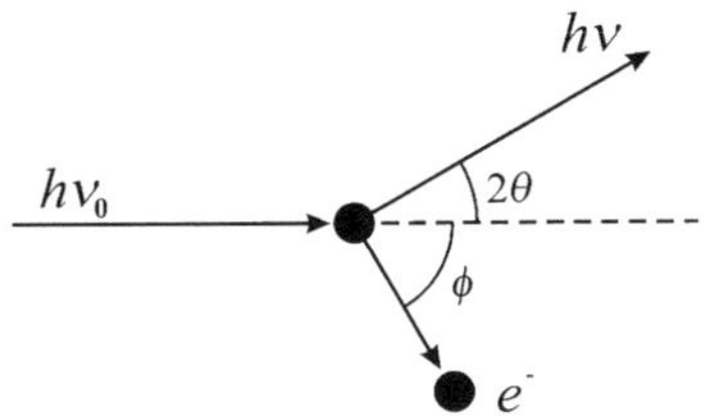

Fig. 3.2 Collision of photon and an electron (Compton scattering)

incident X-ray beam λ_0 before impact and the relationship is given in unit of nm as follows:

$$\Delta\lambda = \lambda - \lambda_0 = \left(\frac{h}{mc}\right)(1 - \cos 2\theta) = 0.002426(1 - \cos 2\theta) \qquad (3.2)$$

The term of $(h/mc) = 0.002426$ nm is called the "Compton wavelength." According to (3.2), the increase in wavelength due to the Compton scattering depends only on the scattering angle 2θ and $\Delta\lambda$ values are estimated to be zero at $2\theta = 0$ and 0.005 nm at $2\theta = 180°$, respectively. The phase of X-rays produced by Compton scattering is not the same as that of the incident X-ray beam, because of the change in wavelength. The Compton modified radiation cannot take part in diffraction because it has no specific phase relation with the incident X-ray beam and therefore cannot produce any interference effect. For this reason, Compton scattering is often called "incoherent scattering."

3.2 Scattering by a Single Atom

If an X-ray beam encounters an atom consisting of the nucleus and a certain number of electrons, each electron produces coherent scattering intensity given by (3.1), the so-called Thomson equation. Since the mass of the nucleus is much larger than that of electron, the X-ray beam cannot oscillate the nucleus to any appreciable extent. The acceleration and deceleration of the nucleus to emit X-ray are not functional. Therefore, when considering the net effect of X-ray scattering from an atom, one needs to take into consideration only scattering by electrons associated with the atom. This is also evident from (3.1) since the square of mass of scattering particle appears in the denominator.

The scattering amplitude of a single atom with atomic number Z containing Z electrons is equal to Z times the scattering amplitude from one electron in the forward direction. Because in direction the scattering angle is zero ($2\theta = 0$), the phases of X-rays scattered by all electrons in one atom are completely coincident, so that the amplitude of the scattered X-rays can be simply added. However, when the scattering angle has nonzero values, there is variation in the phase of X-rays scattered from individual electrons in an atom. In other words, the X-rays scattered by electrons located at, for example, the point A and point B in space, have phase difference because of difference in optical path, when the scattering angle is not zero. As a result, the scattering amplitude of a single atom decreases with increase in θ. The scattering amplitude of an atom also depends on the wavelength of the incident X-rays. For example, at the same scattering angles 2θ, the shorter the wavelength is, the larger the phase difference becomes. This implies that the scattering amplitude becomes relatively small when the shorter wavelength is used.

In order to calculate the scattering amplitude of X-rays for atoms containing more than two electrons, it is necessary to bear in mind that the charge of electrons are not focused at fixed points; rather it is distributed in space like a cloud. The electron density function $\rho(r)$ as a function of distance r away from the nucleus at origin is useful for describing electron distribution. Let us consider that the wave vectors of the incident X-rays and the scattered X-rays are given by $(\boldsymbol{s}_0)$ and $(\boldsymbol{s})$, respectively. The scattered X-rays at the distance r will produce the optical path difference $(\boldsymbol{s}-\boldsymbol{s}_0)\cdot\boldsymbol{r}$ in comparison with the scattered X-rays at origin. If the wavelength of the incident X-rays is given by λ, the scattering amplitude of X-rays irradiated in the direction of $\boldsymbol{s}$ can be described by

$$\rho \exp\left[\frac{2\pi i}{\lambda}(\boldsymbol{s}-\boldsymbol{s}_0)\cdot\boldsymbol{r}\right]\mathrm{d}V \tag{3.3}$$

The amplitude of the coherent scattering from a single electron may be obtained by integrating over the volumes occupied by electron with the help of the phases relevant to $\rho\mathrm{d}V$. Then, the scattering factor f_e per electron in electron unit is obtained

$$f_\mathrm{e} = \int \exp\left[\frac{2\pi i}{\lambda}(\boldsymbol{s}-\boldsymbol{s}_0)\cdot\boldsymbol{r}\right]\rho\mathrm{d}V \tag{3.4}$$

Note that the value of f_e corresponds to the ratio between the amplitude of coherent scattering by a single electron which has distribution and that of X-rays scattered by a single electron whose location is fixed at a point according to the classical view.

Further, the wave vectors of $\boldsymbol{s}_0$ and $\boldsymbol{s}$ are related

$$\boldsymbol{q} = \boldsymbol{s}-\boldsymbol{s}_0 \quad \Rightarrow \quad |\boldsymbol{q}| = q = \frac{2\sin\theta}{\lambda} \tag{3.5}$$

Here, the vector $\boldsymbol{q}$ is frequently called "scattering vector (or wave vector)." This is the vector required to turn the direction of incident X-rays to that of the scattered radiation by angle 2θ, thus, $\boldsymbol{s} = \boldsymbol{s}_0 + \boldsymbol{q}$.

Since the electron density distribution of inner shell is quite likely to be approximated by spherical symmetry, the scattering factors for atom containing more than two electrons can be readily estimated. For example, if the electron density distribution around the nucleus set at origin is given by $\rho = \rho(r)$ as a function of distance r, the scattering factor f_e can be obtained as

$$f_e = \int_0^\infty 4\pi r^2 \rho(r) \frac{\sin 2\pi q r}{2\pi q r} \mathrm{d}r \tag{3.6}$$

The amplitude of the coherent scattering per atom including n electrons is computed by the sum of f_e regarding all electrons using the following equation.

$$f = \sum_j f_{en} = \sum_j \int_0^\infty 4\pi r^2 \rho_j(r) \frac{\sin 2\pi q r}{2\pi q r} \mathrm{d}r \tag{3.7}$$

This f is usually called "X-ray atomic scattering factor" or simply "the atomic scattering factor." Sometimes f is called "the form factor" because it depends on the distribution of electrons around the nucleus. The quantity f provides the efficiency or ability of the coherent scattering per atom and it is defined as a ratio of the amplitude of the wave scattered from one atom to that scattered from one electron under the same condition. Therefore, $f = Z$ for any atom which scatters in the forward direction.

Equation (3.7) implies that f is a function of $(\sin\theta/\lambda)$ for any atom. The electron density distributions in atoms have been provided from the electron wave functions by using several techniques, such as Hatree–Fock and Fermi–Thomas–Dirac approximation and a number of theoretical calculations for the atomic scattering factors of elements were carried out as a function $(\sin\theta/\lambda)$. Such results are compiled in the International Tables for X-ray Crystallography, Vol.C (Kluwer Academic Pub., London, UK, 1999). The atomic scattering factors as a function $(\sin\theta/\lambda)$ are shown in Fig. 3.3 using the results for Al, Fe, and Ag as examples. The f value decreases as $(\sin\theta/\lambda)$ increases.

The variable of Q defined by

$$2\pi q = 2\pi \times \frac{2\sin\theta}{\lambda} = 4\pi\frac{\sin\theta}{\lambda} = Q \tag{3.8}$$

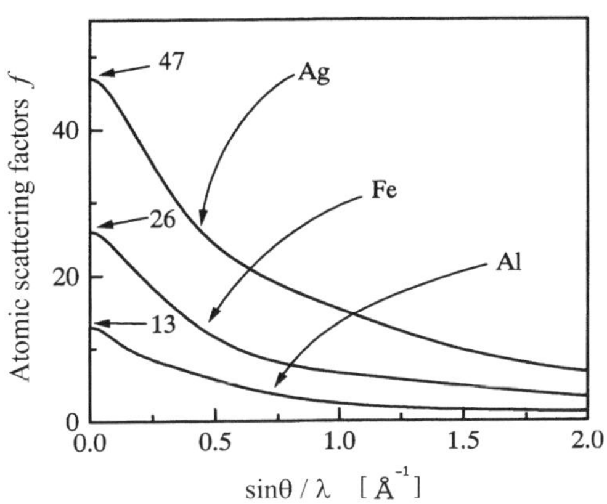

Fig. 3.3 Atomic scattering factors of Al, Fe, and Ag

The variable of Q is often used. It can be obtained from $(2\pi q)$ in (3.7) using (3.5). Equation (3.7) can be rewritten as

$$f = \sum_j f_{en} = \sum_j \int_0^\infty 4\pi r^2 \rho_j(r) \frac{\sin Qr}{Qr} \mathrm{d}r \tag{3.9}$$

It is useful to summarize some important points regarding scattering of X-rays from a single atom. When monochromatic X-ray beam encounters an atom, two scattering processes, coherent and incoherent, simultaneously occur in all directions. The intensity of incoherent (Compton) scattering for light elements (with small atomic number) is found to increase with decreasing atomic number Z. Furthermore, as the quantity of $(\sin\theta/\lambda)$ increases intensity of incoherent (Compton) scattering increases. Thus, the intensities of unmodified (coherent) scattering and modified (Compton) scattering change in opposite ways with Z and $(\sin\theta/\lambda)$. In addition, the sum of the coherent and incoherent scattering intensities is equal to the classical scattering intensity per electron. If i_e represents the intensity of incoherent scattering in electron unit for the completely unpolarized incident X-ray beam, the following equation can be obtained:

$$\begin{aligned} &I_0 \frac{e^4}{m^2c^4R^2}\left(\frac{1+\cos^2 2\theta}{2}\right) f_\mathrm{e}^2 + I_0 \frac{e^4}{m^2c^4R^2}\left(\frac{1+\cos^2 2\theta}{2}\right) i_\mathrm{e} \\ &= I_0 \frac{e^4}{m^2c^4R^2}\left(\frac{1+\cos^2 2\theta}{2}\right) \end{aligned} \tag{3.10}$$

It is note worthy that (3.10) can be simply rewritten in the form, $i_\mathrm{e} = 1 - f_\mathrm{e}^2$.

As previously mentioned, the phase of radiation attributed to the Compton scattering effect has no fixed relation to that of the incident beam, because of the variation in wavelength after collision, suggesting that interference effect cannot be produced by the modified Compton radiation. Therefore, the incoherent scattering intensity $i(M)$ per atom may be given by the simple sum of the incoherent scattering intensity of the respective electrons.

$$i(M) = \sum_j i_{en} = Z - \sum_{j=1}^{Z} f_{en}^2 \tag{3.11}$$

Figure 3.4 shows the calculated results of f and $i(M)$ for Li atom assuming that the electron density distributions for three electrons, two of K shell and one of L shell, are considered to be similar to that of the hydrogen atom. One clearly recognizes the reverse relationship in variation of the intensities of coherent scattering and Compton scattering with respect to $(\sin\theta/\lambda)$. As readily obtained from (3.9), the atomic scattering factor f of Li atom, corresponding to the coherent scattering intensity is represented by $f_\mathrm{Li} = 2f_\mathrm{eK} + f_\mathrm{eL}$, whereas the incoherent scattering intensity $i(M)$ in electron unit is given by the relationship of $i(M) = 3 - 2f_\mathrm{eK}^2 - f_\mathrm{eL}^2$(see (3.11)).

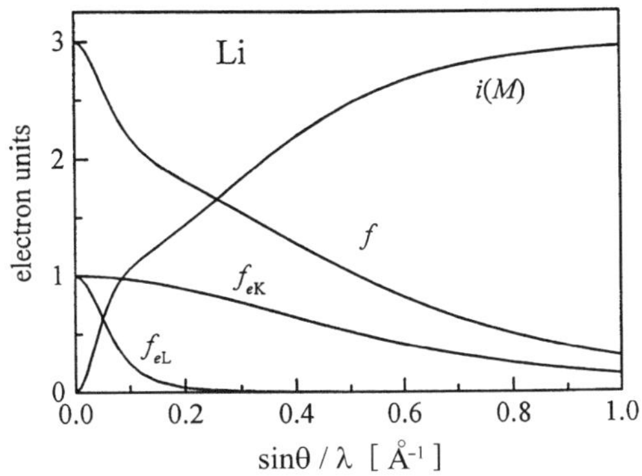

Fig. 3.4 Coherent and incoherent scattering intensities of Li atom calculated based on the assumption that the electron density distributions have spherical symmetry and the interference between electrons in the atom is ignored

3.3 Diffraction from Crystals

Atoms being the constituent of a crystal generates X-rays with the same wavelength as that of the incident X-ray beam by oscillating electrons and the generated X-rays are likely to be the spherical waves centering on respective atoms. This situation is similar to that a wave rolling from one side to the piles lined up on the same line at equal intervals in a pond and propagating to the other side. That is, the diffraction phenomena of X-rays by crystals is attributed to certain phase relations between two or more waves, such as differences in phase produced from the differences in path length of waves and a change in amplitude related to the phase difference. In addition, the most important point to know is that two waves are completely in phase if the difference in path lengths is zero or an integer multiple of wavelength. Some additional details are given below.

The phase of any two waves generally shows deviation of Δ, corresponding to their path difference. Since the phase of two waves is completely coincident (in phase) if the value of Δ is given by an integer multiple of wavelength λ, two waves will combine to form one synthesized wave just like the original one, and its amplitude will be double. The path difference depends on direction of X-ray with respect to the crystal. When the value of Δ is $\lambda/2$, the two waves cancel each other and they are completely out of phase, resulting from the fact that the waves have the same magnitude but opposite amplitude at any point along its path. Situations between these two extreme cases ($\Delta = \lambda$ and $\lambda/2$) are also encountered depending on path difference.

The main target of X-ray diffraction by crystals is to know the particular condition in which the scattered X-rays from atoms and the incident X-rays are completely in phase and reinforce each other to produce a detectable diffraction beam. In other words, we have to find the common relationship that the differences in path length between X-rays scattered from crystals and that of the incident X-rays

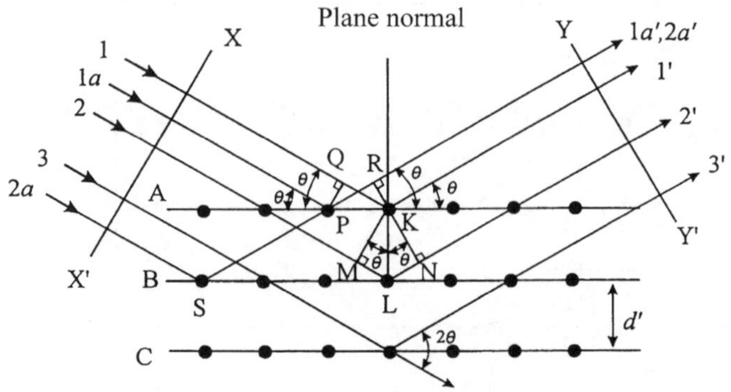

Fig. 3.5 Schematic diagram of diffraction of X-rays by a crystal (Bragg condition)

is an integer multiple of wavelength λ. For this purpose, the most important and familiar method is given by Bragg law which incorporates Bragg angle. In order to facilitate the understanding of Bragg law, Fig. 3.5 is useful. It is also required to remember the following two geometric relationships:

1. The angle between the incident X-ray beam and the normal to the reflection plane is equal to that between the normal and the diffracted X-ray beam. The incident X-ray beam, the plane normal, and the diffracted X-ray beam are always coplanar.
2. The angle between the diffracted X-ray beam and the transmitted one is always 2θ, and this angle is called "the diffraction angle."

If the incident X-rays of wavelength (λ) strike a crystal where all atoms are placed in a regular periodic array with interplanar spacing d', diffraction beam of sufficient intensity is detected only when the "Bragg condition" or "Bragg law" is satisfied

$$2d' \sin\theta = n\lambda \tag{3.12}$$

where n is called the order of reflection and is equal to the number of wavelengths in the path difference between diffracted X-rays from adjacent crystal planes (see Fig. 3.5).

For fixed values of both λ and d', the diffraction occurs at several angles of incidence such as $\theta_1, \theta_2, \theta_3, \ldots$, corresponding to $n = 1, 2, 3, \ldots$. In the first-order reflection ($n = 1$), the path difference between two scattered X-rays denoted by 1' and 2' in Fig. 3.5 is one wavelength. The path difference between X-rays 1' and 3' is two wavelengths, etc. The diffracted X-rays from all atoms in all the planes are considered completely in phase so as to produce the diffracted X-ray beam with appreciable intensity in a particular direction which satisfies the Bragg law. Equation (3.12) may be rewritten as

$$2d \sin\theta = \lambda \tag{3.13}$$

where $d = d'/n$. This form of Bragg law is frequently used.

Table 3.1 Information of plane spacing for seven crystal systems

Cubic	$\frac{1}{d^2} = \frac{h^2+k^2+l^2}{a^2}$
Tetragonal	$\frac{1}{d^2} = \frac{h^2+k^2}{a^2} + \frac{l^2}{c^2}$
Hexagonal	$\frac{1}{d^2} = \frac{4}{3}\left(\frac{h^2+hk+k^2}{a^2}\right) + \frac{l^2}{c^2}$
Trigonal	$\frac{1}{d^2} = \frac{(h^2+k^2+l^2)\sin^2\alpha+2(hk+kl+hl)(\cos^2\alpha-\cos\alpha)}{a^2(1-3\cos^2\alpha+2\cos^3\alpha)}$
Orthorhombic	$\frac{1}{d^2} = \frac{h^2}{a^2} + \frac{k^2}{b^2} + \frac{l^2}{c^2}$
Monoclinic	$\frac{1}{d^2} = \frac{1}{\sin^2\beta}\left(\frac{h^2}{a^2} + \frac{k^2\sin^2\beta}{b^2} + \frac{l^2}{c^2} - \frac{2hl\cos\beta}{ac}\right)$
Triclinic	$\frac{1}{d^2} = \frac{1}{V^2}(S_{11}h^2 + S_{22}k^2 + S_{23}k^2 + 2S_{12}hk + 2S_{23}kl + 2S_{13}hl)$

On the triclinic system, V is the volume of a unit cell and the coefficients are given below.

$$S_{11} = b^2c^2\sin^2\alpha, \qquad S_{12} = abc^2(\cos\alpha\cos\beta - \cos\gamma),$$
$$S_{22} = a^2c^2\sin^2\beta, \qquad S_{23} = a^2bc(\cos\beta\cos\gamma - \cos\alpha),$$
$$S_{33} = a^2b^2\sin^2\gamma, \qquad S_{13} = ab^2c(\cos\gamma\cos\alpha - \cos\beta)$$

In general, the nth order reflection from a certain crystal plane $(h\ k\ l)$ with the interplanar spacing of d could be considered the first-order reflection from a plane $(nh\ nk\ nl)$. Since the $(nh\ nk\ nl)$ plane is parallel to the $(h\ k\ l)$ plane, reflection from $(nh\ nk\ nl)$ plane is equivalent to the first order reflection from planes spaced at the distance $(d = d'/n)$. From such a point of view, 2θ is called the diffraction angle in many cases.

The diffraction angle 2θ of any set of planes $(h\ k\ l)$ can be computed by combining (3.13) with the plane-spacing equations (see Table 3.1) which relate distance between providing the relationship among the distance of adjacent planes to Miller indices and lattice parameters for each crystal system. For example, if the crystal is cubic with the lattice parameter a, the interplanar spacing d and Miller indices $(h\ k\ l)$ are given by

$$\frac{1}{d^2} = \frac{(h^2 + k^2 + l^2)}{a^2} \tag{3.14}$$

By combining the Bragg law with (3.14), one obtains for experiments using the wavelength of λ the following equation.

$$\sin^2\theta = \frac{\lambda^2}{4a^2}(h^2 + k^2 + l^2) \tag{3.15}$$

Equation (3.15) suggests that the diffraction angle, corresponding to diffraction directions, can be determined from the shape and size of the unit cell. This is an important point for structural analysis of substances. The converse is also very valuable, since one can possibly determine an unknown crystal structure by measuring the diffraction angles. In other words, the particular directions of diffracted X-ray beams given by the diffraction angles are related directly to the positions of atoms in the unit cell.

3.4 Scattering by a Unit Cell

As shown in Chap. 2, a crystal is defined as a solid consisting of atoms arranged in a periodic pattern defined by the unit cell. Therefore, when considering the scattering intensity from crystals, it is important to get information of phase differences based on the relationships between the scattering intensity and the atomic positions in one unit cell. Only some essential points are given below. Readers who want to have a deeper knowledge about this field should consult textbooks on structural determination by X-ray crystallography.

The phase difference of X-rays whose path difference is one whole wavelength is 2π radians (360°). Let us consider an atom A is placed at the origin of (000) and atom B at actual coordinates $(x\ y\ z)$ which can be expressed by fractional coordinates $(u\ v\ w)$ where $u = x/a$, $v = y/b$, and $w = z/c$, where a, b, and c are the lattice parameters along each axis. Under these conditions, the phase difference ϕ between the X-rays scattered by atom B and that scattered by atom A at the origin is given by the following equation for the $(h\ k\ l)$ reflection.

$$\phi = 2\pi(hu + kv + lw) \tag{3.16}$$

This relationship is applicable to a unit cell of any shape. If atoms A and B are of different elements, there will be differences not only in phase but also in amplitude. By adding waves scattered by all atoms in a unit cell including the atom at the origin, the resultant wave can be obtained. Each wave can be described by a vector whose length is equal to the amplitude and this vector is inclined to the horizontal-axis at the phase angle. Then, the amplitude and phase of the resultant wave are obtained simply by adding two vectors using the parallelogram law. Such a geometrical solution may be simplified if we use complex numbers to represent the vectors. Complex exponential function can also be employed. The variation in electric field intensity E with time t is frequently expressed in the form

$$E = A\exp(2\pi\, i\, \nu\, t) \tag{3.17}$$

$$= A\cos(2\pi\, \nu\, t) + \mathrm{i}A\sin(2\pi\, \nu\, t) \tag{3.18}$$

where A is the amplitude and ν is frequency. Equation (3.17) is called a complex exponential function and this is also convenient for discussion of wave behavior. The key points are displayed in Figs. 3.6 and 3.7.

A complex number is represented in the form $(a + \mathrm{i}b)$, where a and b are real numbers, whereas $\mathrm{i} = \sqrt{-1}$ is imaginary. A complex number is usually represented in the complex plane and as shown in Fig. 3.6, the real number is plotted as abscissa and the imaginary number as ordinate. Amplitude of the wave is given by A, the length of the vector, and phase by ϕ, the angle between the vector and the horizontal axis. With reference to (3.18) taken together with Fig. 3.6, the wave is now treated by the complex number, $A\ \cos\ \phi + \mathrm{i}\ A\ \sin\ \phi$. Note that these two terms correspond to the horizontal component OM and the vertical one ON of the vector.

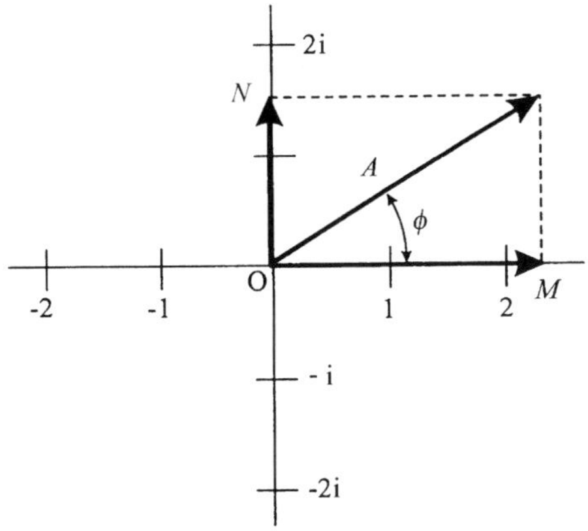

Fig. 3.6 Wave behavior represented by complex numbers

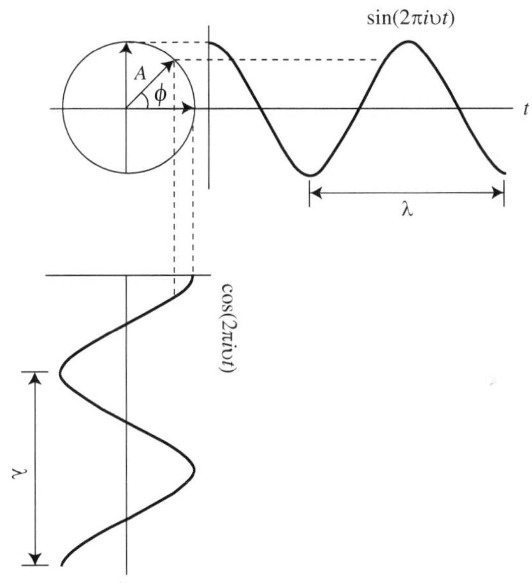

Fig. 3.7 Cosine component is obtained by projecting the wave vector on the real-axis and sine component on the imaginary-axis

In the complex plane, multiplication of a vector by i is equivalent to the rotation of the vector counterclockwise by 90°. Multiplication twice by i, ($i^2 = -1$), rotates the vector through 180°, equivalent to reversing the direction. In other words, when multiplying the horizontal vector 2 in Fig. 3.6 by i converts it into the vertical vector $2i$ along abscissa. Multiplication twice by i converts the horizontal vector 2 into the horizontal vector -2 along the ordinate, pointed in the opposite direction.

Using the power-series expansions of e^{ix}, $\cos x$ and $\sin x$, it can be shown that

$$Ae^{i\phi} = A\cos\phi + iA\sin\phi \tag{3.19}$$

The left hand side of (3.19) is called a complex exponential function. The wave intensity is proportional to the square of the amplitude, so that we have to obtain the square of the absolute value of the wave vector, (A^2). When the wave is described by the complex form, the A^2 value can be obtained by multiplying the complex function by its complex conjugate, (i being replaced by –i). The complex conjugate of $Ae^{i\phi}$ is $Ae^{-i\phi}$, so that $|Ae^{i\phi}|^2 = Ae^{i\phi}Ae^{-i\phi} = A^2$.

Some useful relations, such as $e^{n\pi i} = (-1)^n$, $e^{n\pi i} = e^{-n\pi i}$, where n is any integer, $e^{ix}+e^{-ix} = 2\cos x$ and $e^{ix}-e^{-ix} = 2i\sin x$ are very useful. The amplitude of the scattered wave from each atom in a unit cell is given by the atomic scattering factor of the atom f, and the value of $(\sin\theta/\lambda)$ relevant to diffraction. Since the phase of each wave is given by (3.16) for the $(h\ k\ l)$ reflection using fractional coordinates $(u\ v\ w)$ of the atom, any scattered wave may be expressed in the complex exponential form.

$$Ae^{i\phi} = fe^{2\pi i(hu+kv+lw)} \tag{3.20}$$

Based on this relationship, the sum of the scattered waves from atoms in a unit cell can be computed using the generalized equation.

$$F_{hkl} = \sum_{j=1}^{N} f_j e^{2\pi i(hu_j+kv_j+lw_j)} \tag{3.21}$$

where N is the total number of atoms involved in the unit cell; for example, it is 2 for bcc and 4 for fcc. F defined by (3.21) is called the structure factor or geometrical structure factor and one can obtain the value of F, only if the position and the type of atoms in the unit cell are given. For this reason, (3.21) as well as Bragg condition given by (3.13) are considered very important relationships for analyzing the structure of crystals by X-ray diffraction.

The structure factor F is generally given by a complex number and represents both the amplitude and the phase of the scattered wave obtained from the summation over all atoms in the unit cell. In addition, the absolute value of F represents the amplitude of the resultant scattered wave by adding together waves scattered by individual atoms in the unit cell, based on that of the wave scattered by a single electron. Therefore, the intensity of the resultant wave scattered from all atoms in the unit cell in the direction which satisfies the Bragg law, is simply proportional to $|F|^2$. The value of $|F|^2$ can be obtained by multiplying F of (3.21) by its complex conjugate F^*.

Although there is no theoretical difficulty in calculation of the structure factor given by (3.21), the usefulness and application of (3.21) can be fully appreciated only by working out some actual cases. For example, in the case of the body-centered cell containing two atoms of the same kind, (position of 000 and $\frac{1}{2}, \frac{1}{2}, \frac{1}{2}$) in unit cell, the structure factor is given as follows:

Table 3.2 Relationship between Bravais lattice and reflections

Crystal type	Bravais Lattice type	Reflections possible present	Reflections necessarily absent
Simple	Primitive	Any h, k, l	None
Body-centered	Body-centered	$h + k + l$ even	$h + k + l$ odd
Face-centered	Face-centered	h, k and l unmixed	h, k and l unmixed
Diamond cubic	Face-centered cubic	As fcc, but if all even and $h + k + l \neq 4N$; then absent	h, k and l mixed and if all even and $h + k + l \neq 4N$
Base-centered	Base-centered	h and k both even or both odd* $h + 2k = 3N$ with l even	h and k mixed*
Hexagonal close-packed	Hexagonal	$h + 2k = 3N \pm 1$ with l odd $h + 2k = 3N \pm 1$ with l even	$h + 2k = 3N$ with l odd

* These relationships apply to a cell centered on the C face. If reflections are present only when h and k are unmixed, or when k and l are unmixed, then the cell is centered on the B or A face, respectively.

$$uvw = 000, \frac{1}{2}\frac{1}{2}\frac{1}{2}$$

$$F = f e^{2\pi i \times 0} + f e^{2\pi i\left(\frac{h}{2}+\frac{k}{2}+\frac{l}{2}\right)} = f[1 + f e^{\pi i(h+k+l)}]$$

$$\text{When the number of } (h + k + l) \text{ is even: } F = 2f, F^2 = 4f^2$$

$$\text{When the number of } (h + k + l) \text{ is odd: } F = 0, F^2 = 0$$

The results show that the reflections will be observed for planes such as (110), (200), and (211) whose indices satisfy the condition $(h + k + l)$ is even, but not for the planes (111), (210), (300), etc., because the waves cancel each other being out of phase.

In calculation of the structure factor, information on crystal systems is not required. In the previous example, the given information is only the body-centered cell containing two atoms. No information about the shape of the unit cell, such as cubic, tetragonal, and orthorhombic (see Bravais lattices in Fig. 2.4) was used. The structure factor is completely independent of the shape and size of the unit cell. This also illustrates an important point; any body-centered cell provides the missing reflections when $(h+k+l)$ is an odd number without reference to the specific crystal system. In other words, information about missing reflections provides a clue about the actual atomic arrangement in crystals. The relationship between Bravais lattice and diffraction behavior is summarized in Table 3.2.

The two or more kinds of atoms are included in a unit cell, it is necessary to take into consideration the atomic scattering factor of each atom in calculations using (3.21). They will be further illustrated in Questions 3.12 and 3.13.

3.5 Solved Problems (13 Examples)

Question 3.1 The differential cross-section of the coherent scattering intensity for a free electron is expressed by the following equation with respect to per unit plane angle (radian), if the classical electron radius is set to r_e

$$\frac{d\sigma_e}{d\phi} = \frac{r_e^2}{2}(1+\cos^2\phi)\cdot 2\pi\sin\phi \qquad (m^2/rad)$$

(1) Estimate σ_e which is called the Thomson classical scattering coefficient by the integration about angle.
(2) Express the Thomson classical scattering coefficient in barn unit. (1barn $= 1\times 10^{-28}$ m^2).
(3) When X-rays penetrate an aluminum foil of thickness 1 mm, calculate the probability of the coherent scattering induced from free electrons. The density of aluminum is 2.70×10^6 g/m^3.

Answer 3.1 Figure 1 shows the schematic diagram of the process by which coherent scattering is produced per unit solid angle (steradian) in direction denoted by angle ϕ, when photon strikes a free electron. The relationship of $d\Omega = dA/r^2 = 2\pi\sin\phi d\phi$ is readily established about area (A) denoted by a part of a coaxial cone which is specified by the angle between ϕ and $\phi + d\phi$ on the surface of a sphere of radius r.

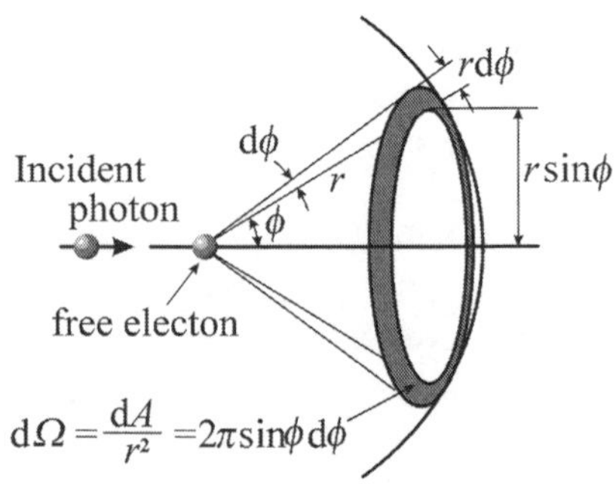

Fig. 1 Gemometric feature found in the cone of diffraction from a free electron

(1) Integration is made from zero to π with respect to the angle ϕ

$$\sigma_e = \int_0^{\pi}\frac{r_e^2}{2}(1+\cos^2\phi)2\pi\sin\phi d\phi = \pi r_e^2\int_0^{\pi}(\sin\phi+\cos^2\phi\sin\phi)d\phi$$
$$= \pi r_e^2\left[-\cos\phi - \frac{\cos^3\phi}{3}\right]_0^{\pi} = \pi r_e^2\left[\frac{4}{3}+\frac{4}{3}\right] = \frac{8}{3}\pi r_e^2$$

where the following relationship is used.

$$t = \cos\phi \quad \rightarrow \quad dt = -\sin\phi d\phi$$

$$\int \cos^2\phi \sin\phi d\phi = \int t^2 dt = \frac{t^3}{3} = \frac{\cos^3\phi}{3}$$

Reference: $I_n = \int \sin^n x dx = -\frac{1}{n}(\sin^{n-1} x \cos x) + \frac{n-1}{n} I_{n-2}$

$$I_1 = \int \sin x dx = -\cos x, \qquad I_0 = \int dx = x$$

(2) The value of the classical electron radius r_e is defined by the following equation using the permittivity of free space $\epsilon_0 = 10^7/(4\pi c^2)$ in electromagnetism.

$$r_e = \frac{e^2}{4\pi\epsilon_0 c^2 m_e} = \frac{e^2}{m_e \times 10^7} = \frac{(1.602 \times 10^{-19})^2}{9.109 \times 10^{-31} \times 10^7} = 2.8719 \times 10^{-15} \quad \text{(m)}$$

where m_e and e are electron rest mass and electron charge respectively. It may be noted for the dielectric constant of vacuum $\epsilon_0 = 8.854 \times 10^{-12}$ (Fm^{-1}) is also employed as electric capacity per unit length. The important point for σ_e obtained by question (1) is that the coherent scattering intensity is proportional to r_e^2. Because of an extremely large mass of the nucleus relative to the electron case, about 1,800 times that of the electron mass, the net effect on the coherent scattering intensity by the nucleus is very small. This is the reason why the scattering caused only by electrons in the atom is considered. The value of σ_e is estimated as follows, using the relation of $1\text{b} = 1 \times 10^{-28}\,\text{m}^2$.

$$\begin{aligned}\sigma_e &= \frac{8}{3}\pi r_e^2 = \frac{8}{3}\pi \times (2.8179 \times 10^{-15})^2 = 66.52 \times 10^{-30}\,(\text{m}^2)\\ &= \frac{66.52 \times 10^{-30}}{1 \times 10^{-28}} = 0.6652 \quad \text{(b)}\end{aligned}$$

Note: The cross-sectional area per unit area is usually given in barn, because the value becomes very small if SI unit is used.

(3) The mass of aluminum per mole (molar mass) is given as 26.98 g in Appendix A.2, and the number of electrons equivalent to the atomic number 13. The number of electrons N_e contained in 1 m^3 from density value is estimated in the following way.

$$N_e = \frac{0.6022 \times 10^{24}}{26.98 \times 10^{-3}} \times 13 \times 2.70 \times 10^3 = 0.783 \times 10^{30} \quad (\text{m}^{-3})$$

Converting the number of electrons of N_e^{film} for aluminum foil with its thickness of 1 mm into the one per unit area (m^{-2}), $N_e = 0.783 \times 10^{27}$ (m^{-2}). Then, the probability P of the coherent scattering intensity due to electrons in this case is found as follows:

$$P = \sigma_e \times N_e^{film} = 66.52 \times 10^{-30} \times 0.783 \times 10^{27} = 0.052$$

Question 3.2 From high-precision scattering experiments, 0.002426 nm is obtained as the wavelength of Compton scattering. Calculate the effective mass of photon m_e using the so-called Einstein relation $E = mc^2$ showing relationship between the mass of photon m and its energy E.

Answer 3.2 X-ray is electromagnetic radiation (photon) of exactly the same nature as visible light except that values of wavelength λ or energy (= frequency ν) are different. The propagation speed of photon c is considered equal to the speed of light in vacuum, 2.998×10^8 m/s.

The following equation is obtained from the relationship $c = h\nu$ by setting Planck constant to h.

$$E = h\nu = \frac{hc}{\lambda}$$

To calculate the effective mass of photon m_e, Einstein relation can be used

$$m_e c^2 = \frac{hc}{\lambda} \quad \rightarrow \quad m_e = \frac{h}{c\lambda}$$

Using $h = 6.626 \times 10^{-34}$ (J · s), $c = 2.998 \times 10^8$ (m/s), and 1 nm = 10^{-9} m.

$$m_e = \frac{6.626 \times 10^{-34}}{2.998 \times 10^8 \times 0.002426 \times 10^{-9}} = 9.110 \times 10^{-31} \quad \text{(kg)}$$

Note: The value of 9.109×10^{-31} kg is also used as electron rest mass. If acceleration voltage exceeds 100 kV, an increase (relativistic) in mass accompanying the variation of speed of photon should be taken into account by using Einstein relation. If the mass of photon and its velocity are set to m_0 and v, respectively, an increase (relativistic) in mass may be estimated from (1.8) in Chap. 1 for the case that acceleration voltage of 200 kV is applied to an electron.

The energy of an electron with its rest mass $m_0 = 9.109 \times 10^{-31}$ (kg) is computed in the following, by coupling with 1(eV) = 1.602×10^{-19} (J).

$$E = m_0 c^2 = \frac{9.109 \times 10^{-31} \times (2.998 \times 10^8)^2}{1.602 \times 10^{19}} = 0.5109 \times 10^6 \text{ (eV)}$$

Since the energy given to an electron is equivalent to 0.2×10^6 (eV) in the case that applied voltage is 200 kV, and the speed v of photon may be estimated as follows.

$$v = c \times \sqrt{1 - \left(\frac{0.5109}{0.2 + 0.5109}\right)^2} = c \times \sqrt{1 - 0.5165} = 0.6953c$$

Because of $\left(\frac{v}{c}\right) = 0.6953$, the increase in mass of photon in this case can be obtained in the following equation:

$$m = \frac{m_0}{\sqrt{1-\left(\frac{v}{c}\right)^2}} = \frac{m_0}{\sqrt{1-(0.6953)^2}} = \frac{m_0}{0.7187} = 1.39m_0$$

Question 3.3 When incident X-rays collide with a free electron, a part of the energy of incident X-rays (photons) is given to the electron as kinetic energy. Accordingly, the energy of the photon after collision is less than the energy before collision. That is, the wavelength of X-rays after collision becomes slightly longer than the wavelength of the incident X-rays and it is called the Compton shift. Answer the following questions related to such incoherent scattering of X-rays.

(1) Obtain the Compton equation for the case where a photon with the energy of $h\nu_0$ (momentum of $h\nu_0/c$) collides with an electron at rest.
(2) Compute the increment of Compton shift in wavelength produced at a scattering angle of 30°.

Here, h is Planck constant, c the speed of light in a vacuum, and ν the frequency.

Answer 3.3

(1) Momentum is usually given by the product mv of mass m and speed v of a desired particle. On the other hand, the momentum of a photon may be expressed by $\frac{h\nu}{c}$ or $\frac{h}{\lambda}$ using some relationships that the kinetic energy of a photon is described by $E = h\nu$, the Einstein relation of $E = mc^2$ and $c = \nu\lambda$, where λ is the wavelength.

The collision between a photon and a free electron is considered an elastic one, as shown in Fig. 1. If the photon having the momentum of $\frac{h\nu_0}{c}$ collides with the electron at rest, the electron is knocked aside and the direction of the photon is deviated through an angle 2θ. Since a part of energy of the incident photon is given to the electron, the momentum of photon $\frac{h\nu}{c}$ after collision is quite likely to become small in comparison with $\frac{h\nu_0}{c}$ before collision.

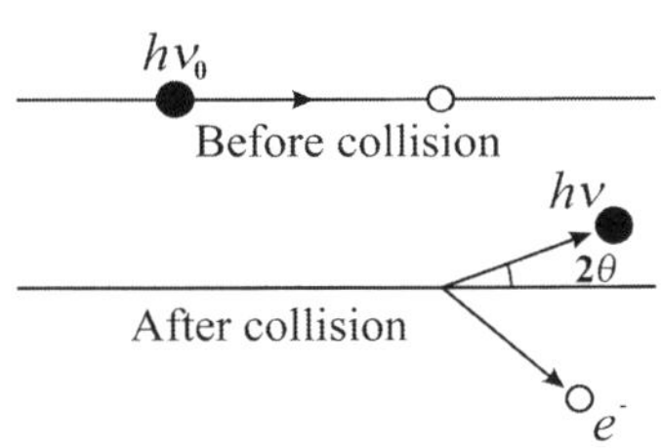

Fig. 1 The elastic collision of photon and an electron (the Compton effect)

It is thought that the law of conservation of energy is satisfied before and after collision. Thus,

$$h\nu_0 + m_0c^2 = h\nu + mc^2 \tag{1}$$

$$mc^2 = h(\nu_0 - \nu) + m_0c^2 = A + B \tag{2}$$

Taking square of both sides of (2) and re-arranging;

$$\begin{aligned}(mc^2)^2 &= \mathrm{A}^2 + \mathrm{B}^2 + 2\mathrm{AB}\\ &= [h(\nu_0 - \nu)]^2 + (m_0c^2)^2 + 2hm_0c^2(\nu_0 - \nu) \qquad (3)\\ &= (h\nu_0)^2 + (h\nu)^2 - 2h^2\nu_0\nu + (m_0c^2)^2 + 2hm_0c^2(\nu_0 - \nu) \qquad (4)\end{aligned}$$

The electron before collisions is at rest, $v_0 = 0$. The momentum of the electron before collision is also zero. Denoting the velocity of electron after collision by v, and invoking the law of conservation of momentum, one obtains

$$\frac{h\nu_0}{c} = \frac{h\nu}{c} + mv \quad \rightarrow \quad mvc = h\nu_0 - h\nu \tag{5}$$

By applying the "law of cosines" ($\mathrm{A}^2 = \mathrm{B}^2 + \mathrm{C}^2 - 2\mathrm{BC}\cos\theta$) to (5),

$$(mvc)^2 = (h\nu_0)^2 + (h\nu)^2 - 2h^2\nu_0\nu\cos 2\theta \tag{6}$$

By subtracting (6) from (4);

$$(mc^2)^2 - (mvc)^2 = (m_0c^2)^2 + 2hm_0c^2(\nu_0 - \nu) - 2h^2\nu_0\nu(1 - \cos 2\theta) \tag{7}$$

The left hand side of (7) may be rewritten as

$$(mc^2)^2 - (mvc)^2 = (mc^2)^2\left[1 - \left(\frac{v}{c}\right)^2\right] = (m_0c^2)^2 \tag{8}$$

Here, we use the relationship with energy in case where an increase in mass of a photon arising from change of its velocity v can be estimated in the following way.

$$mc^2 = \frac{m_0c^2}{\sqrt{1 - \left(\frac{v}{c}\right)^2}} \quad \Rightarrow \quad (mc^2)^2\left[1 - \left(\frac{v}{c}\right)^2\right] = (m_0c^2)^2 \tag{9}$$

Therefore, (7) can be summarized as;

$$2hm_0c^2(\nu_0 - \nu) = 2h^2\nu_0\nu(1 - \cos 2\theta) \tag{10}$$

$$c(\nu_0 - \nu) = \frac{h\nu_0\nu}{m_0c}(1 - \cos 2\theta) \tag{11}$$

On dividing both sides of (11) by $\nu_0\nu$ and using $\lambda_0 = \frac{c}{\nu_0}$ and $\lambda = \frac{c}{\nu}$, we obtain the Compton equation, as follows.

$$\Delta\lambda = \lambda - \lambda_0 = \frac{h}{m_0 c}(1 - \cos 2\theta) \tag{12}$$

(2) Equation (12) suggests that the Compton shift of $\Delta\lambda$ depends only on the scattering angle and it varies from zero in the forward direction ($2\theta = 0$) and to twice the term of $h/(m_0 c)$ in the backward direction ($2\theta = 180°$). It is also noted that the incoherent scattering due to the Compton effect is not always welcome in analyzing the structure of substances, because it increases the background intensity.

Substituting $h = 6.626 \times 10^{-34}$ (J · s), $m_0 = m_e = 9.109 \times 10^{-31}$ (kg), and $c = 2.998 \times 10^8$ (m/s), in (12)

$$\frac{h}{m_e c} = \frac{60626 \times 10^{-34}}{9.109 \times 10^{-31} \times 2.998 \times 10^8} = 0.2426 \times 10^{-11} \quad \text{(m)}$$
$$= 0.0243 \times 10^{-8} \quad \text{(cm)}$$

Next, further substitution of $2\theta = 30°(\cos 30° = 0.866)$ in (12) gives;

$$\Delta\lambda = 0.2426 \times 10^{-11}(1 - 0.866) = 0.0325 \times 10^{-11} \quad \text{(m)}$$
$$= 0.0033 \times 10^{-8} \quad \text{(cm)}$$

Question 3.4 A stream of X-ray quanta with the energy of 200 keV strikes a free electron and the incoherent scattering is produced in a direction 180° relative to the incident beam. The energies of the scattered photon and the recoil electron were found 111.925 and 87.815 keV, respectively. Verify if this incoherent scattering process satisfies the law of conservation of momentum.

Answer 3.4 According to the Einstein relation of $E = mc^2$, the energy is equivalent to the mass and it is also noted that an increase in mass of a quantum is related directly to the change of its speed. If rest mass of an electron 9.109×10^{-31} kg is converted into energy, we obtain 0.5109×10^6 eV = 510.9 keV. Total energy E (it is equivalent to mass) of the electron after collision may be given, in keV, by the sum of the energy of the scattered photon and that of the recoil electron.

$$E + m_0 c^2 = 87.815 + 510.9 = 598.715 \quad \text{keV}$$

The speed of electron v can be computed using (1.8) in the following way:

$$v = c\sqrt{1 - \left(\frac{m_0 c^2}{E + m_0 c^2}\right)^2} = c\sqrt{1 - \left(\frac{510.9}{598.715}\right)^2} = 0.521\, c$$

The energy (equivalent to mass) of the recoil electron is quite likely to increase because of collision and its momentum p may be given by mv and, then we obtain,

$$p = 598.715 \times 0.521\,c = 311.931\,c$$

Since the incoherent scattering is produced at a scattering angle of 180°, corresponding to the extreme backward direction to the incident X-ray beam, the sum of momentum of the incident X-rays and the scattered photon should be compared with the momentum of the recoil electron. For this reason, we obtain the following equation using $p = m_{\text{eff}}c = h\nu/c$.

$$\frac{h\nu_0}{c} + \frac{h\nu}{c} = (200 + 111.925)c = 311.925\,c$$

Although a very small difference of 0.006c is found between the two values, it may be inferred that the law of conservation of momentum is almost satisfied.

Question 3.5 X-rays with energy 51.1 keV produced incoherent scattering by impacting with an electron in the outershell of an atom in a sample. Answer to the following questions.

(1) Estimate the angle (so-called recoil angle ϕ) between the direction of the recoil electron and that of the incident X-ray beam, when the scattering angle 2θ is 20°.
(2) Compute the energy of the scattered photon in this case.

Answer 3.5

(1) In the incoherent scattering process, the recoil angle ϕ is known to correlate with the scattering angles 2θ in the following form. For convenience, Fig. 1 shows the schematic diagram of two angles together with the incident X-ray photon beam.

$$\frac{1}{\tan\phi} = \left(1 + \frac{h\nu}{m_0c^2}\right)\tan\frac{2\theta}{2}$$

In other words, this equation suggests that we can obtain the scattering angle 2θ if the recoil angle ϕ is known or *vice versa*. Similarly, if either the energy $(h\nu_0)$ of the incident X-ray photon beam or the energy $h\nu$ of the scattered photon is given, the energy (E_r) of the recoil electron can be computed. Since the energy of the incident X-ray photon beam is given as 51.1 keV, one obtains

$$\frac{h\nu}{m_0c^2} = \frac{51.1 \times 10^3}{0.5109 \times 10^6} = 0.1$$

Here, we use the information that rest mass of an electron 9.109×10^{-31} kg is equivalent to energy of 0.5109×10^6 eV = 510.9 keV (see Question 3.4).

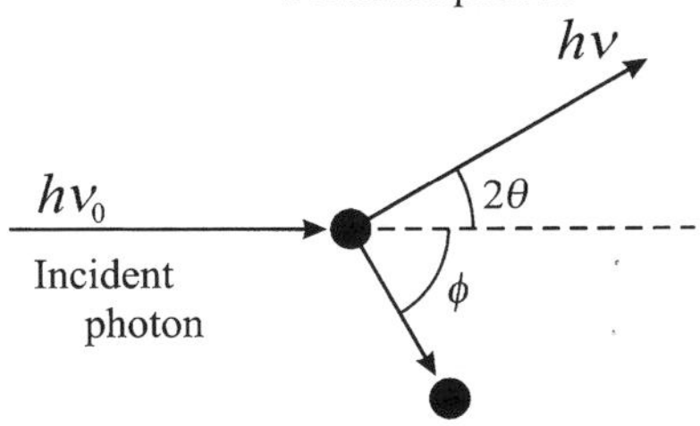

Fig. 1 Schematic diagram for the incoherent scattering process where the incident X-ray photon beam, a scattered photon, and a recoil electron are correlated

On the other hand, from the relationship of $E(\text{keV}) \equiv 1.240/\lambda$ (nm) for the wavelength λ_0 of the incidence X-ray beam, we obtain,

$$\lambda_0 = \frac{1.240}{51.1} = 0.0243 \quad (\text{nm})$$

The recoil angle ϕ can be computed from the value of the scattering angle 20°, as follows:

$$\frac{1}{\tan\phi} = (1 + 0.1)\tan\frac{20}{2}$$

$$\tan\phi = 5.1557 \quad \rightarrow \quad \phi = 79.0^\circ$$

The recoil electron is released in the direction of 79° with respect to the direction of propagation of the incident X-ray beam.

(2) Calculate the Compton shift in the case where the scattering angle is 20°, using (3.2). Since $\cos 2\theta = 0.9397$, one obtains

$$\Delta\lambda = 0.2426 \times 10^{-11}(1 - \cos 2\theta) = 0.2426 \times 10^{-11} \times 0.0603 = 0.0001 \times 10^{-9}\ (\text{m})$$

Therefore, the wavelength of the scattered photon is

$$\lambda = \lambda_0 + \Delta\lambda = 0.0243 + 0.0001 = 0.0244 \quad (\text{nm})$$

The energy of this photon may be estimated from $E = h\nu$,

$$E = h\nu = \frac{1.240}{0.024} = 50.8 \quad (\text{keV})$$

The energy of the recoil electron, E_r, can be computed from the difference in energy between the incident X-ray beam and the scattered photon.

$$E_r = 51.1 - 50.8 = 0.3 \quad (\text{keV})$$

The wavelength of the incident X-ray photon beam with energy of 51.1 keV becomes longer by 0.0001 nm after scattering and the energy of the scattered photon decreases to 50.8 keV. The recoil electron which has the energy of 0.3 keV is released at an angle of 79° from the direction of propagation of the incident X-ray beam.

Question 3.6 Complex number ($A\cos\phi + \mathrm{i}A\sin\theta$) is widely used as an analytic expression for the wave. Using the given diagram of the complex plane, discuss the wave vector with the amplitude and phase. Also confirm the product of a complex number and its complex conjugate is always constant and it is equivalent to the square of the amplitude.

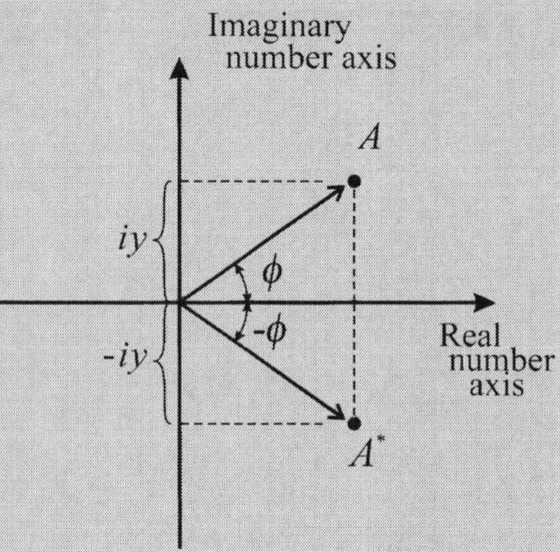

Fig. A Wave vector represented by the complex number

Answer 3.6 A real number can express only one quantity, but a complex number can represent two ingredients (e.g., amplitude and phase of a wave). This particular feature makes the use of vectors much more convenient. A complex number is the sum of a real and an imaginary number and it is usually described by solid dot in the complex plane as shown in the diagram. Here, real numbers are plotted as abscissa and imaginary numbers as ordinates and if the vector drawn from the origin to the solid dot indicates the complex number ($A = x + \mathrm{i}y$), where x and y are real numbers and i is an imaginary number. The length of vector from the origin denoted by $|A|$ corresponds to the amplitude and the phase is given by the angle ϕ between the vector and the axis of real number. Multiplication of a vector by i makes it rotate counterclockwise by 90°. Thus, multiplication by i converts a horizontal vector x to a vertical vector ix.

From the power-series expansions of $\mathrm{e}^{\mathrm{i}x}$, $\cos x$, and $\sin x$, we obtain $\mathrm{e}^{\mathrm{i}x} = \cos x + \mathrm{i}\sin x$. Then, the wave vector is described analytically by either side of the following equation with respect to the complex number A.

$$Ae^{i\phi} = A(\cos\phi + i\sin\phi) \tag{1}$$

$$x = A\cos\phi, \qquad y = A\sin\phi \tag{2}$$

Further, $Ae^{i\phi}$ of the left hand side of (1) is called a complex exponential function.

The complex conjugate of $A = x + iy$ is $A = x - iy$, or $Ae^{-i\phi}$ for $Ae^{i\phi}$, respectively. It is also usually described by A^*. If a wave is described in the complex form, its quantity can be obtained by multiplying the complex expression for the wave by its complex conjugate. As shown in the given diagram of the complex plane, $A = x + iy$ and $A^* = x - iy$ show the so-called mirror symmetry and the following relationship is readily obtained.

$$|A^{i\phi}|^2 = Ae^{i\phi}Ae^{-i\phi} = A^2(e^{i\phi} \cdot e^{-i\phi}) = A^2 \tag{3}$$

This relationship can also be found as follows.

$$A(\cos\phi + i\sin\phi) \cdot A(\cos\phi - i\sin\phi) = A^2(\cos^2\phi + \sin^2\phi) = A^2 \tag{4}$$

$$A \cdot A^* = (x + iy)(x - iy) = x^2 + y^2 = A^2 \tag{5}$$

where the relationship of $i^2 = -1$ is utilized. The operation i in a complex number is equivalent to the square root of -1. For example, when the operation i is additionally applied to the imaginary number $iy \to i(iy) = i^2 y = -y$. Namely, it becomes the real number with the opposite sign by rotating by $\pi/2$ counterclockwise.

The product of a complex number and its complex conjugate becomes the square of the amplitude of the original complex number (see (3)–(5)). This is quite a useful relationship which can be used for calculating the structure factor, because the intensity of a wave is proportional to the square of the amplitude.

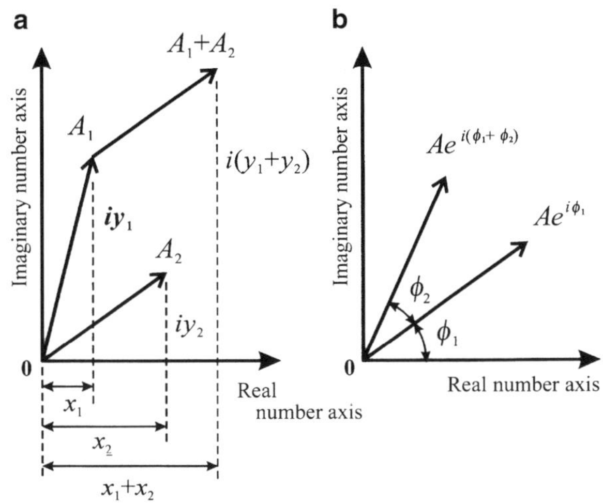

Fig. 1 Addition (**a**) and multiplication (**b**) of complex numbers

Note: The addition of two complex numbers corresponds to the sum of two vectors on a complex plane. The multiplication of two complex numbers is explained by the relationship of two vectors on a complex plane as follows. The rotation by ϕ_2 following the ϕ_1 rotation is equal to the rotation by $(\phi_1 + \phi_2)$. These relationships are illustrated in Fig. 1.

Question 3.7 X-rays belong to the electromagnetic spectrum and its propagation speed is equal to the velocity of light (c). When X-ray wavelength is set to λ, the frequency ν is given by $\nu = c/\lambda$. The cyclic variation in electric field intensity E may be expressed by the following equation, if time and phase are set to t and δ, respectively.

$$E = A\cos 2\pi(\nu t + \delta)$$

Discuss the variation in electric field E and its intensity I in the following two cases of superiomposition of two waves.

(1) Two waves with equal amplitude, but their phases are different.
(2) Two waves with difference in both amplitude and phase.

Answer 3.7 The variations in electric field intensity related to X-rays cannot be directly observed. The meaning of intensity of X-rays may be used in two ways. One is the energy which transits a unit area perpendicular to the direction of propagation of X-rays (wave). Another is the amount proportional to the square of the amplitude of wave obtained as a result of the interference effect of scattered X-rays. The latter is more frequently used in X-ray diffraction crystallography and we usually discuss its relative value. For this purpose, the expression of each wave as a complex exponential function is convenient and the cyclic variation in electric field intensity E can be described as follows:

$$E = A\mathrm{e}^{2\pi i(\nu t+\delta)} = A\cos 2\pi(\nu t + \delta) + \mathrm{i}A\sin 2\pi(\nu t + \delta) \tag{1}$$

where A is the amplitude.

(1) The summation of two waves with equal amplitude and different phase is given in the following equation.

$$E' = A\mathrm{e}^{2\pi i(\nu t+\delta)} + A\mathrm{e}^{2\pi i(\nu t+\delta')} \tag{2}$$

$$= A\mathrm{e}^{2\pi i(\nu t+\delta)}\{1 + \mathrm{e}^{2\pi i(\delta'-\delta)}\} \tag{3}$$

$$= E\{1 + \mathrm{e}^{2\pi i(\delta'-\delta)\}}\} \tag{4}$$

Since the intensity I is proportional to the square of E', we obtain,

$$I = |E'|^2 = |EE^*| = A^2\{1 + e^{2\pi i(\delta' - \delta)}\}^2 \tag{5}$$

$$= A^2 \cdot 2\{1 + \cos 2\pi(\delta' - \delta)\} \tag{6}$$

Here, $x = 2\pi(\delta' - \delta)$ and the following relationships of exponential and trigonometric functions are utilized.

$$1 + e^{ix} = 2 \cdot \left(\frac{e^{i\frac{x}{2}} + e^{-i\frac{x}{2}}}{2} \right) e^{i\frac{x}{2}} = 2\cos\frac{x}{2} \cdot e^{i\frac{x}{2}} \tag{7}$$

$$\cos^2\frac{\alpha}{2} = \frac{1}{2}(1 + \cos\alpha) \tag{8}$$

$$(1 + e^{ix})^2 = 2\cos\frac{x}{2} \cdot e^{i\frac{x}{2}} \cdot 2\cos\frac{x}{2} \cdot e^{-i\frac{x}{2}} \tag{9}$$

$$= 2(1 + \cos x)e^{i\frac{x}{2}}e^{-i\frac{x}{2}} = 2\{1 + \cos 2\pi(\delta' - \delta)\} \tag{10}$$

According to (6), one obtains $I = 4A^2$ if two waves are said to be in phase (= no phase difference) or constructive interference, but $I = 0$ if the phase difference is π (corresponding to the out of phase case or destructive interference).

(2) If two waves differ, not only in amplitude, but also in phase, the resultant E and I are given in the following equation.

$$E = \sum_j A_j e^{2\pi i(\nu t + \delta_j)} = e^{2\pi i\nu t} \sum_j A_j e^{2\pi i\delta_j} \tag{11}$$

$$I = |EE^*| = \left(e^{2\pi i\nu t} \sum_j A_j e^{2\pi i\delta_j} \right) \left(e^{-2\pi i\nu t} \sum_j A_j e^{-2\pi i\delta_j} \right) \tag{12}$$

$$= \left\{ \sum_j A_j^2 + \sum_{j \neq k}\sum A_j A_k e^{2\pi i(\delta_j - \delta_k)} \right\} \tag{13}$$

Here, * indicates the complex conjugate. As shown in (11), we can bundle the term of $e^{2\pi i\nu t}$ related to the frequency of X-rays out of sigma in wave synthesis, so that the $e^{2\pi i\nu t}$ term makes no contribution to scattering intensity. Thus, it is sufficient to consider only amplitude and phase for discussing intensity arising from superposition of waves. For this reason, we can use the following equations instead of E.

$$G = Ae^{2\pi i\delta} \quad \rightarrow \quad I = |GG^*| \tag{14}$$

Equation (13) can also be expressed in the following form when using the relationship $\sum\sum_{j\neq k} = \sum\sum_{j>k} + \sum\sum_{j<k}$.

$$I = \left\{ \sum_j A_j^2 + 2\sum_{j>k}\sum A_j A_k \cos 2\pi(\delta_j - \delta_k) \right\} \tag{15}$$

Since $\sin 2\pi(\delta_j - \delta_k) = -\sin 2\pi(\delta_k - \delta_j)$, all sine terms are zero. On the other hand, $\cos 2\pi(\delta_j - \delta_k) = \cos 2\pi(\delta_k - \delta_j)$, so that cosine terms are twice except for $j = k$.

Note: As shown in Fig. 1, the phase difference between the origin and a position of x at a certain time is considered to be $2\pi\delta$ with respect to a wave which propagates to the x-direction. The phase differences may be given in wavelength measure. Two waves differing in path length by one whole wavelength (λ) will shift in phase by 2π radians or 360°. Therefore, the phase difference is described by the product of the path difference and $(2\pi/\lambda)$ in the following form.

$$\delta = \frac{x}{\lambda} \tag{16}$$

When setting $\boldsymbol{k}$ to a unit vector in the direction of propagation of waves, the phase is considered constant in all positions on a plane perpendicular to $\boldsymbol{k}$. Therefore, the phase difference between the origin and an arbitrary position in a space (see R of Fig. 2) designated by vector $\boldsymbol{r}$ is may be given by

$$2\pi\delta = 2\pi\frac{\boldsymbol{k}\cdot\boldsymbol{r}}{\lambda} \tag{17}$$

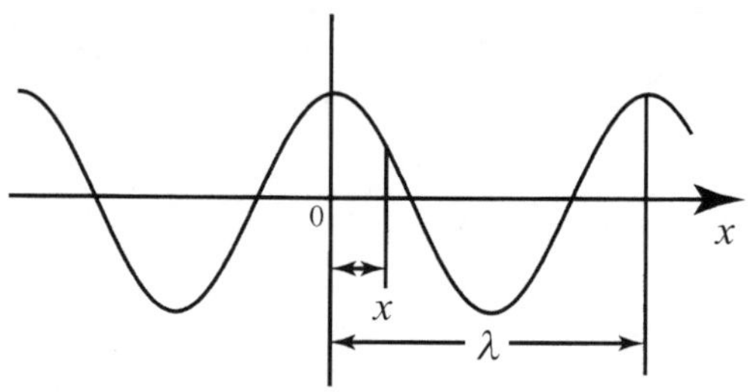

Fig. 1 Phase difference of the wave which propagates to the x-direction

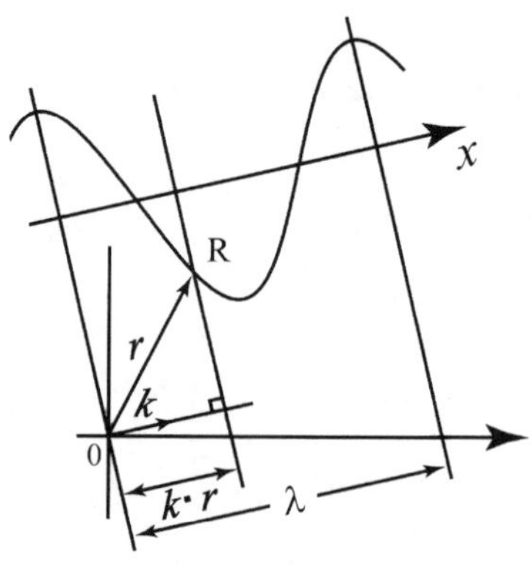

Fig. 2 Phase difference between the origin and an arbitrary position designated by a vector r

For simplification, if the wave vector of $\boldsymbol{K}$ defined by $\boldsymbol{K} = \boldsymbol{k}/\lambda$ is employed, we obtain $\delta = \boldsymbol{K} \cdot \boldsymbol{r}$ and (14) will be rewritten in the following form. Note that the wave vector of $\boldsymbol{K}$ has the length of $1/\lambda$ in the direction of $\boldsymbol{k}$.

$$G = A\mathrm{e}^{2\pi i \boldsymbol{K}\cdot\boldsymbol{r}} \qquad G^* = A\mathrm{e}^{-2\pi i \boldsymbol{K}\cdot\boldsymbol{r}} \tag{18}$$

Question 3.8 There is an atom in which the density distribution of one electron around the nucleus at the distance of $\boldsymbol{r}$ is given by $\rho(\boldsymbol{r})$. Answer the following questions.

(1) Derive the scattering factor f_e when X-rays irradiate this atom.
(2) Derive a generalized equation for the scattering factor of an atom containing Z electrons.

Answer 3.8

(1) The electrons of an atom are known to be situated at different points in space and they may be visualized as points arranged around the nucleus, as shown in Fig. 1 for simplification. Consider the interference of two waves scattered from the electrons at the position of A and B. The two waves scattered in the forward direction ($2\theta = 0$) are said to be "in phase" on a wavefront XX', because these two waves travel exactly the same distance (no path difference) before and after scattering. On the contrary, the other scattered waves in Fig. 1 are not in phase along the wavefront YY', because the path difference ($\overline{\mathrm{AD}} - \overline{\mathrm{CB}}$) is not an integer multiple of wavelength. Partial interference between the wave scattered by electrons A and B produces the result that the net amplitude of the scattered in this direction defined by scattering angle 2θ is less than that of wave scattered in the forward direction. In other words, the scattering amplitude of X-rays decreases, since the phase difference increases with increasing scattering angle 2θ.

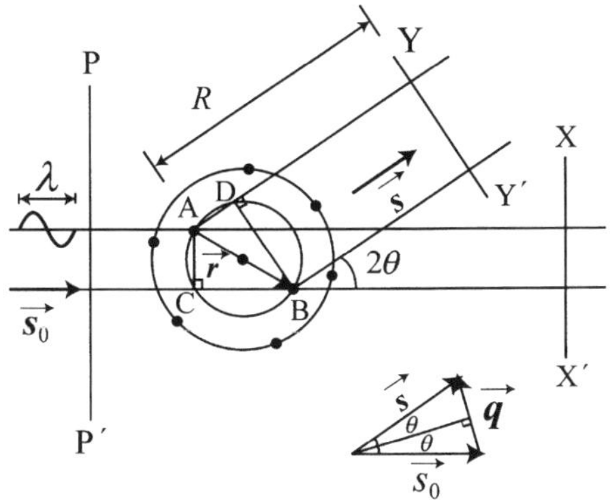

Fig. 1 Scattering of X-rays by an atom containing some electrons arranged as points around the nucleus

The scattered X-rays from A-electron will travel the optical path distance given by $\boldsymbol{s} \cdot \boldsymbol{r}$ before reaching point D. On the other hand, the scattered X-rays from B-electron will be generated just after the incident X-ray beam has traveled the optical path distance $\boldsymbol{s}_0 \cdot \boldsymbol{r}$. For this reason, the phase difference $2\pi\Delta$ (frequently called scattering phase shift) of X-rays scattered from electrons A and B in an atom is given in the following form.

$$2\pi\Delta = 2\pi\frac{(\overline{\mathrm{AD}} - \overline{\mathrm{CB}})}{\lambda} = 2\pi(\boldsymbol{s} \cdot \boldsymbol{r} - \boldsymbol{s}_0 \cdot \boldsymbol{r}) = 2\pi(\boldsymbol{s} - \boldsymbol{s}_0) \cdot \boldsymbol{r} = 2\pi\boldsymbol{q} \cdot \boldsymbol{r} \quad (1)$$

where $\boldsymbol{s}_0$ and $\boldsymbol{s}$ represent the wave number vector of the incident and scattered X-rays, respectively and the absolute value is equal to $1/\lambda$ ($|\boldsymbol{s}_0| = |\boldsymbol{s}| = 1/\lambda$). Equation (1) represents that $(\boldsymbol{s} - \boldsymbol{s}_0) \cdot \boldsymbol{r}$ corresponds to the optical path difference between the scattered wave from the electron at a distance $\boldsymbol{r}$ and that from the electron at the origin. The following relationship is also suggested.

$$\boldsymbol{q} = \boldsymbol{s} - \boldsymbol{s}_0 \quad \Rightarrow \quad |\boldsymbol{q}| = q = \frac{2\sin\theta}{\lambda} \quad (2)$$

This vector $\boldsymbol{q}$ is referred to as the scattering vector. In other words, the scattering vector $\boldsymbol{q}$ is a vector required to turn the incident X-ray wave to direction of 2θ and we obtain the relationship of $\boldsymbol{s} = \boldsymbol{s}_0 + \boldsymbol{q}$. The vector defining the relative locations of electrons A and B is given by $\boldsymbol{r}$. The interference effect of two scattered waves, observed on a wavefront YY' at a distance R which is considerably larger than the distance AB, can be expressed in the following equations.

$$y = \mathrm{e}^{2\pi i(\nu t+\delta)} + \mathrm{e}^{2\pi i(\nu t+\delta+\Delta)} = \mathrm{e}^{2\pi i(\nu t+\delta)}\{1 + \mathrm{e}^{2\pi i(\Delta-\delta)}\} \quad (3)$$

$$= \mathrm{e}^{2\pi i(\nu t+\delta)} \cdot (1 + \mathrm{e}^{2\pi i \boldsymbol{q}\cdot\boldsymbol{r}}) \quad (4)$$

where ν is frequency of the wave and t is time. The 1st term of (4) corresponds to the common phase for a wave which propagates toward a wavefront YY' and the 2nd term is the desired phase effect contributed by the two different scatterers. Nevertheless, we have to note that (4) just covers the case of scattering arising from only two electrons A and B, with A-electron set at the origin. Therefore, if an atom contains n electrons, the interference effect of scattered waves should be considered with respect to the vector $\boldsymbol{r}_j$ showing all positional relationships of n electrons and the generalized equation is as follows:

$$y_n = \mathrm{e}^{2\pi i(\nu t+\delta)} \cdot \sum_{j=1}^{n} f_j \mathrm{e}^{2\pi i \boldsymbol{q}\cdot\boldsymbol{r}_j} \quad (5)$$

where f_j is equivalent to the scattering ability of the jth electron. The 2nd term of (5) gives the scattering amplitude for the case containing n electrons with the scattering ability of f_j.

In quantum mechanics, the electron charge is likely to be distributed in space like a cloud and we use the distribution function $\rho(\boldsymbol{r})$ providing the number of electrons per unit volume as a function of distance. Then, the 2nd term of (5) is given by the following form with respect to f_x called the atomic scattering factor.

$$f_x = \int_{atom} e^{2\pi i \boldsymbol{q}\cdot\boldsymbol{r}} \rho(\boldsymbol{r}) dV \tag{6}$$

Note that $\int \rho(\boldsymbol{r})dV = 1$ for the case of one electron only. When spherical polar coordinates (see Fig. 2) are used for integration about $\boldsymbol{q} \cdot \boldsymbol{r}$, the following relationships are readily found.

$$\boldsymbol{q} \cdot \boldsymbol{r} = qr \cos\beta \tag{7}$$

$$dV = r^2 \sin\beta d\beta d\phi dr \tag{8}$$

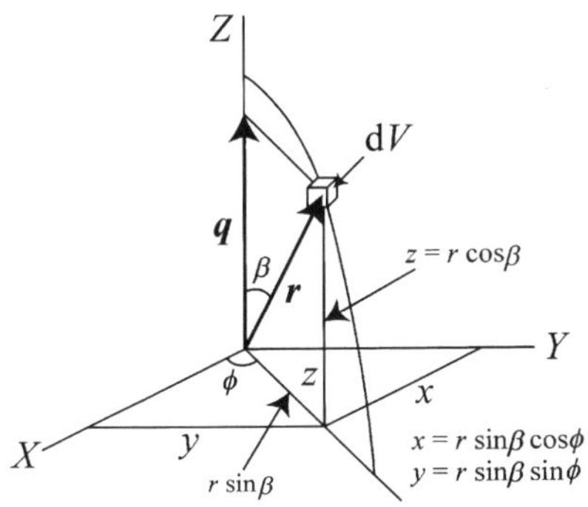

Fig. 2 Variables described by the spherical polar coordinates

By using (7) and (8), we obtain the scattering factor f_e for one electron as follows. It may be worthy of note that f_e given here is defined as a ratio of amplitudes; the ratio of amplitude of the wave scattered by one electron with the distribution function $\rho(\boldsymbol{r})$ to amplitude of the wave scattered by one electron classically localized at the origin (delta function).

$$f_e = \int_{r=0}^{\infty} \int_{\beta=0}^{\pi} \int_{\phi=0}^{2\pi} e^{2\pi i qr \cos\beta} \rho(r) r^2 \sin\beta d\beta d\phi dr \tag{9}$$

$$= 2\pi \int_{r=0}^{\infty} \rho(r) r^2 \int_{\omega=-1}^{1} e^{2\pi i qr\omega} d\omega dr \quad (\omega = \cos\beta) \tag{10}$$

$$= 2\pi \int_{r=0}^{\infty} \rho(r) r^2 dr \left[\frac{e^{2\pi i qr} - e^{-2\pi i qr}}{2\pi i qr} \right] \tag{11}$$

$$= 2\pi \int_{r=0}^{\infty} \rho(r) r^2 dr \times \frac{2i \sin(2\pi qr)}{i 2\pi qr} \tag{12}$$

$$= 4\pi \int_{r=0}^{\infty} \rho(r) r^2 \frac{\sin 2\pi qr}{2\pi qr} dr \tag{13}$$

The following relationships as well as $d\omega = -\sin\beta d\beta$ are also used in this calculation.

By considering the relationships $t = 2\pi i q r$ and $\int e^{tx}dx = \frac{e^{tx}}{t}$, we obtain the following equation.

$$\int_{-1}^{1} e^{tx}dx = \left[\frac{e^{tx}}{t}\right]_{-1}^{1} = \frac{e^{t} - e^{-t}}{t}$$

Note that $e^{ix} - e^{-ix} = 2i\sin x$ can readily found from $e^{ix} = \cos x + i\sin x$ and $e^{-ix} = \cos x - i\sin x$.

(2) With respect to the scattering factor of an atom containing Z electrons, equivalent to the atomic number, we use the approach similar to the f_e case in the following form, if only the electron distributions depend on the distance $\boldsymbol{r}$.

$$f_x = \sum_{j=1}^{Z} f_{e,j} = \sum_{j=1}^{Z} 4\pi \int_{r=0}^{\infty} \rho_j(r) r^2 \frac{\sin 2\pi q r}{2\pi q r} dr \tag{14}$$

$$\sum_{j=1}^{Z} 4\pi \int_{r=0}^{\infty} \rho_j(r) r^2 dr = Z \tag{15}$$

At the condition of $q \to 0$,

$$\frac{\sin 2\pi q r}{2\pi q r} \to 1 \tag{16}$$

Therefore, f_x converges on Z, if the wave vector approaches zero ($q \to 0$), corresponding to scattering in the forward direction (the scattering angle is zero). In other words, the value of f_x is equal to Z times the amplitude of the wave scattered from one electron. The quantity $2\pi q$ found in (13) and (14) is called "wave vector" and the following expressions are also used in many cases.

$$2\pi q = 2\pi \times \frac{2\sin\theta}{\lambda} = 4\pi \frac{\sin\theta}{\lambda} = Q \tag{17}$$

$$f_x = \sum_{j=1}^{Z} \int_{r=0}^{\infty} 4\pi r^2 \rho_j(r) \frac{\sin Qr}{Qr} dr \tag{18}$$

Question 3.9 Hydrogen contains one electron (usually referred to as 1s) in K-shell and its distribution is given by the function of distance r only, as shown in Fig. A. Wave function $\psi_{1s}(r)$ of 1s electron is expressed in the following form in terms of the atomic number Z and Bohr radius r_0.

$$|\psi_{1s}(r)|^2 = \frac{1}{\pi}\left(\frac{Z}{r_0}\right)^3 e^{-2\frac{Z}{r_0}r}$$

Obtain the Fourier transform $F(Q)$ for calculating the atomic scattering factor f_{H} for a hydrogen atom. Note that wave function $\psi_{1s}(r)$ satisfies the following relationship.

$$\int_0^{\infty} 4\pi r|\psi_{1s}(r)|^2 \mathrm{d}r = 1$$

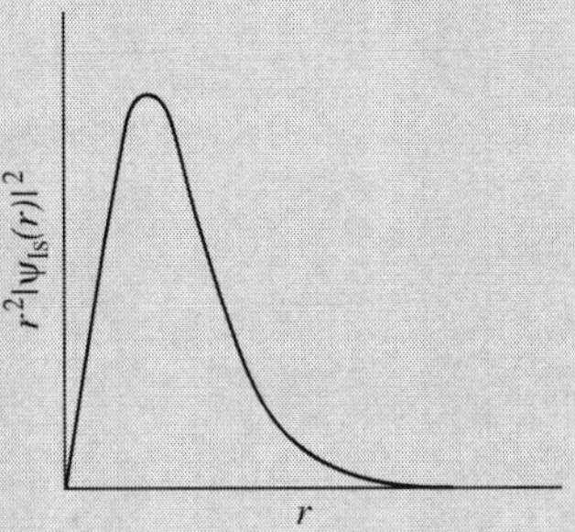

Fig. A Schematic diagram for the radial distribution function of 1s electron of hydrogen

Answer 3.9 We use (3.9) in this calculation.

$$f_{\mathrm{x}} = \sum_{j=1}^{Z} \int_0^{\infty} 4\pi r^2 \rho_j(r) \frac{\sin Q \cdot r}{Q \cdot r} \mathrm{d}r \tag{1}$$

The essential scheme for calculation is to put the given condition, $\rho_1(r) = |\psi_{1s}(r)|^2$, into the integrand of (1) and the integration is simply done. Nevertheless, some key points are shown below, because the integration of (1) is equivalent to the following Fourier transformation. Fourier transform of a function $f(\boldsymbol{r})$ and its inverse Fourier transform $F(\boldsymbol{r})$ are given in the generalized form as follows:

$$\left.\begin{aligned} F(Q) &= \int f(r)\mathrm{e}^{\mathrm{i}Q\cdot r}\mathrm{d}r \\ f(r) &= \int F(Q)\mathrm{e}^{-\mathrm{i}Q\cdot r}\mathrm{d}Q \end{aligned}\right\} \tag{2}$$

where Q corresponds to a vector of reciprocal space (details of Fourier transform are given in other books; for example, An Introduction to X-ray Crystallography, second edition, by M.M.Woolfson, Cambridge University Press, (1997)). Since the integral with respect to angle can be independently treated, if $f(r)$ depends only on distance and directional components are negligible (well considered to be isotropic), we may use the following simplified form.

$$F(Q) = 4\pi \int_0^{\infty} f(r) r^2 \frac{\sin Q \cdot r}{Q \cdot r} \mathrm{d}r \tag{3}$$

With respect to the sphere with radius R, which may be expressed by $f(r) = 1$ in the condition of $r < R$, the following equation is obtained.

$$F(Q) = 4\pi \int_0^R r^2 \frac{\sin Q \cdot r}{Q \cdot r} \mathrm{d}r = \frac{4\pi}{Q} \int_0^R r \sin Qr \mathrm{d}r \tag{4}$$

$$= \frac{4\pi}{Q^3}(\sin QR - QR \cos QR) \tag{5}$$

Here, the following partial integration is used.

$$\int x \sin kx \mathrm{d}x = x \cdot \frac{-\cos kx}{k} - \int 1 \cdot \left(\frac{-\cos kx}{k}\right) \mathrm{d}x$$
$$= -\frac{x \cos kx}{k} + \frac{1}{k}\left(\frac{\sin kx}{k}\right)$$
$$= \frac{1}{k^2}(\sin kx - kx \cos kx)$$

Let us recall the integration of (3) by using $Z/r_0 = 1/t$ for the wave function.

$$\rho_{1s}(r) = f(r) = \frac{1}{\pi}\left(\frac{Z}{r_0}\right)^3 \mathrm{e}^{-2\frac{Z}{r_0}} = \frac{1}{\pi}\left(\frac{1}{t}\right)^3 \mathrm{e}^{-\frac{2}{t}r} \tag{6}$$

The following equation is obtained by substituting (6) for (3) and re-arranging:

$$F(Q) = 4\pi \cdot \frac{1}{\pi}\left(\frac{1}{t}\right)^3 \left(\frac{1}{Q}\right) \int_0^\infty r \cdot \mathrm{e}^{-\frac{2}{t}r} \cdot \sin Qr \mathrm{d}r \tag{7}$$

Equation (7) corresponds to the procedure for estimating the integral value of the following equation.

$$I_1 = \int_0^\infty x \mathrm{e}^{-\alpha x} \sin \beta x \mathrm{d}x \qquad \left(\alpha = \frac{2}{t}, \beta = Q\right) \tag{8}$$

For this purpose, consider the following integrand except for the component x.

$$I_0 = \int_0^\infty \mathrm{e}^{-\alpha x} \sin \beta x \mathrm{d}x \qquad (\alpha > 0) \tag{9}$$
$$= \left[\frac{\mathrm{e}^{-\alpha x}}{-\alpha} \sin \beta x\right]_0^\infty - \int_0^\infty \frac{\mathrm{e}^{-\alpha x}}{-\alpha} \beta \cos \beta x \mathrm{d}x$$
$$= -\frac{1}{\alpha}(0-0) + \frac{\beta}{\alpha}\left\{\left[\frac{\mathrm{e}^{-\alpha x}}{-\alpha} \cos \beta x\right]_0^\infty - \int \frac{\mathrm{e}^{-\alpha x}}{-\alpha}(-\beta \sin \beta x) \mathrm{d}x\right\}$$
$$= -\frac{\beta}{\alpha^2}(0-1) - \frac{\beta^2}{\alpha^2} I_0 \quad \rightarrow \quad \left(1 + \frac{\beta^2}{\alpha^2}\right) I_0 = \frac{\beta}{\alpha^2}$$

Therefore, the value of (9) is obtained as follows.

$$I_0 = \int_0^{\infty} \mathrm{e}^{-\alpha x} \sin \beta x \mathrm{d}x = \frac{\beta}{\alpha^2 + \beta^2} \tag{10}$$

Here, we use the relationship, $\lim_{x \to \infty} \mathrm{e}^{-\alpha x} \sin \beta x = 0$ because $\mathrm{e}^{-\alpha x} = 0 (\alpha > 0)$ if $x = \infty$.

If I_0 is integrated with respect to α, the desired integrand will be $-x\mathrm{e}^{-\alpha x} \sin \beta x$ and it is found close to the form of I_1 in (8). Using this relationship, the value of (8) can be easily calculated.

$$I_1 = -\frac{\mathrm{d}}{\mathrm{d}\alpha} I_0 = -\frac{\mathrm{d}}{\mathrm{d}\alpha} \left(\frac{\beta}{\alpha^2 + \beta^2} \right) = \frac{2\alpha\beta}{(\alpha^2 + \beta^2)^2} \tag{11}$$

Here, we also use that the derivatives of $\alpha^2 + \beta^2 = u$ and $\beta = k$ (constant) are given in the following equation.

$$\left(\frac{k}{u} \right) = -\frac{ku'}{u^2} \quad \rightarrow \quad -\frac{\beta \cdot 2\alpha}{(\alpha^2 + \beta^2)^2}$$

Then, the Fourier transform of the wave function of 1s electron in K-shell is expressed as:

$$F_{1\mathrm{s}}(Q) = 4\pi \cdot \frac{1}{\pi} \cdot \left(\frac{1}{t} \right)^3 \cdot \frac{1}{Q} \times \frac{2 \cdot \left(\frac{2}{t}\right) \cdot Q}{\left\{ \left(\frac{2}{t}\right)^2 + Q^2 \right\}^2} \tag{12}$$

$$= \frac{4^2 \left(\frac{1}{t}\right)^4}{\left\{ \left(\frac{2}{t}\right) + Q^2 \right\}^2} = \frac{4^2 \left(\frac{Z}{r_0}\right)^4}{\left\{ \left(\frac{2Z}{r_0}\right)^2 + Q^2 \right\}^2} \tag{13}$$

Equation (13) provides the scattering factor of a hydrogen atom as a function of Q.

It may be noted that calculations for the scattering factor of various elements have been made in the past as a function of $q = \sin \theta / \lambda$. In this case, (13) is rewritten in the following form by substituting $2\pi q$ for Q.

$$F_{1\mathrm{s}}(q) = \frac{\left(\frac{Z}{r_0}\right)^4}{\left\{ \left(\frac{2Z}{r_0}\right)^2 + (\pi q)^2 \right\}^2} \tag{14}$$

Question 3.10 A hexagonal close-packed cell is known to contain two atoms of the same type located at the positions of (000) and (1/3, 2/3, 1/2). Compute the structure factor F_{hkl}.

Answer 3.10 Fundamentals of the hcp structure are shown in Fig. 1.

$$F_{hkl} = \sum_{j=1}^{2} f_j e^{2\pi i(hu_n+kv_n+lw_n)}$$

$$F_{hkl} = f e^{2\pi i(0+0+0)} + f e^{2\pi i\left(\frac{h}{3}+\frac{2k}{3}+\frac{l}{2}\right)} = f\left[1 + e^{2\pi i\left(\frac{h+2k}{3}+\frac{l}{2}\right)}\right]$$

Set $q = \dfrac{h+2k}{3} + \dfrac{l}{2}$ for further calculation.

$$F_{hkl} = f(1 + e^{2\pi iq})$$

The intensity of scattered waves from the $(h\ k\ l)$ plane is simply proportional to the square of the absolute value of the structure factor and it is obtained by multiplying the F value by its complex conjugate.

$$|F_{hkl}|^2 = f(1 + e^{2\pi iq}) \times f(1 + e^{-2\pi iq}) = f^2(2 + e^{2\pi iq} + e^{-2\pi iq})$$

Here, we use the relationships, $e^{ix} + e^{-ix} = 2\cos x$ and $\cos 2A = 2\cos^2 A - 1$ (double angle of the cosine formula).

$$|F_{hkl}|^2 = f^2(2 + 2\cos 2\pi q) = f^2[2 + 2(2\cos^2 \pi q - 1)] = f^2(4\cos^2 \pi q)$$

Then, the following equation can be obtained.

$$|F_{hkl}|^2 = 4f^2 \cos^2 \pi \left(\frac{h+2k}{3} + \frac{l}{2}\right)$$

According to the characteristic variation of trigonometric function, $\cos x$ is found to be zero if x is $\frac{1}{2}\pi$, $\frac{3}{2}\pi, \ldots$, whereas $\cos x$ is ± 1 if x is $0, \pi, 2\pi, \ldots$ (see Fig. 2). Therefore, $\cos^2 \pi n = 1$ (n is an integer).

When $(h + 2k)$ is described by a multiple of 3, while l is an odd number such as (001) and (111), the term of $\left(\frac{h+2k}{3} + \frac{l}{2}\right)$ gives $\frac{1}{2}\pi$ and $\frac{3}{2}\pi$. In these cases, the scattering intensity is not detected. When l is an even number, while $(h + 2k)$ is a multiple of 3 such as (002) and (112), the value of $q = \left(\frac{h+2k}{3} + \frac{l}{2}\right)$ is an integer and $\cos \pi q = \pm 1, \Rightarrow |F|^2 = 4f^2$. This implies that the scattering intensity is observed in these cases. Nevertheless, it should be kept in mind that all structure factors of such planes have not always the same value as in the following examples.

$$|F_{101}|^2 = 4f^2 \cos^2 \pi \left(\frac{1}{3} + \frac{1}{2}\right) = 4f^2 \cos^2 \left(\frac{5}{6}\pi\right) \rightarrow 3f^2$$

$$|F_{102}|^2 = 4f^2 \cos^2 \pi \left(\frac{1}{3} + \frac{2}{2}\right) = 4f^2 \cos^2 \left(\frac{4}{3}\pi\right) \rightarrow f^2$$

The results for the structure factor of a hexagonal close-packed cell are summarized in Table 1.

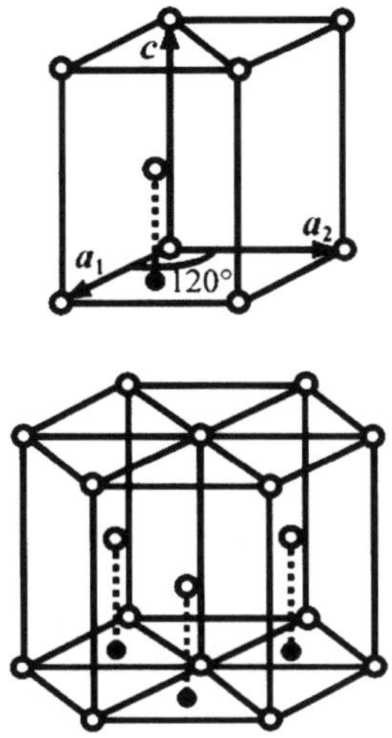

Fig. 1 Fundamentals of the hexagonal close-packed structure

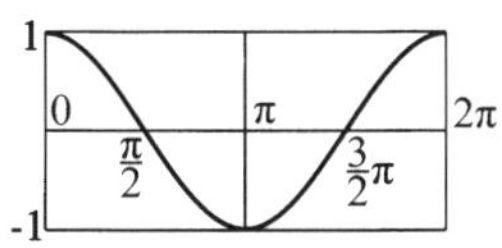

Fig. 2 Variation of cosine function

Table 1 Summary of structure factor of a hexagonal close-packed cell

$h + 2k$	l	$\lvert F\rvert^2$	Examples
$3m$	Odd number	0	001 111 221
$3m$	Even number	$4f^2$	002 110 112
$3m\pm 1$	Odd number	$3f^2$	101 103 201
$3m\pm 1$	Even number	f^2	100 102 200

m : integer

Question 3.11 A unit cell of diamond structure belongs to the cubic crystal system and contains eight atoms of the same type and their positions are as follows.

$$(000)\quad \left(\frac{1}{2}\frac{1}{2}0\right)\ \left(\frac{1}{2}0\frac{1}{2}\right)\ \left(0\frac{1}{2}\frac{1}{2}\right)$$

$$\left(\frac{1}{4}\frac{1}{4}\frac{1}{4}\right)\ \left(\frac{3}{4}\frac{3}{4}\frac{1}{4}\right)\ \left(\frac{3}{4}\frac{1}{4}\frac{3}{4}\right)\ \left(\frac{1}{4}\frac{3}{4}\frac{3}{4}\right)$$

Compute the structure factors F_{hkl} and $|F|^2$.

Answer 3.11

$$\begin{aligned}
F &= f\left[e^{2\pi i(0+0+0)} + e^{2\pi i\left(\frac{h}{2}+\frac{k}{2}+0\right)} + e^{2\pi i\left(\frac{h}{2}+0+\frac{l}{2}\right)} + e^{2\pi i\left(0+\frac{k}{2}+\frac{l}{2}\right)}\right] \\
&\quad + f\left[e^{2\pi i\left(\frac{h}{4}+\frac{k}{4}+\frac{l}{4}\right)} + e^{2\pi i\left(\frac{3h}{4}+\frac{3k}{4}+\frac{l}{4}\right)} + e^{2\pi i\left(\frac{3h}{4}+\frac{k}{4}+\frac{3l}{4}\right)} + e^{2\pi i\left(\frac{h}{4}+\frac{3k}{4}+\frac{3l}{4}\right)}\right] \\
&= f\left[1 + e^{\pi i(h+k)} + e^{\pi i(h+l)} + e^{\pi i(k+l)}\right] \\
&\quad + f\left[e^{\pi i\frac{(h+k+l)}{2}} + e^{\pi i\frac{(3h+3k+l)}{2}} + e^{\pi i\frac{(3h+k+3l)}{2}} + e^{\pi i\frac{(h+3k+3l)}{2}}\right] \\
&= f\left[1 + e^{\pi i(h+k)} + e^{\pi i(h+l)} + e^{\pi i(k+l)}\right] \\
&\quad + f e^{\pi i\left(\frac{h+k+l}{2}\right)}\left[1 + e^{\pi i(h+k)} + e^{\pi i(h+l)} + e^{\pi i(k+l)}\right] \\
&= f\left[1 + e^{\pi i\frac{(h+k+l)}{2}}\right]\left[1 + e^{\pi i(h+k)} + e^{\pi i(h+l)} + e^{\pi i(k+l)}\right]
\end{aligned}$$

At first, let us consider the 2nd term. We find the 2nd term will be zero, if $h\ k\ l$ is a mixture of odd and even numbers and the relationship $e^{n\pi i} = (-1)^n$, n being any integer is taken into account. The relationships $e^{\pi i} = -1$, $e^{2\pi i} = 1$, and $e^{n\pi i} = (-1)^n$, where n is any integer, are also useful for calculating the structure factor. For example, the following results are obtained with respect to planes of (100), (110), and (211), respectively.

$$\begin{aligned}
1 + e^{\pi i(1+0)} + e^{\pi i(1+0)} + e^{\pi i(0+0)} &= 1 - 1 - 1 + 1 = 0 \\
1 + e^{\pi i(1+1)} + e^{\pi i(1+0)} + e^{\pi i(1+0)} &= 1 + 1 - 1 - 1 = 0 \\
1 + e^{\pi i(2+0)} + e^{\pi i(1+0)} + e^{\pi i(1+1)} &= 1 - 1 - 1 + 1 = 0
\end{aligned}$$

On the other hand, the 2nd term is found to be 4 when $h\ k\ l$ is not a mixture of odd and even numbers and then, the structure factor F can be expressed in the following form.

$$\begin{aligned}
F &= 4f\left[1 + e^{\pi i\frac{h+k+l}{2}}\right] \\
|F|^2 &= 16f^2\left[1 + e^{\pi i\frac{h+k+l}{2}}\right]\left[1 + e^{-\pi i\frac{h+k+l}{2}}\right] \\
&= 16f^2\left[1 + 1 + 2\cos\frac{\pi}{2}(h+k+l)\right] = 32f^2\left[1 + \cos\frac{\pi}{2}(h+k+l)\right]
\end{aligned}$$

Here, we use the relationships, $e^0 = 1$ and $e^{ix} + e^{-ix} = 2\cos x$. It is noted that cosine function shows the results: -1 for the condition of an odd multiple of π, $+1$ for a even multiple of π, and zero for an odd multiple of $\pi/2$, respectively. Therefore, the following results are obtained for planes of unmixed $h\ k\ l$.
If $(h + k + l)$ is odd integers, such as (111), (311),

$$\cos x = 0 \quad \rightarrow \quad |F|^2 = 32f^2$$

If $(h + k + l)$ is given by the number of an odd multiple of 2 such as (110), (200)

$$\cos x = -1 \quad \rightarrow \quad |F|^2 = 0$$

If $(h + k + l)$ is given by the number of a even multiple of 2 such as (220), (400),

$$\cos x = +1 \quad \rightarrow \quad |F|^2 = 64f^2$$

Of course, $|F|^2 = 0$ when $h\ k\ l$ is a mixture of odd and even numbers.

Question 3.12 Sodium chloride (NaCl) has a cubic lattice with four Na^+ ions and four Cl^- ions per unit cell. In addition, Na^+ ions occupy the corner of a unit cell as well as the center of the plane, whereas Cl^- ions occupy the center of a cube as well as the midpoint of each edge-line.

This structure can also be produced in the following way; Na^+ ions form fcc lattice and the face-centering translations can reproduce all positions of Cl^- ions, when applied to Cl^- ions located at (1/2, 1/2, 1/2). As a result, any of the Na^+ ions is surrounded by six Cl^- ions and the reverse relationship is recognized. Eight ions in the unit cell of NaCl are given as follows.

$$\mathrm{Na}^+ \quad (000) \quad \left(\frac{1}{2}\frac{1}{2}0\right) \left(\frac{1}{2}0\frac{1}{2}\right) \left(0\frac{1}{2}\frac{1}{2}\right)$$

$$\mathrm{Cl}^- \quad \left(0\frac{1}{2}0\right) \left(\frac{1}{2}00\right) \left(00\frac{1}{2}\right) \left(\frac{1}{2}\frac{1}{2}\frac{1}{2}\right)$$

(1) Compute the structure factor F_{hkl}, assuming that the scattering factors of Na and Cl are expressed by f_{Na} and f_{Cl}, respectively.
(2) Compute the structure factors of planes (111) and (200).

Answer 3.12

$$\begin{aligned}
(1)\ F_{hkl} &= f_{\mathrm{Na}}\left[\mathrm{e}^{2\pi i(0+0+0)} + \mathrm{e}^{2\pi i\left(\frac{h}{2}+\frac{k}{2}+0\right)} + \mathrm{e}^{2\pi i\left(\frac{h}{2}+0+\frac{l}{2}\right)} + \mathrm{e}^{2\pi i\left(0+\frac{k}{2}+\frac{l}{2}\right)}\right] \\
&\quad + f_{\mathrm{Cl}}\left[\mathrm{e}^{2\pi i\left(0+\frac{k}{2}+0\right)} + \mathrm{e}^{2\pi i\left(\frac{h}{2}+0+0\right)} + \mathrm{e}^{2\pi i\left(0+0+\frac{l}{2}\right)} + \mathrm{e}^{2\pi i\left(\frac{h}{2}+\frac{k}{2}+\frac{l}{2}\right)}\right] \\
&= f_{\mathrm{Na}}\left[1 + \mathrm{e}^{\pi i(h+k)} + \mathrm{e}^{\pi i(h+l)} + \mathrm{e}^{\pi i(k+l)}\right] \\
&\quad + f_{\mathrm{Cl}}\left[\mathrm{e}^{\pi i k} + \mathrm{e}^{\pi i h} + \mathrm{e}^{\pi i l} + \mathrm{e}^{\pi i(h+k+l)}\right] \\
&= f_{\mathrm{Na}}\left[1 + \mathrm{e}^{\pi i(h+k)} + \mathrm{e}^{\pi i(h+l)} + \mathrm{e}^{\pi i(k+l)}\right] \\
&\quad + f_{\mathrm{Cl}}\mathrm{e}^{\pi i(h+k+l)}\left[1 + \mathrm{e}^{\pi i(-h-k)} + \mathrm{e}^{\pi i(-h-l)} + \mathrm{e}^{\pi i(-k-l)}\right] \\
&= \left[f_{\mathrm{Na}} + f_{\mathrm{Cl}}\mathrm{e}^{\pi i(h+k+l)}\right]\left[1 + \mathrm{e}^{\pi i(h+k)} + \mathrm{e}^{\pi i(h+l)} + \mathrm{e}^{\pi i(k+l)}\right]
\end{aligned}$$

Here, we use the relationships, $e^0 = 1$ and $\mathrm{e}^{n\pi i} = \mathrm{e}^{-n\pi i}$ (n is an integer).

The 1st term shows the basis of the unit cell, Na at (0 0 0) and Cl at (1/2, 1/2, 1/2), respectively and the 2nd term indicates that this structure can be reproduced by the face-centering translations. Note that the term for translation is the same as the description of F for a fcc lattice. Since the sums $(h + k)$, $(h + l)$, and $(k + l)$ are even integers if h, k, and l are all even or all odd (unmixed), the value of the three exponentials has the value 1. Conversely, the sum of the three exponentials will be -1 for mixed indices. Specifically, the intensity scattered from planes (111) and (200) are observed but the intensity of planes (100) and (210) are not observed. However, even if h, k, and l are unmixed, we have to include the atomic scattering factors for different atoms in the present case. The results of a fcc lattice cannot applied directly to this case.

(2) (111) plane: When h, k, and l are unmixed and $(h + k + l)$ is odd number.

$$F = 4(f_{\mathrm{Na}} - f_{\mathrm{Cl}}) \quad \Rightarrow \quad |F|^2 = 16(f_{\mathrm{Na}} - f_{\mathrm{Cl}})^2$$
$$F_{111} = 16(f_{\mathrm{Na}} - f_{\mathrm{Cl}}) \quad \textit{decrease in intensity due to } (f_{\mathrm{Na}} - f_{\mathrm{Cl}})$$

(200) plane: when h, k, and l are unmixed and $(h + k + l)$ is even number

$$F = 4(f_{\mathrm{Na}} + f_{\mathrm{Cl}}) \quad \Rightarrow \quad |F|^2 = 16(f_{\mathrm{Na}} + f_{\mathrm{Cl}})^2$$
$$F_{200} = 16(f_{\mathrm{Na}} + f_{\mathrm{Cl}}) \quad \textit{increase in intensity due to } (f_{\mathrm{Na}} + f_{\mathrm{Cl}})$$

Note: For computing the structure factor F of ionic crystals, one should use the scattering factor of ions instead of atoms. However, it makes little difference because the scattering factors for atoms are found almost identical to those of ions at higher angle, larger than $(\sin\theta/\lambda) = 3.0$. Only slight difference is found at small angles.

Question 3.13 Structural similarity between an element and a compound is frequently found, as shown by the case of diamond and zinc blende ($\beta-$ZnS) as an example.

(1) Show some points in these two structures, when the atomic positions of Zn and S in the unit cell of $\beta-$ZnS are given in the following.

$$\mathrm{Zn}\quad (000)\quad \left(0\frac{1}{2}\frac{1}{2}\right)\ \left(\frac{1}{2}0\frac{1}{2}\right)\ \left(\frac{1}{2}\frac{1}{2}0\right)$$
$$\mathrm{S}\quad \left(\frac{1}{4}\frac{1}{4}\frac{1}{4}\right)\ \left(\frac{1}{4}\frac{3}{4}\frac{3}{4}\right)\ \left(\frac{3}{4}\frac{1}{4}\frac{3}{4}\right)\ \left(\frac{3}{4}\frac{3}{4}\frac{1}{4}\right)$$

(2) Determine the structure factor of $\beta-$ZnS and indicate the cases where intensity can be observed.

Answer 3.13

(1) The unit cell of both diamond and $\beta-\mathrm{ZnS}$ is expressed by a face-centered cubic lattice as shown in Fig. 1. In diamond, eight atoms per unit cell occupy, (000),$(\frac{1}{4}\frac{1}{4}\frac{1}{4})$, and the translational positions of their face center. Such structural fundamentals are unchanged in ZnS, but Zn occupies the (000) and the translation position of its face center, whereas S occupies $(\frac{1}{4}\frac{1}{4}\frac{1}{4})$ and the translation position of its face center (or the converse).

The structure of zinc blende may be characterized as follows. One of the two elements (e.g., Zn) occupies the corner of unit cell as well as the position of face center and another element (e.g., S) occupies the tetrahedral positions as in diamond. This structural features can also be expressed as superposition of the fcc lattice of Zn with the fcc lattice of S after translating the position by $\frac{1}{4}\frac{1}{4}\frac{1}{4}$ from the Zn lattice.

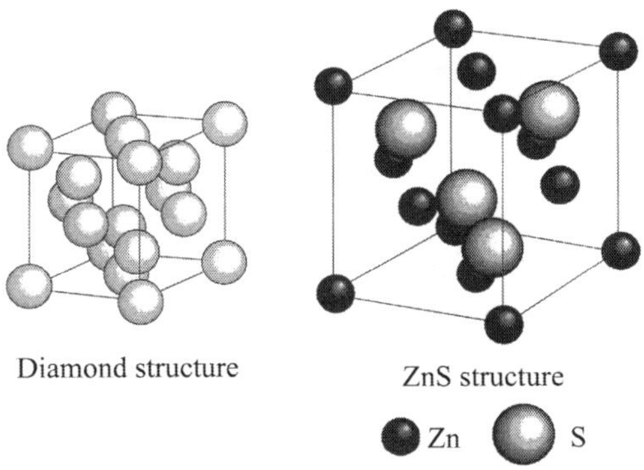

Fig. 1 Relationship between diamond structure and ZnS structure

(2) The structure factor of $\beta-\mathrm{ZnS}$ is calculated as follows:

$$\begin{aligned}
F &= f_{\mathrm{Zn}}\left[\mathrm{e}^{2\pi i(0+0+0)}+\mathrm{e}^{2\pi i(0+\frac{k}{2}+\frac{l}{2})}+\mathrm{e}^{2\pi i(\frac{h}{2}+0+\frac{l}{2})}+\mathrm{e}^{2\pi i(\frac{h}{2}+\frac{k}{2}+0)}\right] \\
&\quad +f_{\mathrm{S}}\left[\mathrm{e}^{2\pi i(\frac{h}{4}+\frac{k}{4}+\frac{l}{4})}+\mathrm{e}^{2\pi i(\frac{j}{4}+\frac{3k}{4}+\frac{3l}{4})}+\mathrm{e}^{2\pi i(\frac{3h}{4}+\frac{k}{4}+\frac{3l}{4})}+\mathrm{e}^{2\pi i(\frac{3h}{4}+\frac{3k}{4}+\frac{l}{4})}\right] \\
&= f_{\mathrm{Zn}}\left[1+\mathrm{e}^{\pi i(k+l)}+\mathrm{e}^{\pi i(h+l)}+\mathrm{e}^{\pi i(h+k)}\right] \\
&\quad +f_{\mathrm{S}}\left[\mathrm{e}^{\pi i\left(\frac{h+k+l}{2}\right)}+\mathrm{e}^{\pi i\left(\frac{h+3k+3l}{2}\right)}+\mathrm{e}^{\pi i\left(\frac{3h+k+3l}{2}\right)}+\mathrm{e}^{\pi i\left(\frac{3h+3k+l}{2}\right)}\right] \\
F &= f_{\mathrm{Zn}}\left[1+\mathrm{e}^{\pi i(h+l)}+\mathrm{e}^{\pi i(h+l)}+\mathrm{e}^{\pi i(h+k)}\right] \\
&\quad +f_{\mathrm{S}}\cdot\mathrm{e}^{\pi i\left(\frac{h+k+l}{2}\right)}[1+\mathrm{e}^{\pi i(k+l)}+\mathrm{e}^{\pi i(h+l)}+\mathrm{e}^{\pi i(h+k)}] \\
&= \left[1+\mathrm{e}^{\pi i(k+l)}+\mathrm{e}^{\pi i(h+l)}+\mathrm{e}^{\pi i(h+k)}\right]\left[f_{\mathrm{Zn}}+f_{\mathrm{S}}\mathrm{e}^{\pi i\left(\frac{h+k+l}{2}\right)}\right]
\end{aligned}$$

The 2nd term represents the basis of the unit cell where Zn is located at (000) and S at $\left(\frac{1}{4}\frac{1}{4}\frac{1}{4}\right)$, respectively. On the other hand, the 1st term is a relationship observed in the description of the structure factor for a fcc lattice. That is, if h k l is a mixture of odd and even numbers, the 1st term will be zero, whereas it becomes 4 for unmixed indices. Thus, we can examine the case where h k l is unmixed using the following equation.

$$F = 4\left[f_{\mathrm{Zn}} + f_{\mathrm{S}}\mathrm{e}^{\pi i\left(\frac{h+k+l}{2}\right)}\right]$$

$|F|^2$, which is proportional to the intensity, can be computed using the complex conjugate as follows:

$$\begin{aligned}|F|^2 &= 4\left[f_{\mathrm{Zn}} + f_{\mathrm{S}}\mathrm{e}^{\pi i\left(\frac{h+k+l}{2}\right)}\right]\cdot 4\left[f_{\mathrm{Zn}} + f_{\mathrm{S}}\mathrm{e}^{-\pi i\left(\frac{h+k+l}{2}\right)}\right]\\ &= 16\left[f_{\mathrm{Zn}}^2 + f_S^2 + 2f_{\mathrm{Zn}}f_{\mathrm{S}}\left(\mathrm{e}^{\pi i\left(\frac{h+k+l}{2}\right)} + \mathrm{e}^{-\pi i\left(\frac{h+k+l}{2}\right)}\right)\right]\\ &= 16\left[f_{\mathrm{Zn}}^2 + f_{\mathrm{S}}^2 + 2f_{\mathrm{Zn}}f_{\mathrm{S}}\cos\frac{\pi}{2}(h+k+l)\right]\end{aligned}$$

Here, we use the relationships of $e^0 = 1$ and $\mathrm{e}^{\mathrm{i}x} + \mathrm{e}^{-\mathrm{i}x} = 2\cos x$.

When it is a even multiple of π, cosine function is +1; when it is an odd multiple of π, -1; and when it is an odd multiple of $\pi/2$, zero. Considering these characteristic features, the following results are obtained for unmixed indices.

$h+k+l$ is odd numbers, $\cos x = 0 \rightarrow |F|^2 = 16(f_{\mathrm{Zn}}^2 + f_{\mathrm{S}}^2)$

$h+k+l$ is a even multiple of 2, $\cos x = +1 \rightarrow |F|^2 = 16(f_{\mathrm{Zn}} + f_{\mathrm{S}})^2$

$h+k+l$ is an odd multiple of 2, $\cos x = -1 \rightarrow |F|^2 = 16(f_{\mathrm{Zn}} - f_{\mathrm{S}})^2$

When h, k, and l are mixed, some are odd and some are even numbers, the intensity is not observed because $|F|^2 = 0$.

Chapter 4
Diffraction from Polycrystalline Samples and Determination of Crystal Structure

There are various methods for measuring the intensity of a scattered X-ray beam (hereafter referred to as diffracted X-ray beam in this chapter) from crystalline materials, and each method has the respective advantage. The most common method is to measure the X-ray diffraction intensity from a powder sample as a function of scattering angle (it is also called diffraction angle) by using a diffractometer. For this reason, several key points of structural analysis will be given with some selected examples on how to obtain structural information of powder samples of interest from measured intensity data using a diffractometer.

4.1 X-ray Diffractometer Essentials

A diffractometer is a precision instrument with two axes (ω and 2θ) of independent rotation. This equipment enables us to obtain the intensity data of a diffracted X-ray beam, as a function of angle, so as to satisfy Bragg's law under the condition of X-rays of known wavelength. The basic design of the diffractometer is illustrated in Fig. 4.1. Three components, X-ray source (F), sample holder (S), and detector (G), lie on the circumference of a circle, as known as the focusing circle. When the position of X-ray source is fixed and the detector is attached on the 2θ-axis, a powder sample in the flat-plate form is usually set on the ω-axis corresponding to the center of the diffractometer. The line focal spot on the target of the X-ray tube is set to be parallel to the diffractometer ω-axis. The main reason for using a flat plate sample is to take advantage of the focusing geometry for effectively collecting the intensity of weak diffracted beams. During the course of measurements, the 2θ-axis rotates two times as much as the ω-axis, and therefore we frequently call it theta two-theta scan. This is to maintain the experimental condition that the angle between the plane of the sample and direction of the incident X-ray beam is equal to that of direction of the diffracted beam, with reference to the direction of propagation of the incident X-ray beam. In other words, the direction of normal to the sample plane should be fit to the direction of the scattering vector $\boldsymbol{q} = \boldsymbol{s} - \boldsymbol{s}_0$ defined by the difference between vector $\boldsymbol{s}_0$ of the incident X-ray beam and vectors of the diffracted X-ray beam $\boldsymbol{s}$. In addition, a circle through the points F (focal spot on the target), S (the center of the

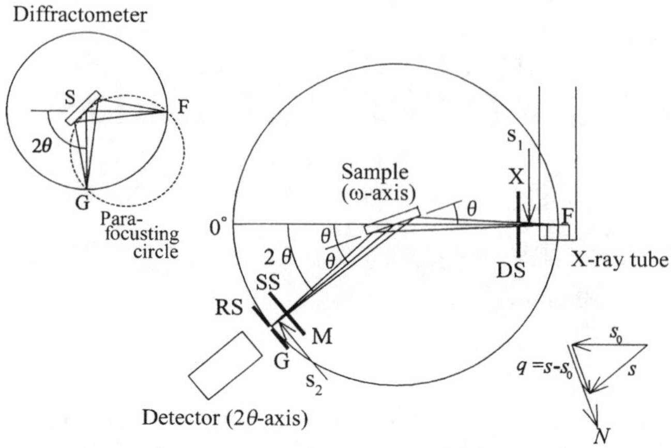

Fig. 4.1 Para-focusing geometry and some essential points of an X-ray diffractometer

diffractometer), and G (the focal point of the diffracted beam) in Fig. 4.1 is named as focusing circle or Rowland circle.

To minimize angular dispersion and to improve spatial resolution for the incident X-ray beam, as well as the diffracted X-ray beam, some slit systems are inserted into the X-ray path. We also use a soller slit, which consists of a set of closely spaced, thin metal plates parallel to the plane of the diffractometer circle and this soller slit is to restrict the perpendicular dispersion of both the incident and the diffracted X-ray beams. As shown in Fig. 4.1, the divergent slits (DS) and the scattering slits (SS) are set to restrict each horizontal dispersion of both incident and diffracted X-ray beams, and the receiving slit (RS) in front of the detector is set to determine the spatial resolution. The important feature of a diffractometer is not only the restriction of dispersion by the DS and SS, but also the focusing of the diffracted X-ray beam from powder samples by the RS. This collimating and focusing principle is called para-focusing. As is seen from Fig. 4.1, the position of RS in front of the detector always matches with a para-focusing spot in the diffractometer, and this makes the intensity measurement effective and the spatial resolution better.

4.2 Estimation of X-ray Diffraction Intensity from a Polycrystalline Sample

Let us consider that the number of atoms is set to be N, the scattering amplitude is A, and the scattered X-rays, which satisfy the so-called Bragg condition, are said to be completely in phase. In this case, the amplitude of diffracted X-rays will be given by NA, so that the total diffraction intensity is $(NA)^2$. That is, the intensity of diffracted X-rays being in phase by satisfying the Bragg condition will be N times the diffraction intensity given by (NA^2), corresponding to the case, where the Bragg condition is not satisfied and no interference is recognized. The value of N is said to be about 10^{22} per gram even in a very small crystal, so that the diffracted X-rays from a powder (crystalline) sample can be measured with sufficient reliability. However, note that the intensity of diffracted X-rays is considerably weak in comparison with the intensity of incident X-rays.

Some selected examples for the "structure factor" required in estimating the intensity of diffracted X-rays are described in Chap. 3. This structure factor F or $|F|^2$ is one way of knowing the relationship between crystal structure and the intensity of the diffracted X-rays from each crystal plane possibly measured. Besides the structure factor, the measured intensity of diffracted X-rays from powder samples contains various components, such as, polarization factor, multiplicity factor, Lorentz factor, absorption factor, and temperature factor. These factors are described in more detail below.

4.2.1 *Structure Factor (see also Chap. 3)*

The structure factor for the (hkl) reflection is given by the following equation:

$$F_{hkl} = \sum_{j=1}^{N} f_j e^{2\pi i(hu_j + kv_j + lw_j)}. \tag{4.1}$$

Here, N represents the total number of atoms in a unit cell, f_j represents the atomic scattering factor of the j – th atom, and $u_j v_j w_j$ are the fractional coordinates for the position of the j – th atom in the corresponding unit cell. Since reflections where $F = 0$ will be zero intensity, we will not observe the diffraction peak and these reflections are called forbidden.

4.2.2 *Polarization Factor*

Thomson's equation described in Chap. 3 (see (3.1)) is estimated from the assumption of a completely unpolarized incident X-ray beam, such as that issuing from an X-ray tube. However, a part of X-rays diffracted from crystals is impossible to have no connection with polarization property. The polarization factor, P, is given in the following form as a function of diffraction angle 2θ for three typical X-ray diffraction experiments with characteristic X-rays being of constant wavelength:

(i) Using a filter only:

$$P = \frac{1 + \cos^2 2\theta}{2}. \tag{4.2}$$

(ii) Using a crystal monochromator in the incident X-ray path:

$$P = \frac{1 + x\cos^2 2\theta}{1 + x}. \tag{4.3}$$

(iii) Using a crystal monochromator in the diffracted X-ray path:

$$P = \frac{1 + x\cos^2 2\theta}{2}. \tag{4.4}$$

With respect to x in both cases of (ii) and (iii), we use $x = \cos^2 2\theta_M$ for an ideally mosaic monochromator crystal or $x = \cos 2\theta_M$ for an ideally perfect monochromator crystal if $2\theta_M$ is twice the Bragg angle of the monochromator crystal.

4.2.3 Multiplicity Factor

Multiplicity factor represents the number of the crystal planes, which have the same spacing and structure factors, but different orientation. For example, in the case of $\{100\}$ planes for cubic crystal, it is six denoted by (100), (010), (001), $(\bar{1}00)$, $(0\bar{1}0)$, and $(00\bar{1})$ and in the case of $\{111\}$ planes, it is eight. Multiplicity factors in various crystal systems are summarized in Table 2.2 in Chap. 2. When crystal orientations in powder samples are completely random, the probability of crystal orientations, which satisfy the so-called Bragg condition is given by the ratio of multiplicity factors, for example, 8–6 for the condition of $\{111\}$ to $\{100\}$.

4.2.4 Lorentz Factor

As mentioned above, a sample usually rotates during measurement with a diffractometer. The intensity of diffracted X-rays is known to be greatest at the exact diffraction angle satisfying the Bragg law. However, the intensity of diffracted X-rays is usually appreciable at angles deviating slightly from the Bragg angle, so that the measured intensity curve as a function of 2θ is given in the form of Fig. 4.2. When all diffracted X-ray beams produced from a crystal sample, as it rotates through the diffraction angle, are caught by a detector, the total energy of the diffracted X-ray beams can be measured. We call this energy the integrated intensity of the corresponding peak, and it is also obtained from the area under the curve of the corresponding peak. That is, the intensity measurement of diffracted X-ray beam from a crystal sample means the measurement of the integrated values of the diffracted intensities produced from volumes with a certain limited size. The volume attributed to the integrated intensity is known to depend on the diffraction angle, so this should be taken into consideration when comparing the intensities diffracted from different crystal planes. Such effects are represented by the so-called Lorentz factor. For example, the Lorentz factor for powder samples is given by the following equation as a function of diffraction angle.

$$[\text{Lorentz factor}] \equiv \frac{1}{\sin^2\theta\cos\theta}. \tag{4.5}$$

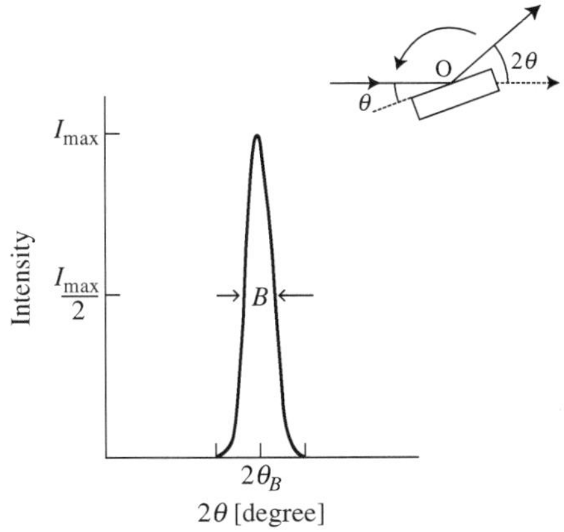

Fig. 4.2 Intensity diffracted from a crystalline sample rotated through the Bragg angle

The Lorentz factor is not only used independently, but also used in many cases as the Lorentz polarization factor by coupling with polarization factor, which is also a function of diffraction angle.

4.2.5 Absorption Factor

Absorption described in Chap. 1 is mainly related to the reduction of X-ray intensity after penetrating a uniform substance. On the contrary, powder samples finely ground are filled up in a sample holder for making a flat plate form to measure the diffracted X-ray beams using a diffractometer. Let us consider the case in which the incident X-ray beam with its intensity I_0 per unit cross-sectional area encounters the powder sample with the angle of γ from the sample surface, and we measure the intensity $\mathrm{d}I$ diffracted by an infinitesimally thin layer characterized by the thickness $\mathrm{d}x$ located at a depth x below the top surface of the sample at the angle of β from the sample surface. The resultant integrated intensity diffracted from such a small volume is given in the following equation.

$$\mathrm{d}I = \frac{I_0}{\sin\gamma}\mathrm{e}^{-\mu x\left(\frac{1}{\sin\gamma}+\frac{1}{\sin\beta}\right)}\mathrm{d}x, \tag{4.6}$$

where μ is the linear absorption coefficient of sample materials.

The total intensity is, therefore, obtained by integrating (4.6) over an infinitely thick sample, and we obtain the following simple form because the relationships of $\gamma = \beta = \theta$ is also allowed in the measurements using a diffractometer.

$$I_D = \frac{I_0}{\sin\theta}\int_0^t \mathrm{e}^{-\frac{2\mu x}{\sin\theta}}\mathrm{d}x = \frac{I_0}{2\mu}\left(1-\mathrm{e}^{-\frac{2\mu t}{\sin\theta}}\right). \tag{4.7}$$

That is, the absorption factor is represented by $\left(1 - e^{-\frac{2\mu t}{\sin\theta}}\right)/2\mu$. Since a sample is usually considered of infinite thickness, it becomes $t \to \infty$, and then we conclude the absorption factor is simply set to $(1/2\mu)$. Therefore, the absorption factor for the measurements with a diffractometer normally used can be negligible as far as handling the relative amounts of intensities, because the absorption factor in cases where the sample of the flat plate form shows infinite thickness is considered to be constant and independent of the diffraction angle. The criterion of judgment that a sample shows infinite thickness can be computed using the following method.

In the diffractometer with the condition of $\gamma = \beta = \theta$, the ratio of the intensity of diffracted X-ray beam for the sample with its thickness t and that of the sample for the semi-infinite plate is given by the following equation with reference to (4.6).

$$G_t = \frac{\int_0^t \frac{I_0}{\sin\theta} e^{-\frac{2\mu x}{\sin\theta}} dx}{\int_0^\infty \frac{I_0}{\sin\theta} e^{-\frac{2\mu t}{\sin\theta}} dx} = 1 - e^{-\frac{2\mu t}{\sin\theta}}$$

$$t = -\frac{\sin\theta}{2\mu} \ln(1 - G_t). \tag{4.8}$$

The value of G_t provides the correlation illustrated in Fig. 4.3 as a function of sample thickness t. For example, as long as the sample thickness is beyond the value of evaluated thickness t when the G_t value is 95%, it may be considered that a sample is of infinite thickness. In the case of silicon powder, if the sample thickness is about 0.5 mm, this criterion is accepted for Cu-Kα radiation.

4.2.6 Temperature Factor

As is well known, atoms in a crystal is not kept at fixed points, they are rather moving around their mean positions by thermal vibration and the amplitude of such

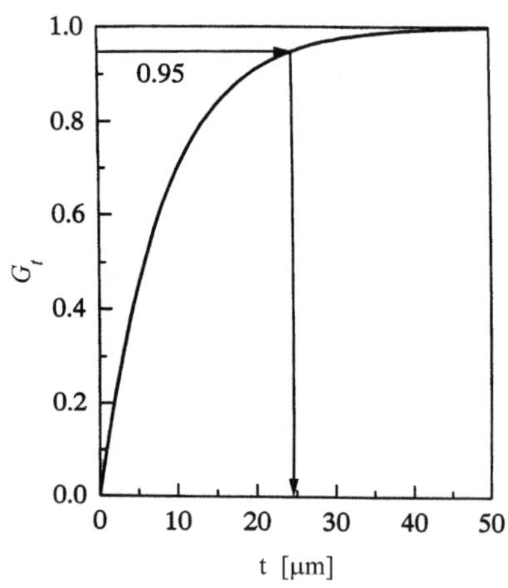

Fig. 4.3 The fraction of the diffracted intensity contributed from a surface layer of depth t to the total diffracted intensity of a sample of semi-infinite thickness

vibration increases with increasing temperature. For example, about 5–6% displacement of an atom from its mean position is estimated at room temperature. Because of this thermal vibration effect of atoms, the reduction in intensity of diffracted X-rays from a crystal sample is detected. The effect of atomic vibrations on the intensity of diffracted X-rays is commonly taken into consideration as a Debye–Waller factor. In practice, for the temperature effect represented as e^{-2M_T} on the intensity of diffracted X-rays, the quantity of M is calculated in the following equations with the atomic scattering factor f.

$$f = f_0 e^{-M_T}$$

$$M_T = 8\pi^2 \langle u^2 \rangle \left(\frac{\sin\theta}{\lambda} \right)^2 = B_T \left(\frac{\sin\theta}{\lambda} \right)^2. \tag{4.9}$$

Here, $\langle u^2 \rangle$ is the mean square displacement of the atom in a direction normal to the diffraction planes. It should be kept in mind that the exact calculation of M_T as a function of temperature is extremely difficult. For this reason, the coefficient B_T in (4.9) is estimated from the measured intensity data at different temperatures and the results are applied to another case.

4.2.7 *General Formula of the Intensity of Diffracted X-rays for Powder Crystalline Samples*

With respect to the intensity of diffracted X-rays for powder crystalline samples, a general formula can be obtained by considering together the factors above mentioned. For example, the intensity I measured by a diffractometer with characteristic radiation monochromated by a filter is given in the following equation.

$$I = |F|^2 p \left(\frac{1 + \cos^2 2\theta}{2 \sin^2\theta \cos\theta} \right) \frac{1}{2\mu} \left(1 - e^{-\frac{2\mu t}{\sin\theta}} \right) e^{-2M_T}. \tag{4.10}$$

Here, F represents structure factor, p is multiplicity factor, and the parenthesis in the 3rd term corresponds to the Lorentz-polarization factor (*LP*). With respect to the *LP*, note that another expression without the numerical value 2, which appears in the denominator is also frequently used. This difference is considered negligible in the intensity calculation, because it is an angular independent constant. If a sample is considered of infinite thickness, the absorption factor is independent of the diffraction angle, being a constant with the value $(1/2\mu)$. Although the temperature factor e^{-2M_T} affects the atomic scattering factor f, it is usually omitted in calculation of the desired intensities, as far as handling the relative amounts of intensities.

4.3 Crystal Structure Determination: Cubic Systems

Analysis of diffraction data is to say "indexing pattern analysis." The crystal structure of metallic elements, which account for about 70% of the periodic table shows the "cubic systems" characterized by the relatively simple atomic arrangement, such as fcc, bcc, hcp, and diamond. Therefore, analysis of diffraction data of metallic elements is not such a difficult task. This is also supported by the following reasons. The first step of structural analysis for the measured X-ray diffraction data is to calculate the position of the scattering angle 2θ, corresponding to the location of the diffraction peaks. Since the current X-ray diffraction experiments are made under computer control, the results are usually examplified by Fig. 4.4, which automatically gives the values of 2θ, d, and (I/I_1). Here, d and (I/I_1) are the plane spacing calculated from Bragg condition using the given wavelength (λ) of X-rays and the relative intensity ratio of the detected peaks with reference to the first peak intensity I_1, respectively.

By combining the Bragg condition (see (3.13)) with the plane spacing for a cubic system, the diffraction peaks with the $\sin^2\theta$ values satisfy the following equation:

$$\frac{\sin^2\theta}{(h^2+k^2+l^2)} = \frac{\sin^2\theta}{S} = \frac{\lambda^2}{4a^2}, \tag{4.11}$$

where $S = h^2 + k^2 + l^2$. Figure 4.4 shows six peaks and their diffraction angles together with the values of plane spacing d are readily obtained. As is clear from (4.11), the sum of the square of plane indices, corresponding to the measured diffraction peaks is always an integer and $\frac{\lambda^2}{4a^2}$ is found a constant for any X-ray diffraction pattern.

There are only four possibilities in cubic systems including diamond lattice and the Miller indices are given as shown in Fig. 4.5. Therefore, the simplest way is by

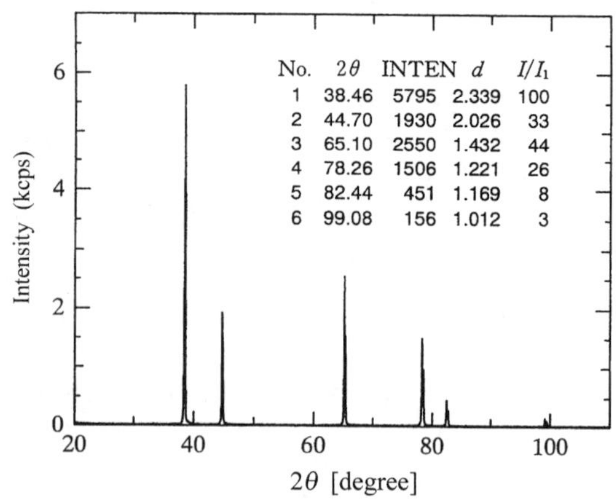

Fig. 4.4 X-ray diffraction pattern of a metal sample with cubic structure obtained by Cu-Kα radiation

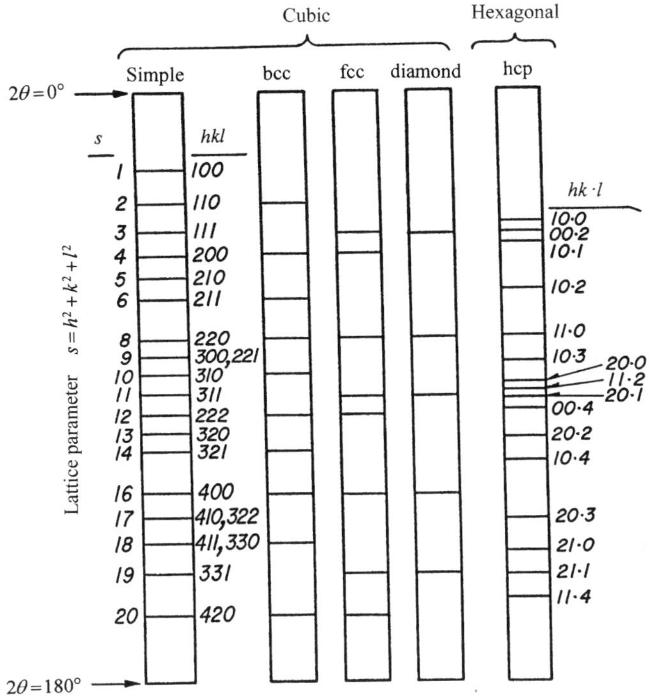

Fig. 4.5 Characteristic sequence of diffraction peaks of four common cubic lattice types together with a hexagonal close-packed lattice

trial and error, as follows. A crystal lattice is assumed to estimate the sum of the square of plane indices (S). Then, the value of $\frac{\sin^2\theta}{S}$ in (4.11) is calculated. When the values are found constant, it can be safely said that the assumed crystal lattice is well accepted. In addition, the lattice parameter (a) can be estimated from $\frac{\lambda^2}{4a^2}$, by applying the wavelength (λ) of the used X-rays. Of course, when a constant value is not obtained, another crystal lattice is assumed and the same calculation is carried out.

As is clear from Fig. 4.5, certain integers such as 7, 15, and 23, will never appear as the value of $S = h^2 + k^2 + l^2$. In other words, when a number such as 7 or 15 appears that means there is an error in analysis. On the contrary, another method using the following equation is also frequently used for analyzing the diffraction data of cubic systems, although the essential points related to (4.11) are unchanged.

$$\frac{4\sin^2\theta}{\lambda^2} = \frac{1}{d^2} = \frac{h^2 + k^2 + l^2}{a^2}. \tag{4.12}$$

4.4 Crystal Structure Determination: Tetragonal and Hexagonal Systems

Indexing pattern analysis of noncubic systems becomes more difficult because the number of unknown parameters increases. With respect to tetragonal systems or hexagonal systems whose plane spacing can be represented by the axial ratios (c/a), a special graphical method called "Hull–Davey chart" has been proposed and it was widely used in the past. For example, the relationship between plane spacing of tetragonal systems and the Bragg condition is given in the following form with two unknown parameters, a and c.

$$\frac{1}{d^2} = \frac{4\sin^2\theta}{\lambda^2} = \frac{h^2+k^2}{a^2} + \frac{l^2}{c^2}. \tag{4.13}$$

We obtain the following equations when taking the logarithm of both sides and slightly rewriting.

$$2\log d = 2\log a - \log\left\{(h^2+k^2) + \frac{l^2}{(c/a)^2}\right\} \tag{4.14}$$

$$\log\sin^2\theta = \log\frac{\lambda^2}{4a^2} + \log\left\{(h^2+k^2) + \frac{l^2}{(c/a)^2}\right\} = -2\log d + \log\frac{\lambda^2}{4}. \tag{4.15}$$

Since the $\log a$ term is canceled out, when taking the difference of the logarithm of d_1 and d_2 assessed with (4.14), to two planes of $(h_1k_1l_1)$ and $(h_2k_2l_2)$, the following equation may be obtained.

$$\begin{aligned} 2\log d_1 - 2\log d_2 = &-\log\left\{(h_1^2+k_1^2) + \frac{l_1^2}{(c/a)^2}\right\} \\ &+\log\left\{(h_2^2+k_2^2) + \frac{l_2^2}{(c/a)^2}\right\}. \end{aligned} \tag{4.16}$$

This equation suggests that the difference between the $2\log d$ values for any two planes depends only upon the axial ratio (c/a) and the indices hkl. Using this fact, Hull and Davey developed in 1921 a devised graphical method (A.H. Hull and W.P. Davey: Phys.Rev., **17** (1921), 549) by constructing the variation of the quantity of $\left\{(h^2+k^2) + \frac{l^2}{(c/a)^2}\right\}$ with c/a. This method was frequently employed when we had no electronic calculator and handbooks of mathematical tables were only available. On the contrary, numerical calculations are easily carried out in recent years, so that an analytical method will be described below using the case of a hexagonal system as an example.

Relationship between plane spacing of hexagonal system and Bragg condition is given by the following form.

$$\frac{4\sin^2\theta}{\lambda^2} = \frac{1}{d^2} = \frac{4}{3}\frac{h^2 + hk + k^2}{a^2} + \frac{l^2}{c^2}. \tag{4.17}$$

Equation (4.17) can be rewritten as follows:

$$\sin^2\theta = \frac{\lambda^2}{4a^2}\left\{\frac{4}{3}(h^2 + hk + k^2)\right\} + \frac{\lambda^2}{4c^2}l^2. \tag{4.18}$$

Lattice parameter a and the axial ratio (c/a) are unique values for a substance of interest. Therefore, when $X = \frac{\lambda^2}{3a^2}$ and $Y = \frac{\lambda^2}{4c^2}$, (4.18) can be rewritten as follows:

$$\sin^2\theta = X(h^2 + hk + k^2) + Yl^2. \tag{4.19}$$

Since hkl is always integer, the value of $(h^2 + hk + k^2)$ should be 0, 1, 3, 4, 7, etc. and the value of l^2 is 0, 1, 4, 9, etc. Keeping these features in mind, we divide the $\sin^2\theta$ values for measured diffraction peaks by the integers, 3, 4, 7, etc. and tabulate the results. Then, we examine the numerical results and look for the common quotient, which is equal to one another or that which is equal to one of the measured $\sin^2\theta$ values. We tentatively get the possible value of X by finding the diffraction peaks corresponding to $(hk0)$ with $l = 0$. The value of X obtained in this process satisfies the following relationship for general indices h, k, and l.

$$Yl^2 = \sin^2\theta - X(h^2 + hk + k^2). \tag{4.20}$$

To obtain the value of Y, we make again the tabulated results by subtracting the value of X, $3X$, and $4X$ from each $\sin^2\theta$ and look for reminders that are in the ratio of 1, 4, 9, etc. because the peaks from the $(00l)$ will have $l^2 = 1, 4, 9, \ldots$ only. Note that the remaining peaks which are neither $(hk0)$ nor $(00l)$ should be assigned as peaks belonging to hexagonal system.

According to this procedure, temporary indexing can be set for several peaks including the values of X and Y. Once these results are obtained, all the measured diffraction peaks can be recalculated and a comparison is made between the calculated $\sin^2\theta$ values and the experimental data. When no inconsistent points can be found, analysis is said "completed". At the same time, the lattice parameters are computed from the relationships of $X = \frac{\lambda^2}{3a^2}$ and $Y = \frac{\lambda^2}{4c^2}$. This analytical method may be best explained by means of a specific example such as Mg and Zn (see Question 4.5).

4.5 Identification of an Unknown Sample by X-ray Diffraction (Hanawalt Method)

There are rather few case, where structural analysis for completely unknown substances are needed, because we usually have some preliminary information, such as possible constituent elements and their volume in a sample. In addition, acomparison of the measured diffraction data with those of a large number of standard substances can be made without any difficulty in a short time by the recent developments of computer and database search software. The powder diffraction database compiles for about 50,000 substances, named by JCPDS cards (Joint Committee on Powder Diffraction Standards) with the cooperation of many societies in U.S.A. Canada, United Kingdom, and France. Then, we may identify an unknown substance of interest by searching the diffraction data so as to find one, which exactly matches the pattern of the unknown substance compiled in the database. This procedure was originally proposed by Hanawalt *et al.* (J.D. Hanawalt, H.W. Rinn and L.K. Frevel: Ind. Eng. Chem. Anal. Ed., **8** (1936)244.; **10**(1938)457.) using the concept that the powder diffraction pattern of a substance is characteristic of that substance, like a fingerprint and the measured powder diffraction pattern is quite likely to be reproduced by the simple sum of those of pure substances if a sample contains two components or more. In the Hanawalt method, each substance is characterized by the d-values of the three strongest diffraction peaks. The essential points of this method [referred to as Hanawalt method] are described below.

The values of d_1, d_2, and d_3 for three strongest diffraction peaks coupled with their relative intensities (I/I_1) are used to search the corresponding pattern in the database. Analysis is completed if we find the exact match between the diffraction data of interest and those of the standard materials. Of course, when applying the Hanawalt method, it is also important to acquire information in advance about type of elements contained in a sample and the composition ratio of the elements contained, for example, by fluorescent X-ray analysis.

A comparison was manually carried out using the index tables called "Search Manual," where substances are listed in order of d-values of the three strongest peaks, d-spacings and intensities are given for a total eight peaks of each substance. Another manual where substances are listed alphabetically by name is also available. However, the automatic search by computer is now widely employed instead of manual searching. When the automatic search by applying the Hanawalt method is made by computer together with the automatic calculation of d values as output of an automatic diffractometer, it should be kept in mind "what work should be carried out" or "what you are going to do".

The d values and their relative intensities are known to be affected by several factors, for example, in cases where a sample includes some water molecules, a sample may react with oxygen, water vapor, carbon dioxide, etc. in the atmosphere or a sample may be received with preferred orientation. Such situation sometimes leads to the erroneous identification of a desired sample by accidentally matching with the reference data in the automatic search operation, so that attention is again strongly suggested.

When using the Hanawalt method, we have to keep in mind that a certain experimental uncertainty is included and some reference data are not so reliable even if it is compiled very recently. Do not forget that some materials show a very similar diffraction profile, as exemplified by Au and Ag, Si, and ZnS or a ferrite and a spinel compound, etc. In other words, identification of an unknown substance by the Hanawalt method is very effective, but it is also important to know this method within the relative comparison only.

If the unknown is a single component, the search procedure is relatively straightforward, whereas the analysis of mixture becomes more difficult, since a d-value from one component is superimposed on a d-value from another and the three strongest peaks in the pattern of unknown substance consist of such mixed d-values. It may be safely said that the maximum components contained in one sample is limited to three to analyze using the Hanawalt method. This limitation may be solved in the near future by the development of the next generation-type software, which enables us to show an almost identical performance of the search procedure by human eyes.

Details of the JCPDS cards have been described in several monographs (e.g., B.D. Cullity, Elements of X-ray Diffraction, 2nd edition, Addison-Wesley, Reading Massachusetts, (1978)) and are not duplicated here. Nevertheless, Questions 4.4 $\sim$ 4.9 in this chapter facilitate to readers understanding of "how to use the Hanawalt method for determining the structure of unknown samples." Of course, careful sample preparation and good experimental arrangement will eliminate unrelated diffraction data. Some points which should be taken into account when analyzing data using the Hanawalt method are summarized as follows:

(1) If the particle size of a powder crystalline sample is larger than several 10 μm, good reproducibility of the intensity pattern is not always confirmed and one can frequently find a difference more than several 10% for every measurement.
(2) Keep in mind the used wavelength of X-rays for measurement (for example: Cu-Kα or Mo-Kα), because some differences may be found in the relative intensities, depending upon the wavelength due to anomalous dispersion factors.
(3) The reverse order of the relative intensities may be found in a sample, which is easily received under the preferred orientation, like clay minerals.
(4) When impurities are included in a sample, a certain shift toward the lower angle in the peak positions is quite likely to appear. (Such shift usually becomes distinct in the peak detected at higher angles.)
(5) If the goniometer center or its zero point is found to be eccentric, the position of the diffraction peaks is readily deviated from the JSPDS reference values. Badly aligned slits may also be an origin to produce similar deviation.
(6) When the diffraction peak, which is not identified by the JCPDS data, is observed, careful judgment should be given by considering some factors, such as impurity contamination and presence of a solid solution phase or an ordered phase (super-lattice).

4.6 Determination of Lattice Parameter of a Polycrystalline Sample

The structure of a sample becomes clear by obtaining information of the shape and size of a unit cell from the angular positions of the measured diffraction data, and the correct Miller indices are also assigned to respective peak called "indexing the diffraction pattern." Then, we can estimate the lattice parameters. Since the lattice parameters are structure sensitive, it is known that many substances show volume change of about 0.001% with variation of one degree in temperature.

There is no complicated procedure in determining the lattice parameters. The sufficiently reliable parameter values are obtainable from the measurements, as precisely as possible, with respect to the d – values of any particular set of lattice planes whose index is proved. In this case, we use the Bragg law to determine the d – value. On the contrary, for example, the lattice parameter a of a cubic substance is simply proportional to the d – values. Then, the accuracy in determining the d – value (or lattice parameter a) depends on precision in $\sin\theta$, which is a derived quantity, not on precision in θ, which is a measured quantity. The value of $\sin\theta$ changes very slowly with θ in the close vicinity of 90° from which it may be suggested that the uncertainty in measurement of $\sin\theta$ decreases with increasing the value of θ.

The following equation can be obtained by differentiating the Bragg law with respect to θ.

$$\frac{\Delta d}{d} = \frac{\Delta\lambda}{\lambda} = -\Delta\theta\cot\theta. \tag{4.21}$$

Equation (4.21) suggests that when θ is brought as much as possible close to 180°, the fractional error in d as denoted by $\Delta d/d$, equivalent to resolution, approaches zero, so that we should use the diffraction peaks as close to the value of $2\theta = 180°$, as possible to calculate the best precision and the true value of the lattice parameter. Namely, the value of highly precise d can be obtained. However, note that the diffraction peaks detected at the higher angle side are affected more or less by a temperature factor. In addition, since measurement of a diffraction peak at $2\theta = 180°$ is physically impossible, the extrapolation of the lattice parameter versus certain functions of θ, as producing a straight line, is generally employed. Various methods for such extrapolation are proposed. For example, the method of Cohen using $\cos 2\theta$ or the method of Nelson–Riley, which utilizes the function given by $\frac{1}{2}\left\{\frac{\cos^2\theta}{\sin\theta} + \frac{\cos^2\theta}{\theta}\right\}$ is widely used with respect to a cubic system. Figure 4.6 shows the results using the method of Nelson–Riley, as an example.

Precise lattice parameter measurements of a powder crystalline sample is affected by various factors such as misalignment of the diffractometer parts, displacement of the flat sample from the diffractometer axis, vertical divergence of the incident X-ray beam, etc. One simple way for checking the deviation from an ideal case is to use the "internal standard method", where the uncertainty being determined, is compared with the diffraction data from a standard substance, mixed with the sample of interest. High-purity silicon powder or tungsten powder, whose lattice parameter is already measured precisely, is widely used as the standard substance.

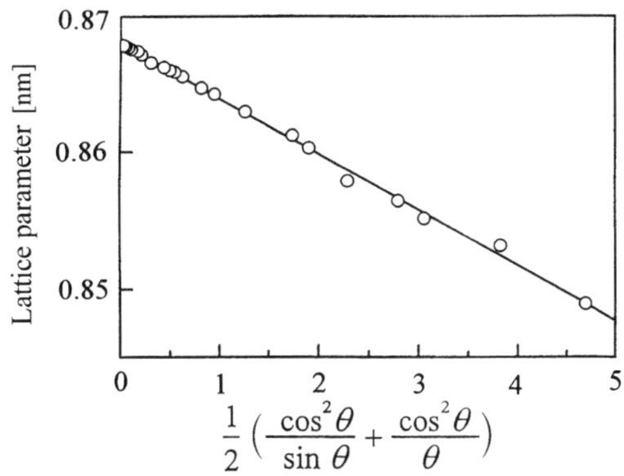

Fig. 4.6 Extrapolation of measured lattice parameters using the Nelson–Riley method

Nevertheless, considering many factors, it may be safely said to readily determine the lattice parameter within the range of ±0.1% of an error, whereas it is very difficult to determine the value within the error range of ±0.01%. For later purpose, it is required to utilize the particular equipment only for the precision measurements of the lattice parameter equipped with four crystal monochromators.

4.7 Quantitative Analysis of Powder Mixtures and Determination of Crystalline Size and Lattice Strain

X-ray diffraction analysis is used not only to identify the phase of unknown substance, as well as the estimation of the lattice parameters, but also to determine the concentration of that phase in the mixture. The peak profile is also employed to estimate the particle size of very small crystals called "crystallites" in a powder sample. The essential points of these applications are described below.

4.7.1 Quantitative Determination of a Crystalline Substance in a Mixture

Quantitative X-ray diffraction analysis is based on the fact that the intensity of the diffraction pattern of a desired crystalline substance depends on the concentration of that phase in a mixture. The relationship between intensity and concentration is not always linear, but it is found to be possible when we focus on a particular peak of the desired substance with reference to the case of that substance alone. Therefore, the diffraction peak corresponding to the specific plane of the desired crystalline substance should be observed at the fully separated angle from those of other ingredients and its integrated intensity enables us to provide an index of the amount. Of course, it is desirable to determine the concentration from the integrated intensity for one peak and more.

When carrying out the diffraction intensity measurements using a diffractometer under the $\theta - 2\theta$ operation, the integrated intensity I_{ij}, of the specific diffraction peak i of the desired crystalline substance j to be determined, is given by the following equation.

$$I_{ij} = \mathrm{K}\left[\frac{P_{ij} F_{ij}^2 (LP)}{V_{cj}^2}\right]\frac{V_j}{2\mu}. \tag{4.22}$$

The meaning of each symbol is as follows. K: constant independent of the quantity and type of sample, V_{cj}: volume of a unit cell of the crystalline substance j, P_{ij} and F_{ij}: multiplicity factor and the structure factor of the diffraction peak i of the substance j, respectively, (LP): Lorentz polarization factor and V_j: volume ratio of crystalline substance j, and μ: the average linear absorption coefficient of a sample. If the used wavelength of X-rays and the specific diffraction peak employed for quantification are set and the portion employed as a constant is collectively stipulated with R_{ij}, (4.22) can be simplified in the following form.

$$I_{ij} = \frac{R_{ij} V_j}{\mu}. \tag{4.23}$$

Let us set the density of the crystalline substance j to ρ_j and that of other substance to ρ_{M}, while the volume ratio and average linear absorption coefficients of a substance other than the desired crystalline substance j are set to V_{M} and μ_{M}, respectively. Then, by using the weight fraction (or weight ratio) w_j, (4.23) is rewritten in the following form.

$$I_{ij} = \frac{R_{ij} w_j}{\rho_j} \Bigg/ \left[w_j\left(\frac{\mu_j}{\rho_j} - \frac{\mu_{\mathrm{M}}}{\rho_{\mathrm{M}}}\right) + \frac{\mu_{\mathrm{M}}}{\rho_{\mathrm{M}}}\right]. \tag{4.24}$$

If the average mass absorption coefficient of a sample is given by (μ^*/ρ), (4.24) becomes

$$I_{ij} = \frac{R_{ij} w_j}{\rho_j} \Bigg/ \left(\frac{\mu^*}{\rho}\right). \tag{4.25}$$

As a result, the integrated intensity I_{ij} of the specific diffraction peak i of the desired crystalline substance j can be subject to information, proportional to the weight fraction w_j or in inverse proportion to the average mass absorption coefficients (μ^*/ρ) of a sample.

If the chemical composition and the average mass absorption coefficient of the desired sample are available, by applying other techniques such as fluorescent X-ray analysis, a calibration curve can be prepared from the measurements on a set of synthetic samples containing known concentrations of the crystalline substance of interest, so that the integrated intensity of the specified diffraction peak of the crystal substance j, is given as a function of concentration, in advance. When the chemical compositions and the average mass absorption coefficient of the desired sample are unavailable, we can solve such difficulty by using the so-called "internal standard method" by adding a known reference substance such as NaCl and CaF_2 to the sample at a certain known weight fraction. Examples are shown in Questions 4.14 ∼ 4.17.

4.7.2 Measurement of the Size of Crystal Grains and Heterogeneous Distortion

The term "particle size" or "grain size" is used when the size of individual crystals is less than about 0.1 μm (100 nm), because a particle of a real crystalline powder sample generally consists of many fine units called "crystallites," which can be considered as a single crystal, as illustrated in Fig. 4.7. Here, it should be suggested not to confuse the grain size of a crystalline powder sample and the size of crystallites. Although the size of crystallites may be the same as that of grain size in some cases, they are essentially different physical quantities. When saying "the size of a crystal" in X-ray diffraction analysis, it usually refers to the "size of crystallites" concerning a factor, which makes a diffraction peak broad. If the size of crystallites in a crystalline powder sample becomes small less than 0.005 μm (5 nm), the measured diffraction peak profile clearly deviates from that of the same sample with the standard size of 0.5–10 μm in diameter.

Uniform distribution in all directions of crystallites in a sample is also required to provide the diffraction peak profile to enable us to give sufficiently reliable results. However, if the diameter of crystallites in a sample becomes small, for example, less than 0.1 μm, the measured diffraction profiles broaden out so as to say "peak broadening." This is attributed to the fact that the periodic region in the atomic

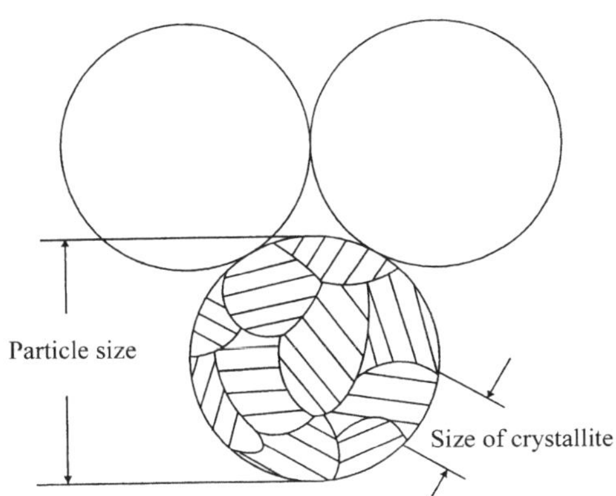

Fig. 4.7 Schematic diagram for grain size and crystallite of a crystalline powder sample

arrangements producing the same diffraction peak profile is limited. Peak broadening is also detected if the sample is deformed plastically and then the lattice planes become distorted so as to change the spacing of any particular set from one grain to another or from one portion of a grain to another. Such nonuniform strain is introduced, for example, by mechanical grinding. Therefore, it is necessary to take these two factors into consideration in the X-ray structural analysis, using the measured diffraction peak profile for a sample consisting of fine particles.

Setting the wavelength of the incident X-rays to λ, plane spacing to d and Bragg angle to θ_B, we obtain the generalized equation called "Bragg law."

$$\lambda = 2d \sin \theta_B. \tag{4.26}$$

Usually, so-called destructive interference is not perfect if the incident X-rays cover the Bragg condition only slightly different from θ_B. This is particularly true when the number of planes completely satisfying the Bragg condition is not sufficient to be involved by reducing the particle size of a crystalline powder sample. That is, the phase difference, being slightly out of harmony with Bragg law for waves scattered by the spacing of any particular plane set, is correlated with the grain size of crystallites. The details should be referred to other monographs on X-ray diffraction, the certain width B_r of a diffraction peak is generally observed in the angular region near $2\theta_B$ as shown in Fig. 4.8. Note that the X-ray scattering intensity, in the angular region between a slightly smaller angle $2\theta_1$ near $2\theta_B$ and a slightly larger angle $2\theta_2$, is not zero, as the size of crystallites decreases. Such peak profile is characterized by the distribution with values intermediate between zero and the maximum at $2\theta_B$ as illustrated in Fig. 4.8.

Since θ_1 and θ_2 are very close to θ_B, the relationship of $\theta_1 + \theta_2 = 2\theta_B$ is recognized so that the approximation of $\sin\{(\theta_1 - \theta_2)/2\} = \{(\theta_1 - \theta_2)/2\}$ is also well accepted. The peak width B_r is usually defined in radians at the intensity equal to half the maximum intensity called the integral width for a peak. Nevertheless, we may use an angular width B in terms of 2θ as a measure of a peak width B_r. Namely, if $B_r = (2\theta_1 - 2\theta_2)/2 = \theta_1 - \theta_2$ is taken into consideration, as the peak width (see Fig. 4.9), the following equation will be obtained.

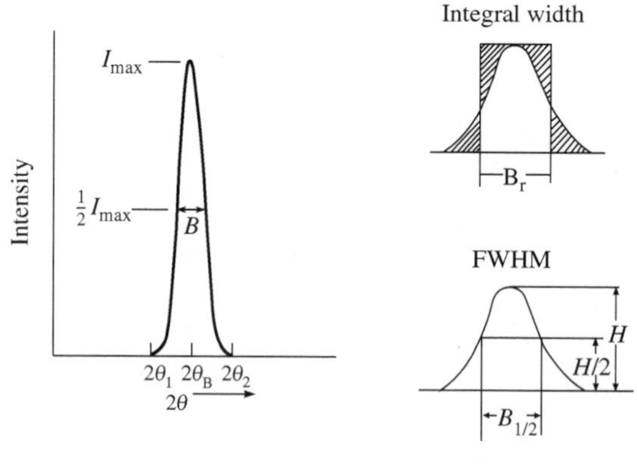

Fig. 4.8 Schematic diagram for X-ray diffraction peak profile of a fine crystalline sample

Fig. 4.9 Change in peak profile, peak position, and its width arising from lattice strain

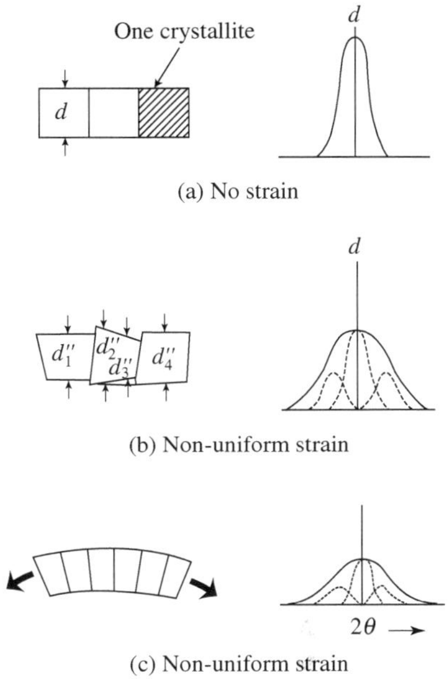

$$2t\left(\frac{\theta_1-\theta_2}{2}\right)\cos\theta_B = \lambda \tag{4.27}$$

$$t = \frac{\lambda}{B_r\cos\theta_B}. \tag{4.28}$$

Here, we use path-difference equations at the Bragg condition together with the entire thickness of the crystalline sample, t, in comparison with the distance between adjacent planes. On the contrary, by using $B_{1/2}$ of FWHM (full width at half its maximum intensity) instead of the integral width for the corresponding peak, the following equation called "Scherrer's equation" is widely used for estimating the size of crystallites from measured diffraction peak profile.

$$t = \frac{0.9\lambda}{B_{1/2}\cos\theta_B}. \tag{4.29}$$

The value of t in (4.28) and (4.29) represents the diameter of crystallites, perpendicular to the plane, which corresponds to the measured diffraction peak. If the diameter of crystallites becomes small, for example, down to about $t = 0.05\,\mu\text{m}$ (50 nm), the peak broadening can be appreciable by making the B_r value the order of 4×10^{-3} radians (0.2 degrees). However, we have to mention that all diffraction peaks have a measureable width arising from some factors such as the divergence of the incident beam and width of the X-ray source (not an infinitesimal thin line spot), even if adjustments of optical alignment for diffractometer are carried out perfectly. In other words, the width given in (4.28) and (4.29) refers to the extra peak broadening due to the particle size effect alone. For this reason, when calculating the mean value of the size of crystallites from the measured width B_{obs}, it is desirable, in advance, to determine the value of width (B_{i}) of peak broadening caused by factors except for the particle size effect, frequently called instrumental broadening factor.

Specifically, the shape of a diffraction peak is approximated by Gaussian distribution, and the value of the width B_{r} related only to change in size of crystallites may be calculated using the following equation.

$$B_{\mathrm{r}}^2 = B_{\mathrm{obs}}^2 - B_{\mathrm{i}}^2. \tag{4.30}$$

For this purpose, we frequently use α – quartz crystalline particles with 25 μm in diameter as a standard sample. In this case, α – quartz crystalline particles are fully annealed at 1,073 K and slowly cooled down to room temperature to remove any distortion and strain in crystallites.

On the contrary, when a polycrystalline sample, which consists of many crystal grains, is plastically deformed, slip is known to occur in each grain and the crystal grain varies its shape, such as being flattened and elongated in the direction of rolling. The variation in shape of any one crystal grain should be related not only to the forces applied to the sample as a whole, but also to the fact that each crystal grain retains contact on its boundary surfaces with all neighbors. As a result of the restraint by neighbors, a plastically deformed crystal grain in a solid substance usually has regions of the lattice left in an elastically bent or twisted. Then, the diffraction peak profile of such polycrystalline sample should be affected, more or less, by some residual strain. This makes the diffraction peak broad.

When all crystal grains receive a uniform strain, a plane spacing changes uniformly and then a certain shift in the position of the diffraction peak is observed. However, when applying a non-uniform strain, the different size of distortion depending on the place, to crystal grains, the plane spacing varies at random and then the resultant diffraction peak shows a certain width. Such situation is illustrated in Fig. 4.9. In order to estimate the peak broadening due to the effect of particle size, as well as the effect of nonuniform (inhomogeneous) strain, the Hall method as described in the following equation is commonly used.

$$B_{\mathrm{r}} \cos\theta = \frac{\lambda}{\varepsilon} + 2\eta \sin\theta, \tag{4.31}$$

where B_{r} is the peak width (integral width) resulting from the size of crystallites (ε) and 2η is the amount equivalent to inhomogeneous strain.

Specifically, the integral width (see Fig. 4.9) showing the width of rectangular which is the same as peak heights and has same area is computed for more than two diffraction peaks and then the graphic relation between $B_{\mathrm{r}} \cos\theta$ and $\sin\theta$ is drawn. Analysis is often carried out using $B_{1/2}$ of FWHM instead of the integral width. Since a linear correlation is usually observed between these two quantities, a slope of the line gives the (2η) value for inhomogeneous strain and an intersection with the $B_r \cos\theta$ axis corresponds to the reciprocal of the size of crystallites (ε), as easily seen from (4.31). It is also suggested for this approach to use the measured diffraction data of the planes belonging to the same set, such as (111) and (222) or (200) and (400) as much as possible, because the size of crystallites frequently appears to depend on the planes to be used.

Under the present experimental conditions, the size of the crystallites can be computed within the error of about ±several percent when using the Scherrer method or the Hall method. This is true for the cases where the size of crystallites in a sample of interest lies in the range about 0.005 μm (5 nm) below 0.05 μm. However, it should also be kept in mind that the value of particle size obtained from the diffraction peak profile is not always in agreement with those found by other techniques, such as electron microscopy and a laser particle size analyzer.

4.8 Solved Problems (18 Examples)

Question 4.1 If an X-ray diffraction pattern for a powder sample of Cr (crystal system: body-centered cubic and lattice parameter $a = 0.2884$ nm) is measured using a conventional diffractometer, we will obtain six diffraction peaks corresponding to (110), (200), (211), (220), (310), and (222) planes. Compute the Lorentz-polarization factor using Cu-Kα radiation $\lambda = 0.1542$ nm.

Answer 4.1 By combining the Bragg law with the plane spacing equation for the cubic system, the following relation is readily obtained

$$\sin\theta = \frac{\lambda}{2a}\sqrt{h^2 + k^2 + l^2}. \tag{1}$$

On the contrary, the Lorentz-polarization factor (*LP*) is given in the following form.

$$LP = \frac{1 + \cos^2 2\theta}{2\sin^2\theta\cos\theta}. \tag{2}$$

Note that another expression without the numerical value 2, which appears in the denominator of (2), is also frequently utilized. By applying (110), (200), (211), (220), (310), and (222) to (hkl) in (1) together with $\lambda = 0.1542$ nm, we obtain $\sin\theta$ and θ from which the Lorentz-polarization factor (*LP*) can be computed using (2). The results are summarized in Tables 1 and 2.

Table 1 Calculated results of (1) for $\sin\theta$ and θ

	hkl	$h^2+k^2+l^2$	$\sqrt{h^2+k^2+l^2}$	$\sin\theta$	θ	2θ
1	110	2	1.4142	0.3781	22.22	44.44
2	200	4	2	0.5347	32.32	64.64
3	211	6	2.4495	0.6548	40.90	81.80
4	220	8	2.8284	0.7561	49.12	98.24
5	310	10	3.1623	0.8454	57.71	115.42
6	222	12	3.4641	0.9254	67.73	135.46

Table 2 Calculated results of (2) for the LP factor

	$\sin\theta$	$\sin^2\theta$	$\cos\theta$	$\cos 2\theta$	$\cos^2 2\theta$	$1+\cos^2 2\theta$	$2\sin^2\theta\cos\theta$	LP
1	0.3781	0.1430	0.9257	0.7140	0.5098	1.5098	0.2648	5.70
2	0.5347	0.2859	0.8451	0.4283	0.1834	1.1834	0.4832	2.45
3	0.6548	0.4288	0.7559	0.1426	0.0203	1.0203	0.6483	1.57
4	0.7561	0.5717	0.6545	−0.1433	0.0205	1.0205	0.7484	1.36
5	0.8454	0.7147	0.5342	−0.4293	0.1843	1.1843	0.7636	1.55
6	0.9254	0.8564	0.3790	−0.7128	0.5081	1.5081	0.6492	2.32

Question 4.2 With respect to a powder sample of Cr, compute the temperature factor at room temperatures (293 K) as a function of $\frac{\sin\theta}{\lambda}$ by applying the Debye approximation to the evaluation for the effect of thermal vibration.

Answer 4.2 The effect of thermal vibration of atoms on the intensity of diffracted X-rays is commonly taken into consideration as a Debye–Waller factor and such effect, in practice, represented by e^{-2M_T} is introduced in the following equation through the atomic scattering factor f.

$$f = f_0 e^{-M_T}, \tag{1}$$

where f_0 corresponds to the atomic scattering factor without thermal vibration. It may be noteworthy that the temperature effect responds to multiply the intensity computed from f_0^2 by e^{-2M_T}, since the intensity is proportional to f^2.

M_T involves the amplitude u of thermal vibration, as well as the diffraction angle 2θ in the following form.

$$M_T = 8\pi^2 < u^2 > \left(\frac{\sin\theta}{\lambda}\right)^2 = B_T \left(\frac{\sin\theta}{\lambda}\right)^2. \tag{2}$$

Here, $< u^2 >$ is the mean-square of the displacement to a direction perpendicular to the diffracted plane. Although it is difficult to calculate the absolute values of $< u^2 >$ as a function of temperature, the following simple equation is proposed and widely used to evaluate the thermal vibration effect of atoms by using the Debye approximation.

$$B_T = \frac{6h^2}{mk_B}\frac{T}{\Theta^2}\left\{\phi(x) + \frac{x}{4}\right\}. \tag{3}$$

Here, T is the absolute temperature, m is the mass of the vibrating atom, h and k_B are Planck constant and Boltzmann constant, respectively, and Θ the so-called Debye characteristic temperature of substance given by the absolute temperature.

In addition, $\phi(x)$ in case of $x = \frac{\Theta}{T}$ is tabulated in some textbooks (e.g., B.D. Cullity, Elements of X-ray Diffraction, 2nd Edition, Addison-Wesley, (1978); see also Appendix A.6). We are likely to obtain from (3) that the reduction in peak intensity due to the temperature factor will be certainly detected for substances, such as Pb and Bi whose characteristic Debye temperature is relatively low. This is particularly true for the peaks observed at higher angles.

The value of m in (3) is equivalent to the quotient when the atomic weight M of the substance divided by Avogadro's number N_A. At this time, the atomic scattering factors compiled, for example in the International Tables for X-ray Crystallography, are given as a function of $\frac{\sin\theta}{\lambda}$ in unit of Angstrom, so that (3) can be rewritten as follows:

$$\frac{6h^2}{mk_B} = \frac{6N_A h^2}{Mk_B} = \frac{6 \times (0.6022 \times 10^{24}) \times 10^3 \times (6.626 \times 10^{-34})^2}{M \times (1.3806 \times 10^{-23}) \times 10^{-20}}$$
$$= \frac{1.15 \times 10^4}{M}. \tag{4}$$

Since kg is used for the Planck constant and Boltzmann constant given by SI unit, we have to take molar mass of substance per kg and $1\text{A} = 10^{-8}\text{cm} = 10^{-10}$ m into consideration. Note that angstrom denoted by Å is also used in comparison with the electric current A.

The Debye characteristic temperature of Cr is found to be 485 K from Appendix A.2 and then the value of x is estimated as follows:

$$x = \frac{485}{293} = 1.66. \tag{5}$$

From Appendix A.6, we obtain $\phi(x) = 0.660$ at $x = 1.66$ and put these values to (3) together with 51.996 g for the atomic weight of Cr per mole (see Appendix A.2).

$$B_T = \frac{1.15 \times 10^4}{51.996} \times \frac{293}{(485)^2} \times \left(0.660 + \frac{1.66}{4}\right) = 0.296. \tag{6}$$

Consequently, we obtain the relationship of $\mathrm{e}^{-2M_T} = \mathrm{e}^{-0.592\left(\frac{\sin\theta}{\lambda}\right)^2}$ with respect to the temperature factor of a powder sample of Cr. The results are listed in Table 1, and they are also illustrated in Fig. 1.

Table 1 Calculated results of temperature factor of Cr

$\frac{\sin\theta}{\lambda}$	0	0.1	0.2	0.3	0.4	0.5	0.6
$\left(\frac{\sin\theta}{\lambda}\right)^2$	0	0.01	0.04	0.09	0.16	0.25	0.36
e^{-2M_T}	1	0.99	0.98	0.95	0.91	0.86	0.81

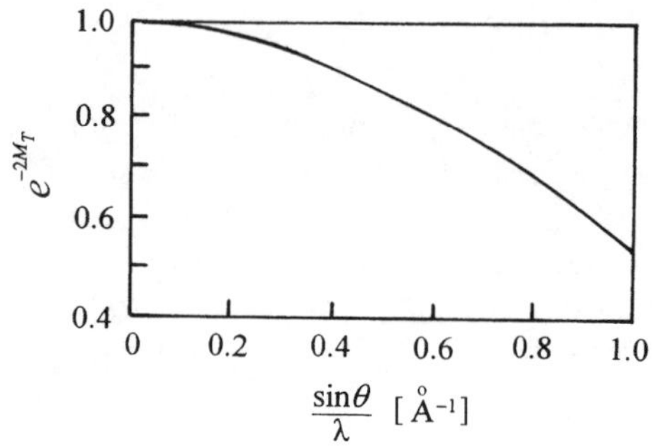

Fig. 1 Variation of temperature factor of Cr as a function of $\sin\theta/\lambda$

Question 4.3 With respect to a powder samples of Cr (crystal system: body-centered cubic, lattice parameter $a = 0.2884$ nm), compute the intensities of several peaks possibly detected when Cu-Kα radiation ($\lambda = 0.1542$ nm) is used.

Answer 4.3 The general formula of the intensity of diffracted X-rays for a powder crystalline sample is given in the following.

$$I = |F|^2 p \left(\frac{1 + \cos^2 2\theta}{2 \sin^2 \theta \cos \theta} \right) \frac{1}{2\mu} \left(1 - e^{-\frac{2\mu t}{\sin\theta}} \right) e^{-2M_T}. \tag{1}$$

Here, F is structure factor, p multiplicity factor, the parenthesis in the 3rd term Lorentz- polarization factor (LP), and e^{-2M_T} the temperature factor. The values of LP and e^{-2M_T} have been obtained in the results of Questions 4.1 and 4.2. With respect to the absorption factor, it supposes that the usual case with a sample thickness is considered of infinite thickness and it is omitted in this calculation.

The atomic scattering factor f_{Cr} is taken from the compiled data in Appendix A.3 by considering the $\sin\theta/\lambda$ values, whereas the multiplicity factor p is readily obtained from the results of Table 2.2 in Chap. 2. In addition, we also take the relationship of $|F|^2 = 16f^2$ for a body-centered cubic into consideration (see Chap. 3.4). The results are summarized in Table 1.

However, the temperature factor has been included using the results of Question 4.2 and the results are summarized in Table 2. In the column at the right end of Table 2, the experimental values of the intensity ratio of peaks are also given for comparison.

Table 1 Calculated results of the peak intensities for powder sample of Cr possibly detected using Cu-Kα radiation (without the temperature factor)

	hkl	S	$\sin\theta$	$\frac{\sin\theta}{\lambda}$ (A^{-1})	f_{Cr}	$\lvert F\rvert^2$	p	LP	I_{cal}	I_{cal}/I_1
1	110	2	0.3781	0.245	15.8	999	12	5.70	6.83×10^4	100
2	200	4	0.5347	0.347	13.2	697	6	2.45	1.02×10^4	15
3	211	6	0.6548	0.425	11.7	548	24	1.57	2.06×10^4	30
4	220	8	0.7561	0.490	10.6	449	12	1.36	0.73×10^4	11
5	310	10	0.8454	0.548	9.8	382	24	1.55	1.43×10^4	21
6	222	12	0.9254	0.606	9.2	339	8	2.32	0.63×10^4	9

$S = h^2 + k^2 + l^2$

Table 2 Calculated results of the peak intensities for powder sample of Cr possibly detected when the temperature factor is included

	e^{-2M_T}	I'_{cal}	I'_{cal}/I_1	Exp.
1	0.97	6.63×10^4	100	100
2	0.93	0.95×10^4	14	16
3	0.90	1.85×10^4	28	30
4	0.87	0.64×10^4	10	18
5	0.84	1.20×10^4	18	20
6	0.81	0.51×10^4	8	6

Question 4.4 X-ray measurements using Cu-Kα radiation ($\lambda = 0.1542$ nm) provide the diffraction patterns of **Figs. A**, **B**, and **C** for three metallic samples, which are known cubic system. The relevant numerical data for three cases are summarized in Tables. By applying the fundamental equation obtained by combining the Bragg law with the plane spacing equation for the cubic system to three cases, index the pattern and compute the lattice parameter.

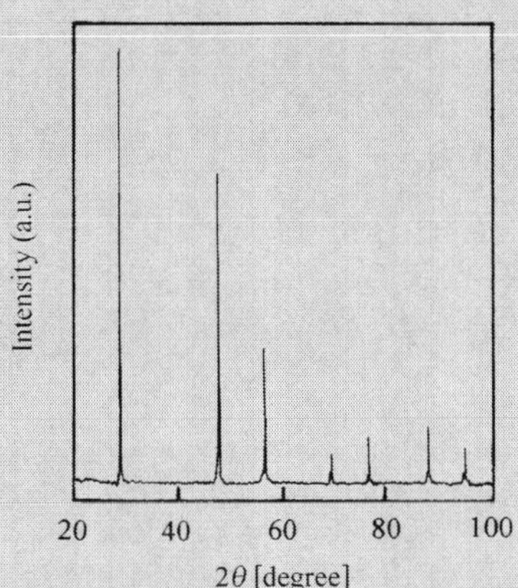

Fig. A Diffraction pattern of sample A

Table A Diffraction data of sample A

	2θ	d(A)	I/I_0
1	28.41	3.142	100
2	47.33	1.921	57
3	56.11	1.639	28
4	69.08	1.360	7
5	76.34	1.248	11
6	88.03	1.110	13
7	94.95	1.046	5

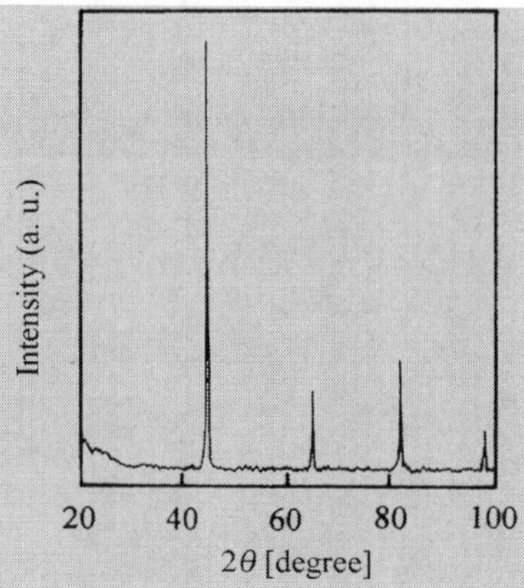

Fig. B Diffraction pattern of sample B

Table B Diffraction data of sample B

	2θ	d(A)	I/I_0
1	44.51	2.036	100
2	51.90	1.762	43
3	76.45	1.246	22
4	93.02	1.063	19
5	98.50	1.018	7

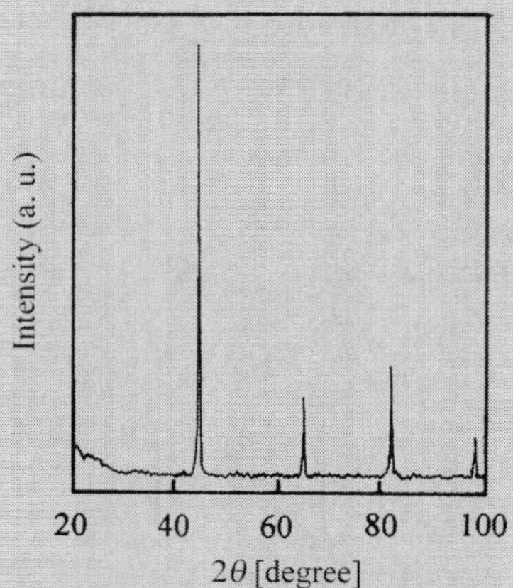

Fig. C Diffraction pattern of sample C

Table C Diffraction data of sample C

	2θ	d(A)	I/I_0
1	44.40	2.041	100
2	64.59	1.443	20
3	81.76	1.178	26
4	98.31	1.019	7

Answer 4.4 The following equation can be given for a cubic system.

$$\frac{\sin^2\theta}{(h^2+k^2+l^2)} = \frac{\sin^2\theta}{S} = \frac{\lambda}{4a^2}. \tag{1}$$

Here, λ is the wavelength of used X-rays, a is a lattice parameter, and hkl is Miller indices. The sum of the square of plane indices ($S = h^2 + k^2 + l^2$) corresponds to the measured diffraction peak is always an integer and $\frac{\lambda^2}{4a^2}$ is also a constant for any X-ray diffraction pattern. Therefore, the process of indexing the pattern is to find a set of integers, which yields a constant quotient when divided one by one into the measured $\sin^2\theta$ values. In these trial computations, the plane indices of

(111), (200), (220) etc. are input if a face-centered cubic lattice is assumed, next (110), (200), (211), etc. are input if a body-centered cubic lattice is assumed. For this reason, to minimize the trial computation, you are suggested to watch carefully the measured X-ray diffraction pattern of three samples.

With reference to the guideline of Fig. 4.5, as for the pattern of Sample C, the diffraction peaks have appeared at equal intervals and has the characteristics features of body-centered cubic. From such point of view, we could recognize the characteristic feature of face-centered cubic in Sample B, where two diffraction peaks are first observed at relatively near angles and another diffraction peak appears in a slightly isolated angle position. However, the peak positions observed in Sample A appear to differ from the features of Samples B and C. Therefore, we may be sure that our trial calculation is done by assuming fcc for Sample B and bcc for Sample C.

Table 1 (Sample B) Example of trial calculation on the assumption of fcc

	2θ	$\sin\theta$	$4\sin^2\theta/\lambda^2$	$*$	$h^2+k^2+l^2$	hkl	a (nm)
1	44.51	0.3787	24.1258	1	3 (1.00)	111	0.3526
2	51.9	0.4376	32.2141	1.34	4 (1.33)	200	0.3524
3	76.45	0.6188	64.4157	2.67	8 (2.67)	220	0.3524
4	93.02	0.7255	88.5454	3.67	11 (3.67)	311	0.3525
5	98.50	0.7576	96.5542	4	12 (4.00)	222	0.3525

$*$The value in the column of $4\pi\sin^2\theta/\lambda^2$ divided by 24.1258.

Table 2 (Sample C) Example of trial calculation on the assumption of bcc

	2θ	$\sin\theta$	$4\sin^2\theta/\lambda^2$	$*$	$h^2+k^2+l^2$	hkl	a (nm)
1	44.40	0.3778	24.0113	1	2 (1.00)	110	0.2886
2	64.59	0.5343	48.0244	2	4 (2.00)	200	0.2886
3	81.76	0.6545	72.0627	3	6 (3.00)	211	0.2885
4	98.31	0.7565	96.2740	4	8 (4.00)	220	0.2883

$*$The value in the column of $4\pi\sin^2\theta/\lambda^2$ divided by 24.0113.

From these results of Tables 1 and 2, Sample B has an fcc structure and its lattice parameter is 0.3525 nm and Sample C has a bcc structure and its lattice parameter is 0.2885 nm. It may be mentioned Ni, $fcc, a = 0.35238$ nm and Cr, $bcc, a = 0.28839$ nm compiled in Appendix A.9 for reference.

Next, the trial calculations on the assumption of fcc or bcc were made for Sample A, but we could not obtain good results of indexing the pattern, as shown in the column of Table 3 marked by $*$. Then, a trial calculation was further carried out, when assuming diamond structure for Sample A. The results are summarized in the right-hand side columns of Table 3.

When assuming diamond structure for Sample A, the coincidence with the values marked by $*$. and the square of plane indices $h^2+k^2+l^2$ has been confirmed as seen in Table 3. The resultant lattice parameter is 0.5437 nm, which is in good agreement with the value of Si, $a = 0.54309$ nm cited in Appendix A.9.

Table 3 (Sample A) Example of trial calculation on the assumption of diamond structure

	2θ	$\sin\theta$	$4\sin^2\theta/\lambda^2$	$*$	$h^2+k^2+l^2$	hkl	a (nm)
1	28.41	0.2454	10.1307	1	3 (1.00)	111	0.5442
2	47.33	0.4014	27.1048	2.68	8 (2.67)	220	0.5433
3	56.11	0.4703	37.2084	3.67	11 (3.67)	311	0.5437
4	69.08	0.5670	54.0826	5.34	16 (5.33)	400	0.5439
5	76.34	0.6180	64.2493	6.34	19 (6.33)	331	0.5438
6	88.03	0.6948	81.2102	8.02	24 (8.00)	422	0.5436
7	94.95	0.7370	91.3748	9.02	27 (9.00)	511	0.5436

$*$The value in the column of $4\pi\sin^2\theta/\lambda^2$ divided by 10.1307.

Question 4.5 X-ray measurement using Cu-Kα radiation ($\lambda = 0.1542$ nm) provides the diffraction pattern (see Fig. A) for a certain metallic sample and the relevant numerical data are summarized in Table A. With reference to the guideline of Fig. 4.5, the characteristic features of fcc, bcc, and diamond structures are not found in the diffraction pattern presently measured. Therefore, let us perform the procedure of indexing the pattern by assuming a hexagonal system as another typical structure of metallic elements and compute the lattice parameter.

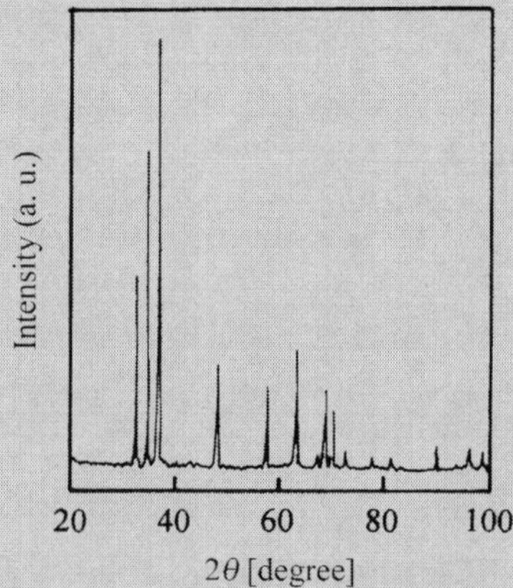

Fig. A Diffraction pattern of a sample

Table A Diffraction data of a sample

	2θ	d(A)	I/I_0
1	32.16	2.7836	26
2	34.37	2.6695	40
3	36.61	2.4548	100
4	47.80	1.9030	18
5	57.38	1.6060	10
6	63.07	1.4741	15
7	67.36	1.3903	2
8	68.64	1.3675	12
9	70.02	1.3439	8
10	72.53	1.3034	2
11	77.85	1.2271	2
12	81.50	1.1811	2

Answer 4.5 The hexagonal unit cell is known to be characterized by lattice parameters a and c, and the plane spacing equation coupled with the Bragg law is given as follows.

$$\frac{4\sin^2\theta}{\lambda^2} = \frac{1}{d^2} = \frac{4}{3}\frac{h^2+hk+k^2}{a^2} + \frac{l^2}{c^2}. \tag{1}$$

We also obtain the following simple form, when setting with $X = \lambda^2/3a^2$ and $Y = \lambda^2/4c^2$,

$$\sin^2\theta = X(h^2+hk+k^2) + Yl^2. \tag{2}$$

Note that permissible values of $(h^2 + hk + k^2)$ are 1, 3, 4, 7, 9, etc. At first, we will try to look for the $(hk0)$-type reflections ($l = 0$) and a constant value X are tentatively obtained. Once the value of X is known, we can estimate the value of Y in the following equation by rewriting the terms in (2).

$$Yl^2 = \sin^2\theta - X(h^2+hk+k^2). \tag{3}$$

Next, we subtract from each $\sin^2\theta$ value the values of X, $3X$, and $4X$ and $7X$, etc, and look for reminders Yl^2, which are in the ratio of 1, 4, 9, etc. because l is integers. They are likely to be the $(00l)$-type reflections and then we can find the value of Y. Finally, all the peaks may be indexed from a combination of X and Y values.

The results obtained by applying this procedure to the measured 12 diffraction peaks are summarized in Table 1.

Table 1 Trial calculation on the assumption of hexagonal structure

	2θ	$\sin\theta$	$\sin^2\theta$	$\frac{\sin^2\theta}{3}$	$\frac{\sin^2\theta}{4}$	$\frac{\sin^2\theta}{7}$	$\sin^2\theta - X$	$\sin^2\theta - 3X$	$\sin^2\theta - 4X$
1	32.16	0.277	[0.0767]	0.0256	0.0192	0.0110	0		
2	34.37	0.2955	0.0873	0.0291	0.0218	0.0125	0.0105		
3	36.61	0.3141	0.0987	0.0329	0.0247	0.0141	0.0219		
4	47.80	0.4051	0.1641	0.0547	0.041	0.0234	(0.0873)		
5	57.38	0.4801	0.2305	[0.0768]	0.0576	0.0329	0.1537	0	
6	63.07	0.5230	0.2735	0.0912	0.0684	0.0391	0.1967	0.0431	
7	67.36	0.5542	0.3071	0.1024	[0.0768]	0.0439	0.2303	0.0767	0
8	68.64	0.5638	0.3179	0.106	0.0795	0.0454	0.2411	(0.0875)	0.0107
9	70.02	0.5737	0.3291	0.1097	0.0823	0.0470	0.2523	0.0987	0.0219
10	72.53	0.5915	0.3499	0.1166	0.0875	0.0500	0.2731	0.1195	0.0427
11	77.85	0.6283	0.3948	0.1316	0.0987	0.0564	0.3180	0.1644	(0.0876)
12	81.50	0.6528	0.4261	0.1420	0.1065	0.0609	0.3493	0.1957	0.1189
								$X = 0.0768$	

When watching the columns of $\sin^2\theta$, $\frac{\sin^2\theta}{3}$, and $\frac{\sin^2\theta}{4}$ in Table 1, the value of 0.0768 will be a possible common quotient denoted by a box. This result implies that the 5th peak and the 7th peak correspond to $h^2+hk+k^2 = 3$ and $h^2+hk+k^2 = 4$, i.e., (110) and (200), respectively. Then, we can estimate the lattice parameter as follows:

$$X = 0.0768 = \frac{\lambda^2}{3a^2} \rightarrow a = 0.3212\,\text{nm}.$$

It may be added that the value of 0.0768 is not found in the column $\frac{\sin^2\theta}{7}$, but the diffraction pattern shows a peak at $2\theta = 94.31°$. Therefore when a similar calculation is done, $\sin^2\theta = 0.5377 \rightarrow \sin^2\theta/7 = 0.0768$ is obtained.

Next, $\sin^2\theta - X$, $\sin^2\theta - 3X$, and $\sin^2\theta - 4X$ are calculated using $X = 0.0768$ and the results are listed in the column at the right end of Table 1. The values of $0.0873 \sim 0.0876$ denoted by parentheses are considered as a common term and it may be applicable to the results in the column of $\sin^2\theta$.

Since a diffraction peak of (001) plane does not appear in a hexagonal system by the extinction rule, the 2nd peak is assumed to be as (002), $l = 2$. Therefore, we can estimate the values of Y and the lattice parameter c using $Yl^2 = 4Y$.

$$4Y = 0.0875 \rightarrow Y = 0.0219$$

$$Y = \frac{\lambda^2}{4c^2} = 0.0219 \rightarrow c = 0.5210\,\text{nm}.$$

In addition, $l = 4 \rightarrow (004)$ corresponds to $0.0219 \times 16 = 0.3504 = \sin^2\theta$. It may safely be said that the 10th peak ($\sin^2\theta = 0.3499$) corresponds to the case of $l = 4 \rightarrow (004)$, although there is difference in detail of the present calculations. Thus, four diffraction peaks are temporarily indexed. Final check on their reliability will be made in the usual manner by comparing the measured and calculated $\sin^2\theta$ values and in this process the indices are assigned to all peaks. Recalculated results are summarized in Table 2.

Table 2 Recalculated results of the present sample by considering a hexagonal structure with $X = 0.0768$, $Y = 0.0219$

	2θ	$\sin\theta$	$\sin^2\theta$	$X + Y$	$\sin^2\theta$cal	$h^2 + hk + k^2$	h and k	l	
1	32.16	0.2770	0.0767	X+0Y	0.0768	1	1, 0	0	100
2	34.37	0.2955	(0.0873)	0X+4Y	0.0876	0	0, 0	2	(002)
3	36.61	0.3141	0.0987	X+Y	0.0987	1	1, 0	1	101
4	47.80	0.4051	0.1641	X+4Y	0.1644	1	1, 0	2	102
5	57.38	0.4801	0.2305	3X+0Y	0.2304	3	1, 1	0	110
6	63.07	0.5230	0.2735	X+9Y	0.2739	1	1, 1	3	103
7	67.36	0.5542	0.3071	4X+0Y	0.3072	4	2, 0	0	200
8	68.64	0.5638	0.3179	3X+4Y	0.3180	3	1, 1	2	112
9	70.02	0.5737	0.3291	4X+Y	0.3291	4	2, 0	1	201
10	72.53	0.5915	(0.3499)	0X+16Y	0.3504	0	0, 0	4	(004)
11	77.85	0.6283	0.3948	4X+4Y	0.3948	4	2, 0	2	202
12	81.50	0.6528	0.4261	X+16Y	0.4272	1	1, 1	4	104
13	90.43	0.7098	0.5038	4X+9Y	0.5043	4	2, 0	3	203
14	94.31	0.7333	0.5377	7X+0Y	0.5376	7	2, 1	0	210
15	96.85	0.7481	0.5597	7X+Y	0.5595	7	2, 1	1	211
16	99.24	0.7618	0.5803	3X+16Y	0.5808	3	1, 1	4	114

Summary: $a = 0.3212$ nm, $c = 0.5210$ nm, and $c/a = 1.622$.

For indexing the pattern, it may be again mentioned that it is important to predict a possible crystal structure to some extent from features of the measured X-ray diffraction pattern, for example with reference to the guideline of Fig. 4.5. In the X-ray diffraction pattern of a hexagonal system, you may find the feature characterized by three peaks at equal intervals in a relatively low angle region and some other peaks appear in a higher angle region. However, the relative intensity ratio of the first three peaks depends on materials of interest. That is, no common relation exists in the hexagonal system.

The substance presently analyzed is considered to be Mg from the lattice parameters and their ratio of c/a, which agree well with the results of Mg: $a = 0.32095$ nm, and $c/a = 1.6235$ compiled in the Appendix A.9.

Question 4.6 When measuring the X-ray diffraction pattern of magnesium oxide (MgO) powder sample by Cu-Kα radiation ($\lambda = 0.1542$ nm), ten diffraction peaks were obtained in the scattering angle (2θ) as shown below. Index the pattern and compute the lattice parameter by referring to MgO having the NaCl-type structure.

2θ(degree): 36.93, 42.91, 62.30, 74.64, 78.64, 94.06, 105.75, 109.78, 127.29 and 143.77.

Answer 4.6 Since NaCl structure is also classified into a cubic system, the following equation can be used.

$$a = d \times \sqrt{h^2 + k^2 + l^2}. \qquad (1)$$

On the contrary, the following particular relationships are given for the structure factor of NaCl structure. (See Question 3.12.)

(1) When the Miller indices hkl are unmixed and $(h + k + l)$ is odd: Example (111) plane,

$$|F|^2 = 16(f_{\mathrm{Na}} - f_{\mathrm{Cl}})^2 \Rightarrow 16(f_{\mathrm{Mg}} - f_{\mathrm{O}})^2.$$

Table 1 The diffraction data and the result of indexing for MgO powder sample

	2θ	d(nm)	hkl	$h^2 + k^2 + l^2$	$\sqrt{h^2 + k^2 + l^2}$	a(nm)
1	36.95	0.2433	111	3	1.732	0.4214
2	42.91	0.2108	200	4	2	0.4216
3	62.30	0.1490	220	8	2.828	0.4214
4	74.64	0.1272	311	11	3.317	0.4219
5	78.64	0.1217	222	12	3.464	0.4216
6	94.06	0.1054	400	16	4	0.4216
7	105.75	0.0967	331	19	4.359	0.4215
8	109.78	0.0942	420	20	4.472	0.4213
9	127.29	0.0860	422	24	4.899	0.4213
10	143.77	0.0811	511	27	5.196	0.4214

$d = \frac{\lambda}{2 \sin \theta}$ ($\lambda = 0.1542$ nm)

(2) When the Miller indices hkl are unmixed and $(h + k + l)$ is even: Example (200) plane,

$$|F|^2 = 16(f_{\mathrm{Na}} + f_{\mathrm{Cl}})^2 \Rightarrow 16(f_{\mathrm{Mg}} + f_{\mathrm{O}})^2.$$

(3) When the Miller indices hkl are mixed (for example (100) (210) etc.), the intensity is not observed due to the extinction rule.

The results of indexing are summarized in Table 1 and the mean value of the lattice parameter is estimated to $a = 0.4215$ nm.

Question 4.7 When measuring the X-ray diffraction pattern of potassium chloride (KCl) powder sample by Cu-Kα radiation ($\lambda = 0.1542$ nm), twelve diffraction peaks were obtained with the scattering angle (2θ) as shown below. Index the pattern and compute the lattice parameter by referring to KCl having NaCl-type structure.

2θ(degree): 24.48, 28.35, 40.50, 47.92, 50.18, 58.66, 66.39, 73.54, 87.68, 94.58, 101.51, and 108.65.

Answer 4.7 We can attempt the procedure, similar to the previous MgO case, without any difficulty. Nevertheless, the following point should be taken into account. The atomic number of the components are K $= 19$ and Cl $= 17$, respectively, so that the value of $(f_{\mathrm{K}} - f_{\mathrm{Cl}})^2$ becomes small. Then, if the Miller indices hkl are unmixed and $(h + k + l)$ is odd, such as (111), the extinction rule definitely suggests that the diffraction intensity should be observed, but only quite weak intensity is expected. In other words, since some of these weak diffraction peaks are not always observed, depending on the experimental condition, careful analysis for indexing the pattern should be made (refer to Question 3.12). Note that the Miller indices of (111), (311), (331), (511), (531), (533), etc. are included in this category.

Table 1 The diffraction data and the result of indexing for KCl powder sample

	2θ	d(nm)	hkl	$h^2 + k^2 + l^2$	$\sqrt{h^2 + k^2 + l^2}$	a(nm)
1	24.48	0.3637	111	3	1.732	0.6299
2	28.35	0.3148	200	4	2	0.6296
3	40.50	0.2228	220	8	2.828	0.6300
4	47.92	0.1899	311	11	3.317	0.6299
5	50.18	0.1818	322	12	3.464	0.6298
6	58.66	0.1574	400	16	4	0.6296
7	66.39	0.1408	420	20	4.472	0.6297
8	73.54	0.1288	422	24	4.899	0.6271
9	87.68	0.1113	440	32	5.657	0.6296
10	94.58	0.1049	600	36	6	0.6294
11	101.51	0.0996	620	40	6.325	0.6300
12	108.65	0.0949	622	44	6.633	0.6295

Using these fundamental information, the result of indexing is summarized in Table 1. $a = 0.6298$ nm is obtained, as mean value of the lattice parameter. As a result, we found that the diffraction peaks corresponding to (331), (511), and (531) were not detected. The reason is shown below using the (331) plane as an example.

The value of $(h^2 + k^2 + l^2)$ corresponding to the (331) plane is estimated to 19 and it may be assigned to a peak observed at $2\theta = 66.39°$, denoted by the (420) plane in Table 1. However, it gives $\sqrt{h^2 + k^2 + l^2} = 4.359$ and $a = 0.6137$ nm. The difference from those of other calculations is found rather significant. While, if assuming that a peak observed at $2\theta = 66.39°$ is assigned to the (420) plane, we obtain the reasonable results by finding $h^2 + k^2 + l^2 = 20$ and $\sqrt{h^2 + k^2 + l^2} = 4.472 \rightarrow a = 0.6297$ nm. For this reason, it may safely be concluded that the diffraction peak corresponding to the (331) plane does not appear in the present measurements of KCl powder sample.

Question 4.8 When measuring the X-ray diffraction pattern of an unknown powder sample possibly consisting of a single phase by Cu-Kα radiation ($\lambda = 0.1542$ nm), the diffraction pattern (see Fig. A) and the relevant numerical peaks were obtained as listed in Table A. Identify this unknown sample and compute the lattice parameter by applying the Hanawalt Search manual.

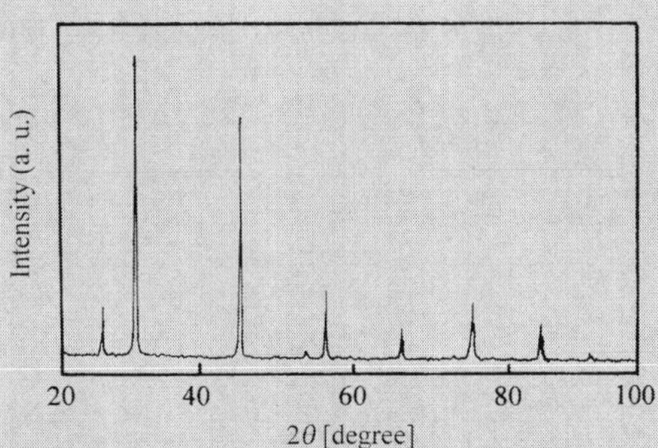

Fig. A Diffraction pattern of an unknown sample

Table A Diffraction data of unknown sample

	2θ	d (A)	I/I_1
1	27.44	3.251	15
2	31.73	2.820	100
3	45.54	1.992	60
4	53.67	1.708	5
5	56.46	1.630	25
6	65.56	1.424	10
7	75.87	1.253	20
8	84.01	1.152	10
9	91.98	1.072	5

Answer 4.8 The Hanawalt method is to search a well-matched pair between the measured diffraction pattern with that of a standard substance compiled in the database using the d – values of three strongest diffraction peaks together with their relative intensity ratios (I/I_1) as an indicator.

In the experimental results of Table A, the 2nd peak with $d = 2.820$ A is found the strongest and taking this peak as basis, the 3rd and the 5th peaks will serve as a set of three strongest diffraction peaks as listed in Table 1.

Table 1 A set of three strongest diffraction peaks

2θ(degree)	d(A)	I/I_1
31.7	2.820	100
45.5	1.992	60
56.5	1.630	25

When investigating this combination of three d – values of 2.82, 1.99, and 1.63 using the Hanawalt Search Manual, we can find the information given in Table 2.

Table 2 Data taken from the Hanawalt Search Manual

Spacing and intensity								Substances	File number	Fish number
2.82_9	**1.99_9**	**2.26_x**	1.61_9	1.51_9	1.49_9	3.57_8	2.66_8	$(ErSe_2)Q$	19-443	1-106-F6
2.82_x	**1.99_6**	**1.63_2**	3.26_1	1.26_1	1.15_1	1.41_1	0.89_1	NaCl	5-0628	1-18-F8
2.82_4	**1.99_4**	**1.54_x**	1.20_4	1.19_4	2.44_3	5.62_2	4.89_2	$(NH_4)_2WO_2Cl_4$	22-65	1-145-D12
2.82_x	**1.99_8**	**1.26_3**	1.63_2	1.15_2	0.94_1	0.89_1	1.41_1	$(BePd)_2C$	18-225	1-90-D1

x: strongest peak and the number of subscript is the intensity ratio (Ex. 9 → 90% and 4 → 40%).

As for the combination of the three d – values of 2.82, 1.99, and 1.63, good agreement with the NaCl case is deemed and further the compiled reference data suggest that the d – values of 3.26 (as 4th peak in Table 2) and 1.26 (as 5th peak in Table 2) are not badly assigned to the experimental d – values of 3.251 and 1.253, corresponding to the 1st and 7th peaks, respectively, although we find an inversion of their relative intensity ratios. Thus, the present unknown sample is quite likely to be sodium chloride (NaCl). Then, if we pick up the JCPDS card of the file number 5-0628 from database, information displayed in Fig. 1 is acquired. The value of $d = 3.26$ cited in the 4th column of I/I_1 at the upper left of the JCPDS card shows the largest d – value observed in this substance, i.e., the diffraction peak observed at lowest angle. With respect to this point, good agreement with the measured d – value of 3.25 is also recognized. Accordingly, the sample has been judged to be NaCl through this process.

Since NaCl is a cubic system, the lattice parameter is estimated using the plane spacing equation of $\frac{1}{d^2} = \frac{h^2+k^2+l^2}{a^2}$, as shown in Table 3.

On the contrary, there are also points to note. Intensity ratios of X-ray diffraction pattern change easily depending on the experimental conditions of the sample

Table 3 Example of sample which hypothesized NaCl

d(nm)	hkl	$h^2+k^2+l^2$	a (nm)	
0.3251	111	3	0.5631	
0.2820	200	4	0.5640	
0.1992	220	8	0.5634	
0.1708	311	11	0.5665	
0.1630	222	12	0.5646	$a = 0.5643$ nm

5-0628

d	2.82	1.99	1.63	3.258	NaCl	★
I/I_1	100	55	15	13	Sodium Chloride	Halite

				d Å	I/I_1	hkl
Rad. Cu	λ 1.5405	Filter		3.258	13	111
Dia.	Cut off	Coll.		2.821	100	200
I/I_1		d corr. abs.?		1.994	55	220
Ref. Swanson and Fuyat, NBS Circular 539, Vol. II,				1.701	2	311
		41(1953)		1.628	15	222
Sys. Cubic		S.G. O_H^5 - Fm3m		1.410	6	400
a_0 5.6402	b_0	C_0	A C	1.294	1	331
α	β	γ	Z 4	1.261	11	420
Ref. Ibid.				1.1515	7	422
ε α	nωβ 1.542	ε γ	Sign	1.0855	1	511
2V	Dx 2.164 mp	Color		.9969	2	440
Ref. Ibid.				.9533	1	531
An ACS reagent grade sample recrystallized				.9401	3	600
twice from hydrochloric acid.				.8917	4	620
X-ray pattern at 26℃				.8601	1	533
				.8503	3	622
				.8141	2	444
Replaces 1-0993, 1-0994, 2-0818						

Fig. 1 JCPDS card of NaCl

preparation of a powder sample, scanning speed of diffractometer and others. In fact, a diffraction peak with a relatively low intensity, which appears near $d = 1.294 \Rightarrow 73.1°$, is registered in the JCPDS card, but it is not detected in the present measurement. This suggests strong requirements of preliminary information such as types of the elements contained and their fractions before identifying an unknown sample as well as indexing the pattern.

Question 4.9 When measuring the X-ray diffraction pattern of an unknown powder sample by Cu-Kα radiation ($\lambda = 0.1542$ nm), the diffraction pattern (Fig. A) and the relevant numerical peaks were obtained as listed in Table A. Identify this unknown sample and compute the lattice parameter by applying the Hanawalt Search manual.

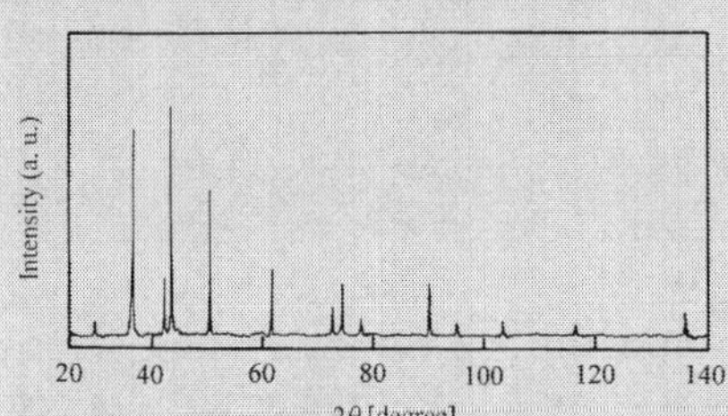

Fig. A Diffraction pattern of an unknown sample

Table A Diffraction data of unknown sample

	2θ	d(A)	I/I_1		2θ	d(A)	I/I_1
1	29.63	3.015	5	8	74.01	1.281	20
2	36.38	2.470	75	9	77.71	1.229	6
3	42.38	2.133	30	10	90.25	1.088	20
4	43.36	2.087	100	11	95.21	1.044	3
5	50.42	1.810	55	12	103.61	0.981	5
6	61.77	1.502	25	13	116.64	0.906	4
7	72.30	1.307	10	14	136.18	0.831	8

Answer 4.9 In the experimental results of Table A, the 4th peak with $d = 2.087$ A is found the strongest and taking this peak as basis, the 2nd and the 5th peaks will serve as a set of three strongest diffraction peaks as listed in Table 1.

Table 1 A tentative set of three strongest diffraction peaks

2θ	d(A)	I/I_1
43.36	2.087	100
36.38	2.470	75
50.42	1.810	54

When investigating this combination of three d – values of 2.09, 2.47, and 1.81 using the Hanawalt Search Manual, we can find the combination of 2.09 and 2.47 in the compiled database which regards $d_1 = 2.09$ as the strongest one and the 2nd strongest peak with $d_2 = 2.47$. However, the combination of the 3rd strongest peak corresponding to $d_3 = 1.81$ is not found. This implies a potential that the present unknown sample contains more than one substance.

For this reason, the diffraction peak of $d_1 = 2.087$ and that of $d_2 = 2.470$ should be assigned to the strongest diffraction peak of a different substance, respectively, and further consideration is made by combining the d – values of $d_3 = 1.810$ or $d_4 = 2.133$, $d_5 = 1.502$, etc. For example, the identification is examined in the combination of 2.09-1.81 or 2.09-2.13. As a result, good agreement with the Cu case (File number 4-0836) is deemed by finding the combination of 2.09-1.81-1.28 (see Table 2). For readers convenience, information of the JCPDS card number 4-0836 is given in Fig. 1.

Table 2 Data (I) taken from the Hanawalt Search Manual

Spacing and intensity								Substances	File number
2.09_x	**1.81_5**	**3.62_2**	1.28_2	1.09_1	2.56_1	1.62_1	1.05_1	$(AlNi_3C_{0-5})4.5C$	29-58
2.09_x	**1.81_4**	**1.28_2**	1.09_1	1.04_1	0.83_1	0.81_1	0.00_1	$(Co_2GeC_{0-25})4.25C$	29-475
2.09_x	**1.81_5**	**1.28_2**	1.09_2	0.83_1	0.81_1	1.04_1	0.90_1	Cu	4-0836
2.07_x	**1.81_5**	**1.99_4**	4.44_3	3.19_2	2.84_x	2.71_2	1.58_2	$NaSn_2F_5$	15-619
2.07_x	**1.81_6**	**1.77_3**	1.27_x	1.28_3	1.27_3	3.22_2	3.82_1	$(Ni_2V)6P$	17-715

A peak with $d = 2.09$ is to be observed at the lowest angle and the 4th strongest peak with the d –values of 1.09 compiled in the JSPDS card is reasonably equivalent to the 10th peak ($d = 1.088$) in Table A. Based on these results, further indexing of the pattern should be made for residual peaks by excluding the diffraction peaks assigned to Cu (the JCPDS card number 4-0836) from the measured 14 diffraction peaks. The resultant seven peaks are listed as in Table 3. Note that it is necessary to recalculate the intensity ratios of I/I_1 with respect to a new set of three strongest peaks and the results are summarized in Table 4.

4-0836

d 4-0846	2.088	1.808	1.278	2.088	Cu	★
I/I_1 4-0836	100	46	20	100	Copper	

Rad. Cu λ 1.5405 Filter Ni
Dia. Cut off Coll.
I/I_1 d corr. abs.?
Ref. Swanson and Tatge, JC Fel. Reports, NBS, (1949)

Sys. Cubic (F.C.) S.G. O_H^5 - Fm3m
a_0 3.6150 b_0 C_0 A C
α β γ Z 4
Ref. Ibid.

$\varepsilon\alpha$ $n\omega\beta$ $\varepsilon\gamma$ Sign
2V Dx8.936 mp Color
Ref. Ibid.

Johnson and Matthey-spec. sample, annealed at 700℃ in vacuum.
At 26℃
To replace 1-1241, 1-1242, 2-1225, 3-1005, 3-1015, 3-1018

d Å	I/I_1	hkl	d Å	I/I_1	hkl
2.088	100	111			
1.808	46	200			
1.278	20	220			
1.0900	17	311			
1.0436	5	222			
0.9038	3	400			
.8293	9	331			
.0883	8	420			

Fig. 1 JCPDS card of Cu

Table 3 New diffraction data set of an unknown sample

	2θ	d(A)	I/I_1		2θ	d(A)	I/I_1
1	29.63	3.015	5→7	7	72.30	1.307	10→13
2	36.38	2.470	75→100	9	77.71	1.229	6→8
3	42.38	2.133	30→40	12	103.61	0.981	5→7
6	61.77	1.502	25→33				

Table 4 A new set of three strongest diffraction peaks

2θ	d(A)	I/I_1
36.38	2.470	100
42.38	2.133	40
61.77	1.502	33

When investigating this combination of $d-$values of 2.47, 2.13, and 1.50 using the Hanawalt Search Manual, we can find information given in Table 5.

According to the data of Table 5, a possible candidate is (CaN)8F or Cu_2O showing the 4th strongest peak with $d_4 = 1.29$. However, we find the appreciable difference in the 5th peak; $d_5 = 0.98$ for (CaN)8F and $d_5 = 3.02$ for cuprous oxide (Cu_2O), respectively. The intensity ratios of the measured peaks with $d-$values of 3.02, 1.23, and 0.98 show almost the same magnitude, so that the order of the intensity ratio is considered not necessarily a key factor in a comparison with those of the reference data. Thus, it may safely be concluded on the present case that the measured $d-$values of 3.02, 1.23, and 0.98 are rather attributed to Cu_2O. To make sure

Table 5 Data (II) taken from the Hanawalt Search Manual

Spacing and intensity								Substances	Files No./number
2.48_x	**2.14_4**	**1.51_4**	1.29_2	0.98_1	1.24_1	0.00_1	0.00_1	(CaN)8F	16-116
2.47_x	**2.14_4**	**1.51_2**	1.29_2	3.02_1	1.23_1	0.98_1	0.96_1	Cu_2O	5-0667
2.46_x	**2.14_5**	**2.24_6**	1.23_x	1.37_a	2.09_6	1.42_6	1.35_6	(Ta, Co)13R	21-270
2.51_5	**2.13_5**	**2.23_x**	2.09_0	2.98_3	2.47_5	1.33_5	0.80_5	$(Ru_{0-23}W_{0-22}B_{0-55})8O$	24-994
2.51_6	**2.13_6**	**2.21_x**	1.27_0	2.29_7	1.33_5	4.85_4	1.81_4	$(Re_2P)12O$	17-391

05-0667 Wavelength=1.5405

Cu20

Copper Oxide

Cuprite, syn

Rad: CuKal λ: 1.5405 Filter: Ni d-3p:

Cut off: Int.: Diffract. 1/Icor.:

Ref: Swanson, Fuyat. Natl. Bur. Stand. (U.S.) Circ. 539, II. 23 (1953)

Sys.: Cubic S.G.: Pn3m (224)

a: 4.2696 b: c: λ: C:

α: β: γ: Z: 2 mp:

Rel: Ibid.

Dx: 6.106 Dm: SS/FOM: F_{13}=56(.0117, 20)

Color. Violet red
Pattern taken at 28 C. Sample prepared at NBS, Gaithereburg, Maryland, USA by sintering CuCl abd Na2 CO2 at ~800 C, then leaching with water and drysing. Spectroscopic analysis:<1% Ca, Si:<0.1% Al, Mg:<0.01% Ag, B, Ba, Fe,Ti; <0.001% Mn, Pb, Sn, Opaue mineral optical data on specimen from Liskeard, Cornwall, England, UK. Pattern reviewed by Martin, K., McCarthy, G., North Dakota State University, Fargo, North Dakota, USA. ICDD Grant-in-Aid(1990). Agrees well with experimental and calculated patterns. Additional weak reflection [indicated by brackets] was observed. Ag2 0 type. PSC: cP6. Mwt: 143.09. Volume[CD]: 77.83.

d Å	Int	h	k	l
3.0200	9	1	1	0
2.4650	100	1	1	1
2.1350	37	2	0	0
1.7430	1	2	1	1
1.5100	27	2	2	0
1.3502	1	3	1	0
1.2870	17	3	1	1
1.2330	4	2	2	2
1.0674	2	4	0	0
0.97950	4	3	3	1
.95480	3	4	2	0
.87150	3	4	2	2
.82160	3	5	1	1

Fig. 2 JCPDS card of Cu_2O (information from CD-ROM)

this conclusion, the JCPDS card number 5-667 illustrated in Fig. 2 is checked. As a result, the peaks with weak intensity corresponding to $d = 1.74$, 1.35, and 1.07 are not observed, but all seven diffraction peaks listed in Table 3 can be reasonably assigned to the reference data of Cu_2O.

Summarizing the present results, the unknown sample is identified as a mixture of Cu and Cu_2O from the given experimental data. As shown in this example, if knowing in advance that the sample includes copper and oxygen by applying fluorescent X-ray analysis, the existence of Cu, CuO, Cu_2O, etc. should be taken into account. Or if getting to know that Cu is contained in a sample, it is desirable to keep in mind as a potential that copper oxide or copper sulfide may coexist.

Question 4.10 Explain the conditions for the precise lattice parameter measurements of a powder crystalline sample using the diffractometer with characteristic X-ray radiation.

Answer 4.10 In the Bragg law of $2d \sin\theta = \lambda$, where λ is a constant when utilizing characteristic X-ray radiation. We can estimate the lattice parameter from the spacing d if we measure the Bragg angle. Note that it is related to $\sin\theta$, not θ. Thus, the precision in d or the lattice parameter depends on precision in θ, not on the measured quantity, which is precision in θ. This situation is fortunate because the $\sin\theta$ value changes very slowly with θ in the close vicinity of 90 degrees. Namely, the variation of $\sin\theta$ becomes very small in high angle region, as shown in Fig. 1. A comparison is demonstrated in two cases, where θ is near 80° and near 10°, respectively. This result clearly suggests that to obtain the precise lattice parameter, we should utilize the diffraction peaks measured in the high angle region such as θ near 90° or the scattering angle 2θ near 180°.

A similar guideline is readily obtained as follows, when differentiating the Bragg law $2d \sin\theta = \lambda$ with respect to angle θ.

$$\sin\theta = \frac{\lambda}{2d} \Rightarrow \cos\theta \cdot \Delta\theta = -\frac{\lambda}{2d^2}\Delta d = -\sin\theta\frac{\Delta d}{d}. \tag{1}$$

It follows

$$\frac{\Delta d}{d} = -\cot\theta \cdot \Delta\theta \qquad \left(\because \cot\theta = \frac{\cos\theta}{\sin\theta}\right). \tag{2}$$

With respect to the relationship between the plane spacing d and the lattice parameter a, the following equation is given in the cubic system.

$$a = d\sqrt{h^2 + k^2 + l^2} \Rightarrow \frac{\Delta a}{a} = \frac{\Delta d}{d}. \tag{3}$$

From (2) and (3), we obtain

$$\frac{\Delta a}{a} = \frac{\Delta d}{d} = -\cot \cdot \Delta\theta. \tag{4}$$

As for (4), to reduce the fractional error of the lattice parameter ($\Delta a/a$) as much as possible, it is desirable for the value of $\cot\theta$ to become small down to zero. That is, if θ approaches 90° (the diffraction angle 2θ approaches 180°), the value of $\cot\theta$ will also approach zero. The key factor of the precise lattice parameter measurements is again recognized to use the diffraction peaks having 2θ values as near to 180 as possible. In addition, when using the diffraction peak in the high angle region, it is important to minmize the error related to the angle measurement, $\Delta\theta$ (or $\Delta 2\theta$), as much as possible (see Fig. 2).

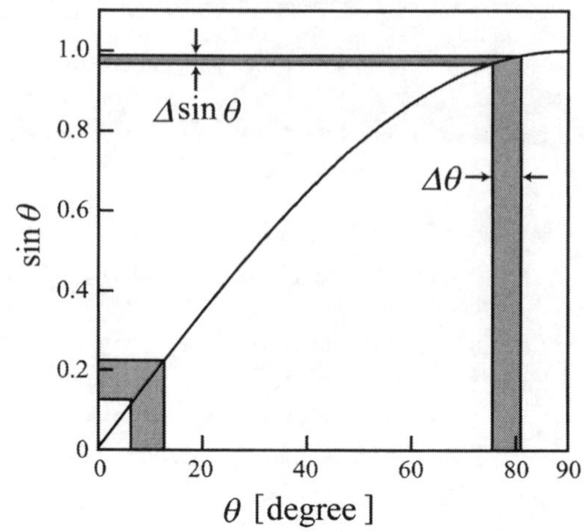

Fig. 1 Variation of $\sin\theta$ with θ

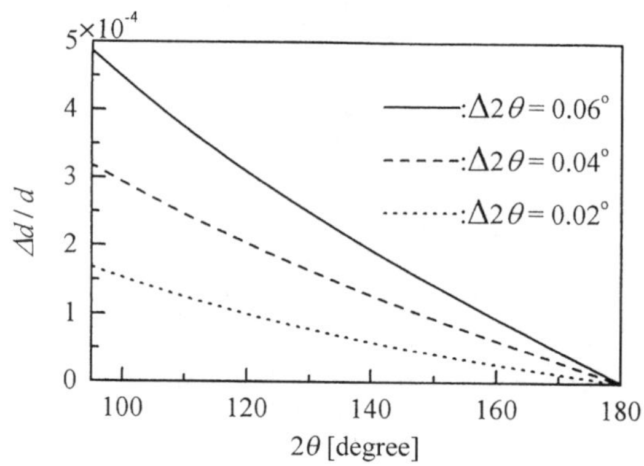

Fig. 2 Relationship between the error in the measured diffraction angle 2θ and the error in d-value

Question 4.11 Four typical characteristic X-rays commonly are utilized for X-ray diffraction analysis and their wavelengths are summarized in Table A.

Table A The wavelength of typical characteristic X-ray (unit: nm)

Elements	$K\alpha$ (weighted average) *	$K\alpha_2$ Strong.	$K\alpha_1$ Very strong.	$K\beta_1$ Weak.
Fe	0.1937355	0.193998	0.1936042	0.175661
Co	0.1790260	0.179285	0.1788965	0.162079
Cu	0.1541838	0.154439	0.1540562	0.1392218
Mo	0.0710730	0.071359	0.0709300	0.0632288

* $K\alpha_1$ was averaged with $K\alpha_2$, weighing twice of $K\alpha_2$.

When an X-ray diffraction pattern for a powder sample of tungsten (crystal system: body-centered cubic and lattice parameter $a = 0.31648\,\text{nm}$) is measured using these four characteristic radiations, compute the angles of the possibly detected diffraction peaks corresponding to (220), (310), and (222) planes.

Answer 4.11 From the plane spacing equation of a cubic system coupled with the Bragg condition, the following equation is given.

$$\sin\theta = \frac{\lambda}{2a}\sqrt{h^2+k^2+l^2}. \tag{1}$$

For the (220) plane, $\sqrt{h^2+k^2+l^2} = \sqrt{8} = 2.8284$. Then, the given values of the wavelengths listed in Table A and the lattice parameter $a = 0.31648\,\text{nm}$ are put into (1), the diffraction angles 2θ can be obtained as shown in Table 1.

Table 1 Possible diffraction angles for a peak corresponding to the (220) plane of tungsten

Elements	Kα	Kα_2	Kα_1	Kβ_1
Fe	119.93	120.20	119.79	103.43
Co	106.26	106.48	106.15	92.81
Cu	87.10	87.28	87.01	76.94
Mo	37.03	37.19	36.96	32.82

Similarly, we can estimate the diffraction angles 2θ with respect to the (310) plane as shown in Table 2 and those for the (222) plane as shown in Table 3, respectively.

Table 2 Possible diffraction angles for a peak corresponding to the (310) plane of tungsten

Elements	Kα	Kα_2	Kα_1	Kβ_1
Fe	150.89	151.81	150.59	122.71
Co	126.87	127.2	126.7	108.14
Cu	100.76	100.99	100.65	88.14
Mo	41.6	41.77	41.51	36.83

For (310) plane, $\sqrt{h^2+k^2+l^2} = \sqrt{10} = 3.1623$.

Table 3 Possible diffraction angles for a peak corresponding to the (222) plane of tungsten

Elements	Kα	Kα_2	Kα_1	Kβ_1
Fe	*	*	*	148.04
Co	156.92	157.75	156.52	125.01
Cu	115.09	115.39	114.94	99.27
Mo	45.78	45.98	45.68	40.49

For (222) plane, $\sqrt{h^2+k^2+l^2} = 3.4641$. $*$: Since it does not match with the conditions of (1), no diffraction peak can be observed.

It may be safely suggest that the experimental uncertainty of the angle measurements is the order of ±0.02° and the repeated-reproducibility of angle is the order of ±0.005° at most for a commercial diffractometer. Considering these experimental uncertainties together with the results of Tables 1–3, we can appreciable difference in angle of a peak detected in the high angle region when the combination of Kα_1 with not only Kβ_1 but also Kα_2. In other words, the measurements of the diffraction peaks appeared in the high angle region using the characteristic X-ray radiations of different wavelengths are effective to obtain the lattice parameter with sufficient reliability. For readers' convenience, we estimate the possible differences in the diffraction angle for the (321) and (400) planes of tungsten, if the characteristic radiations of copper are employed, the results are summarized in Table 4.

Table 4 The possible diffraction angles for the peaks corresponding to the (321) and (400) planes of tungsten using the characteristic radiation of copper

	$\sqrt{h^2+k^2+l^2}$	Kα	Kα_2	Kα_1	Kβ_1
321	3.7417	131.41	131.83	131.20	110.77
400	4	154.00	154.83	153.59	123.24

Question 4.12 The so-called Kα doublet of characteristic X-ray radiation is often used for the precise measurements of a lattice parameter. The effect of this Kα doublet on the diffraction angle becomes more prominent in a peak detected in the high angle region. Explain the reason and its separation (Kα_1 and Kα_2).

Answer 4.12 There are several radiations in the K-set, but usually only the three strongest radiations are used in normal diffraction work. They are Kα_1, Kα_2, and Kβ_1. In addition, the Kα_1 and Kα_2 components are known to have wavelengths so close together and they are not always resolved as separate radiations. However, if resolved, we call it the Kα doublet.

As shown in Fig. 1 of the X-ray spectrum for Mo, the Kα radiation consists of a mixture of Kα_1 and Kα_2, where 0.07093 nm for Kα_1 and 0.07136 for Kα_2, respectively (see Question 4.11). With respect to Cu – Kα radiation, there is about 0.0004 nm difference between Kα_1(wavelength of 0.154056 nm) and Kα_2(0.154439 nm).

As easily understood from the Bragg condition, $2d\sin\theta = \lambda$, for a peak with the relatively large d – value (at the small scattering angles 2θ), the difference between Kα_1 and Kα_2 is difficult to detect. Such situation is as follows, using the case where the d – value is 0.3 nm or 0.1 nm.

$$\sin\theta_1 - \sin\theta_2 = \frac{1}{2d}(\lambda_1 - \lambda_2). \tag{1}$$

In case of $d = 0.3$ nm

$$\sin\theta_1 - \sin\theta_2 = \frac{0.0004}{2 \times 0.3} = 6.7 \times 10^{-4} \qquad \theta_1 - \theta_2 \cong 0.038°. \tag{2}$$

In case of $d = 0.1$ nm

$$\sin\theta_1 - \sin\theta_2 = \frac{0.0004}{2 \times 0.1} = 0.02 \qquad \theta_1 - \theta_2 \cong 1.15°. \tag{3}$$

In other words, with respect to the diffraction peaks measured in the scattering angles 2θ from 90° to 160°, corresponding to the planes with small d – values, the effect attributed to the difference between Kα_1 and Kα_2 is clearly observed.

Nevertheless, the Kα_1 and Kα_2 components are not always resolved as separate peaks completely. In many cases, it is observed as a peak having partial overlap, which is separated to some degree. We have to use an additional method for estimating the scattering angles and the essential points of such method are given below, although recent X-ray diffraction equipment includes computer software for this purpose.

Figure 2 shows a diffraction peak as partially separated with relatively good resolution and in this case, the estimation of the scattering angles is not a difficult task. Considering that a diffraction peak is attributed to each wavelength and two components by Kα_1 and Kα_2 can reasonably be separated and then the middle-point of full width of half maximum intensity is used as a scattering angle.

With respect to a devised method for applying to an example of Fig. 3, where the resolution of two components is not sufficient, the Rachinger method is frequently employed (for details, see other textbooks, e.g., H.P. Klug and L.E. Alexander: *X-ray Diffraction Procedures, 2nd Edition*, John-Wiley & Sons, New York, (1973)). The Rachinger method is based on the fact that the intensity ratio of Kα_1 and Kα_2 is described by 2 : 1, as well as the difference $\Delta\lambda$ of these two wavelengths is constant. That is, the diffraction peak profile by Kα_2 is considered the same as the Kα_1 case, and its intensity is $\frac{1}{2}$ and its diffraction angle is shifted only by $\Delta 2\theta_r = 2\tan\theta \times \frac{\Delta\lambda}{\lambda_{K\alpha_1}}$. Then, we may obtain the following equation.

$$I(2\theta) = I\alpha_1(2\theta) + \frac{1}{2} I\alpha_1(2\theta + \Delta 2\theta_r), \tag{4}$$

where $I(2\theta)$ is the diffraction peak profile by Kα_1 and $I(2\theta)$ is the measured diffraction peak profile. In practice, the profile fitting the measured diffraction peak profile is carried out by a computer software using the relationship of (4) to estimate the values of the scattering angles by Kα_1 and Kα_2.

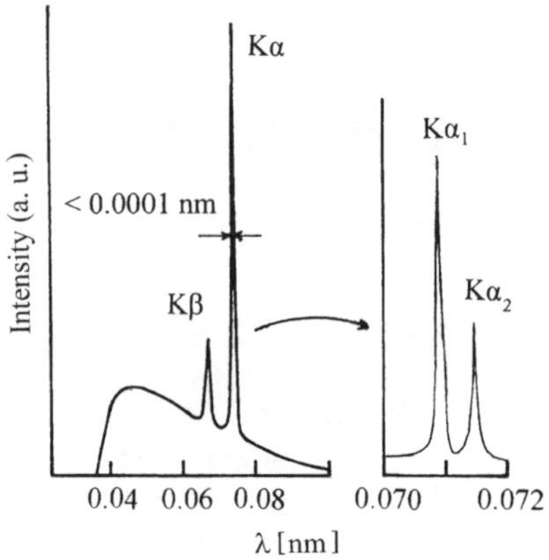

Fig. 1 Spectrum of X-rays produced from Mo target with resolved doublet

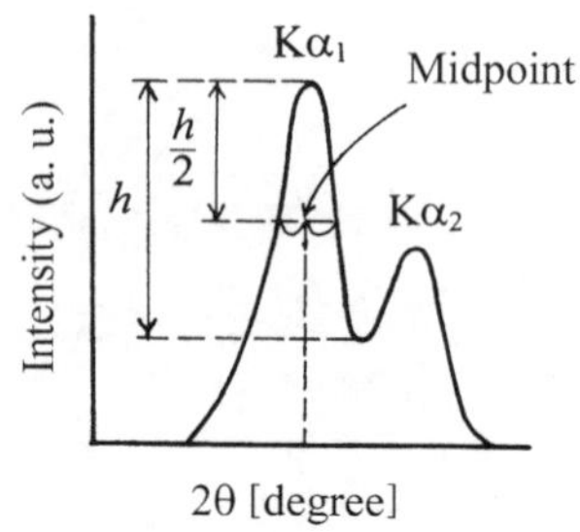

Fig. 2 Separation for a peak partially separated with relatively good resolution

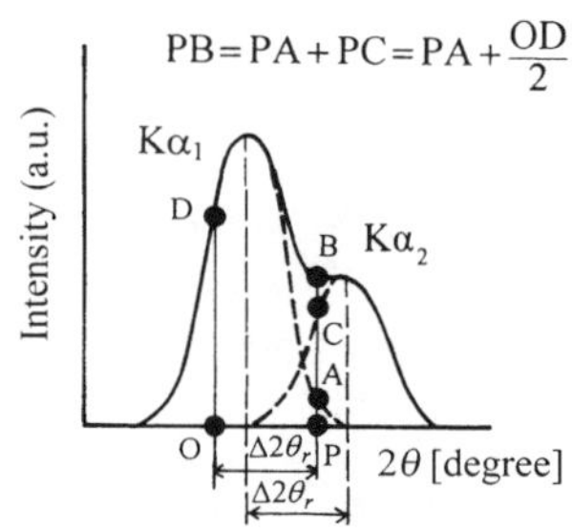

Fig. 3 Separation for a peak partial overlapped with two components using the Rachinger method

Question 4.13 In the diffraction experiments for magnesium oxide (MgO: NaCl type structure) utilizing two characteristic X-rays, the diffraction peaks corresponding to four lattice planes in the high angle region were obtained and they are summarized in Table A. Compute the lattice parameter using the extrapolation methods.

Table A Diffraction data of magnesium oxide sample (2θ/degrees)

Characteristic X-ray	λ(nm)	*hkl*			
		400	420	422	511
Cu-Kα	0.15418_4	94.17	109.9	127.44	144.04
Cu-Kβ_1	0.13922_2	82.85	95.37	108.12	118.44

Answer 4.13 The lattice parameter a for a cubic system can be computed from the measured diffraction data using the following equation. To use the extrapolation method, the value of $\sin^2\theta$ is simultaneously calculated and the results are given in Table 1.

$$a = \frac{\lambda}{2\sin\theta}\sqrt{h^2+k^2+l^2}. \tag{1}$$

Table 1 Calculated results using (1)

	hkl	$\sqrt{h^2+k^2+l^2}$	θ	a(nm)	$\sin\theta$	$\sin^2\theta$	$\sin^2\theta'$	$\sin\theta'$	a'(nm)
1	400	4	47.08_5	0.4210	0.7324	0.5364			
2	400	4	41.42_5	0.4209	0.6616	0.4377	(0.5369)	(0.7327)	(0.4209)
3	420	4.4721	54.95	0.4211	0.8187	0.6703			
4	420	4.4721	47.68_5	0.4209	0.7395	0.5469	(0.6707)	(0.8190)	(0.4210)
5	422	4.8990	63.72	0.4212	0.8966	0.8039			
6	422	4.8990	54.06	0.4212	0.8096	0.6555	(0.8039)	(0.8966)	(0.4212)
7	511	5.1962	72.02	0.4211	0.9512	0.9048			
8	511	5.1962	59.22	0.4210	0.8591	0.7381	(0.9052)	(0.9514)	(0.4210)

Until the 1970s, one of the main tools for X-ray diffraction analysis was the film method using a Debye–Scherrer camera. In this method, only the back-reflection region is suitable for precise lattice parameter measurement and its schematic diagram is given in Fig. 1. For a Debye–Scherrer camera, film shrinkage ($\Delta S'$), incorrect camera radius (ΔR), and off-centering (Δx) of a sample in the camera are considered as possible sources of error in the measured values related to θ and the following equations have been proposed for these factors.

$$\frac{\Delta d}{d} = -\cot\theta\cdot\Delta\theta \tag{2}$$

$$\frac{\Delta d}{d} = -\frac{\sin\theta}{\cos\theta}\cdot\Delta\theta = \frac{\sin\phi}{\cos\phi}\cdot\Delta\phi \tag{3}$$

$$\frac{\Delta d}{d} = \frac{\sin\theta}{\cos\theta}\left[\left(\frac{\Delta S'}{S'} - \frac{\Delta R}{R}\right)\phi + \frac{\Delta x}{R}\sin\phi\cos\phi\right]. \tag{4}$$

Here, the relationships of $\phi = 90° - \theta$, $\Delta\phi = -\Delta\theta$, $\sin\phi = \cos\theta$, and $\cos\phi = \sin\theta$ are utilized. In the back-reflection region, we usually choose small value of ϕ, so that the approximation of $\sin\phi \approx \phi$ and $\cos\phi \approx 1$ is well accepted. For this reason, the first term of (4) may be replaced by $\sin\phi\cos\phi$. The following simplification can be made.

$$\frac{\Delta d}{d} = \left(\frac{\Delta S'}{S'} - \frac{\Delta R}{R} + \frac{\Delta x}{R}\right)\sin^2\phi. \tag{5}$$

The quantities contained within parentheses of (5) are constant which will be determined by the experimental condition. Therefore, the fractional error of the d – value is found to be directly proportional to $\sin^2\phi$ or $\cos^2\theta$, where K is a constant.

$$\frac{\Delta d}{d} = \mathrm{K}\sin^2\phi = \mathrm{K}\cos^2\theta \qquad (\because \sin\phi = \cot\theta). \tag{6}$$

Therefore, the following equation is established for a cubic system as an example.

$$\frac{\Delta d}{d} = \frac{\Delta a}{a} = \frac{a - a_0}{a_0} = \mathrm{K}\cos^2\theta. \tag{7}$$

The values of lattice parameter a computed from one set of the diffraction data are plotted as a function of $\cos^2\theta$ and a straight line will be obtained. The true value of the lattice parameter, a_0, can be calculated by extrapolating this straight line to $\cos^2\theta \to 0$. In addition, by considering the relationship of $\sin^2\theta = 1 - \cos^2\theta$, various a − values computed from the measured diffraction data may be plotted against $\sin^2\theta$ and a straight line will be extrapolated to $\sin^2\theta \to 1$. This extrapolation method is also widely used.

Figure 2 shows the results of magnesium oxide in Table 1, plotted as a function of $\sin^2\theta$ using the calculation result in Table 1, from which the lattice parameter of $a_0 = 0.4212$ nm is obtained.

The diffraction peaks by Kα and Kβ radiations are produced from one d −value, so that the technique of converting the value of $\sin\theta$, is frequently employed. This is simply corresponding to the angle by Kα (or Kβ) and calculated for the other by Kβ (or Kα). For this purpose, we use the following equation readily obtained by the Bragg law.

$$\sin^2\theta' = \sin^2\theta_\beta \times \frac{\lambda^2_{\mathrm{K}\alpha}}{\lambda^2_{\mathrm{K}\beta}}. \tag{8}$$

For the convenience of readers, the examples of recalculated results by using (8) are included in the column on the right-hand side of Table 1.

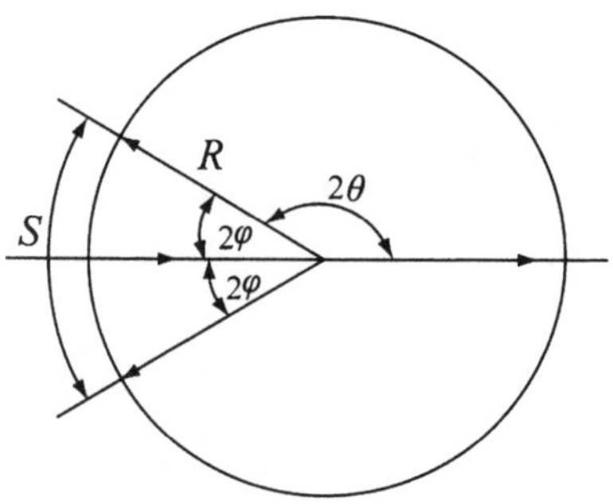

Fig. 1 Schematic diagram of the back-reflection region in Debye–Scherrer camera

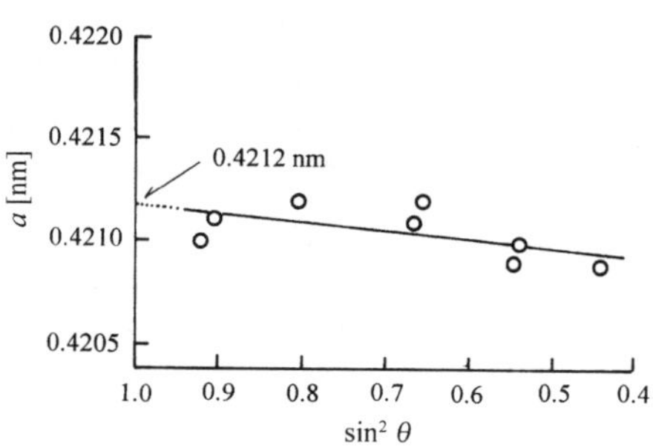

Fig. 2 The computation of true value of the lattice parameter a_0 by extrapolation to $\sin^2\theta$

Question 4.14 Some methods are known to be proposed for quantitative analysis of powder mixtures using the integrated intensity of a specific diffraction peak and the relative merit and demerit of each method have already been discussed. The so-called "direct comparison method" is widely used to determine phase proportions in a mixture of phases, because the reference peak is from another phase in the mixture. Namely, this method does not require a sample of the pure phase. For example, this method is employed to determine the amount of retained austenite phase in a quenched steel sample and its usefulness and validity has been confirmed. Obtain fundamental equations for the direct comparison method with respect to a case, where two crystalline substances are mixed. (Reference: The original paper of a direct comparison method is given as follows. B.L. Averbach and M. Cohen: Trans. AIME, Vol. **176** (1948), p.401.)

Answer 4.14 Integrated intensity I of the diffracted peak from a single phase powder sample measured by a diffractometer (usually : joule-sec^{-1} and m^{-1}) is given in the following form if a sample of the flat plate form keeps the condition of infinite thickness.

$$I = \mathrm{K}_0 \cdot \left[|F|^2 \cdot p \cdot \left(\frac{1+\cos^2 2\theta}{2\sin^2\theta \cdot \cos\theta} \right) \right] \frac{1}{\Omega^2} \frac{\mathrm{e}^{-2M_T}}{2\mu}. \tag{1}$$

Here, F represents structure factor, p multiplicity factor, the term within parentheses Lorentz polarization factor (LP), μ a linear absorption coefficient, and Ω the volume of unit cell of the constituent crystalline substance in a sample. On the contrary, K_0 covers the whole factors, such as Thomson scattering, cross-sectional area of the incident X-ray beam, scattering angle 2θ determined by diffractometer radius, etc. It also includes the amount independent of the absorption coefficient of the sample. Although the temperature factor e^{-2M_T} affects on the atomic scattering factors f, it is usually omitted in calculation of the desired intensities, as far as handling the relative amounts of intensities.

Let us consider a binary system containing two phases of α and β. If w represents the weight fraction of each component and its density is ρ, the mass absorption coefficient of the mixture is described as follows.

$$\frac{\mu_\mathrm{m}}{\rho_\mathrm{m}} = w_\alpha \left(\frac{\mu_\alpha}{\rho_\alpha} \right) + w_\beta \left(\frac{\mu_\beta}{\rho_\beta} \right). \tag{2}$$

The weight of a unit volume of a binary system is ρ_m, and then, for example, the weight of α – component in the mixture can be expressed by $w_\alpha \rho_\mathrm{m}$. The volume fraction c_α of α–component in the mixture may be given by the following equation.

$$c_\alpha = \frac{w_\alpha}{\rho_\alpha} \rho_\mathrm{m}. \tag{3}$$

A similar expression is given for c_β. In addition, let us consider that a specific diffraction peak attributed to α – component is found completely resolved as a separate peak from any diffraction peak produced by β – component. In such case, the intensity of (1) can be simply expressed as I_α. However, we have to take the volume fraction c_α in the mixture into account, as well as the value of μ_m related to absorption of the mixture. That is, the intensity of α – component in the mixture can be handled by an independent constant K_1 with two values c_α and μ_m, as described in the following form.

$$I_\alpha = \mathrm{K}_1 \frac{c_\alpha}{\mu_\mathrm{m}}. \tag{4}$$

The constant K_1 can be estimated from the experimental condition, similar to K_0 in (1). However, the determination of its absolute value is certainly difficult and it is usually unknown. When obtaining the intensity ratio of I_α with respect to the intensity of a specific diffraction peak of another component, the K-value cancels out and disappears. By using this particular feature, we can determine the concentration of α – component.

The intensity I can be rewritten in a very simple equation, if terms other than K_0 of (1) are expressed by R in the following form.

$$R = \left[|F|^2 \cdot p \cdot \left(\frac{1 + \cos^2 2\theta}{2 \sin^2 \theta \cdot \cos \theta} \right) \right] \frac{\mathrm{e}^{-2M_T}}{\Omega^2} \tag{5}$$

$$I = \frac{\mathrm{K}_0 R}{2\mu}. \tag{6}$$

Only R is a value which depends on the scattering angle 2θ and hkl or $|F|^2$. From the relationships of (4) and (6), the intensity of the specific diffraction peak for each component is given by the following form, respectively.

$$I_\alpha = \frac{\mathrm{K}_0' R_\alpha c_\alpha}{2\mu_\mathrm{m}}, \qquad I_\beta = \frac{\mathrm{K}_0' R_\beta c_\beta}{2\mu_\mathrm{m}}. \tag{7}$$

If the ratio between I_α and I_β is obtained, both K_0' and $2\mu_\mathrm{m}$ will be cancelled.

$$\frac{I_\beta}{I_\alpha} = \frac{R_\beta c_\beta}{R_\alpha c_\alpha}. \tag{8}$$

Since the ratio of R_β / R_α can be computed from (5), if I_α and I_β are measured with reasonable reliability, the ratio of c_β / c_α is readily obtained. Once the ratio of c_β / c_α is obtained, the value of c_α (or c_β) can be estimated from the well-known relationship of $c_\alpha + c_\beta = 1$ in a binary system.

This method is convenient because of no standard substance. However, it should be kept in mind that we have to measure the integrated intensity of a peak attributed to α – component is found completely resolved as a separate peak from any diffraction peak produced by β – component. In other words, if the desired peaks of two components are so close together and they are not resolved, the direct comparison method is no longer an effective tool for quantitative analysis of crystalline powder mixtures.

Question 4.15 When the diffraction experiment with a diffractometer and Cu-Kα radiation ($\lambda = 0.1542$ nm) was made on a mixed sample of Al powder including Si, four clearly separated diffraction peaks were obtained as shown in Fig. A. Preliminary analysis suggests that they are assigned as diffraction peaks corresponding to (111) and (220) planes of Al and (111) and (200) planes of Si. Then, the integrated intensities of these four peaks were carefully measured. The results are summarized in Table A. Calculate the volume fraction of Al and Si using the direct comparison method.

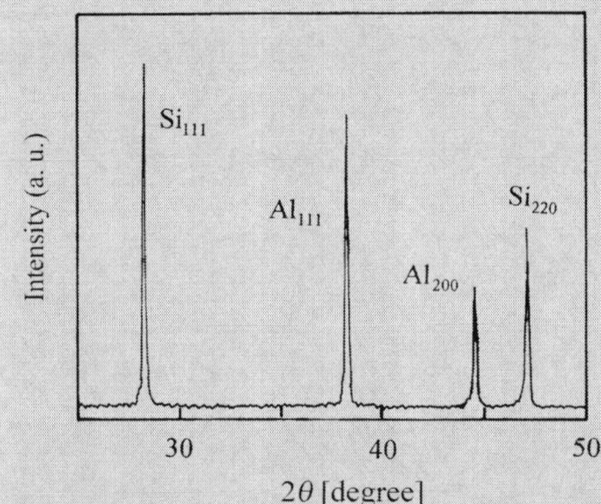

Fig. A Diffraction pattern of Al powder sample containing Si

Table A Four peaks measured for Al powder sample containing Si

	2θ(degree)	Integrated intensity I	hkl
1	28.41	180.3	Si (111)
2	38.46	216.2	Al (111)
3	44.73	93.1	Al (200)
4	47.33	118.4	Si (220)

Answer 4.15 In the direct comparison method, the integrated intensity and the volume fraction of each component, Al and Si in the present case, are as follows.

$$\frac{I_{\text{Si}}}{I_{\text{Al}}} = \frac{R_{\text{Si}} c_{\text{Si}}}{R_{\text{Al}} c_{\text{Al}}} \tag{1}$$

$$R = \left[|F|^2 \cdot p \cdot \left(\frac{1 + \cos^2 2\theta}{2 \sin^2 \theta \cdot \cos \theta} \right) \right] \frac{\text{e}^{-2M_T}}{\Omega^2}, \tag{2}$$

where F is structure factor, p multiplicity factor and Ω the volume of unit cell. The value of R_{Al} or R_{Si} can be calculated from (2).

(i) Component of Al

Fundamental information of aluminum is as follows: face-centered cubic (a = 0.4049 nm), density 2.70×10^6 g/m^3, molar mass 26.982 g, and Debye characteristic temperature $\Theta = 428$ K. Let consider the measurements at room temperature

293 K(20°C), and then we compute the temperature factor as listed in Table 1 (refer to Question 4.2). Here, we use the results of $x = \frac{\Theta}{T} = \frac{428}{293} = 1.46 \rightarrow \phi(x) = 0.693$.

$$B_T = \frac{1.15 \times 10^4}{M} \times \frac{T}{\Theta^2} \times \left\{\phi(x) - \frac{x}{4}\right\}$$
$$= \frac{1.15 \times 10^4}{26.982} \times \frac{293}{(428)^2} \times \left(0.693 + \frac{1.46}{4}\right) = 0.721\,. \quad (3)$$

$$e^{-2M_T} = e^{-2B_T\left(\frac{\sin\theta}{\lambda}\right)^2}. \quad (4)$$

On the contrary, the Lorentz polarization factor (*LP*) is computed using (5) and the results are listed in Table 2.

$$LP = \frac{1 + \cos^2 2\theta}{2\sin^2\theta\cos\theta}. \quad (5)$$

Table 1 Temperature factor for diffraction peaks of Al

	2θ(degree)	$\frac{\sin\theta}{\lambda}$	$\left(\frac{\sin\theta}{\lambda}\right)^2$	$-2B_T\left(\frac{\sin\theta}{\lambda}\right)^2$	e^{-2M_T}
Al(111)	38.46	0.214	0.0458	−0.0660	0.936
Al(200)	44.73	0.247	0.0610	−0.0880	0.916

Table 2 Lorentz polarization factor for diffraction peaks of Al

	2θ(degree)	$\sin\theta$	$\sin^2\theta$	$\cos\theta$	$\cos 2\theta$	$\cos^2 2\theta$	$1+\cos^2 2\theta$	$2\sin^2\theta\cos\theta$	LP
Al(111)	38.46	0.3294	0.1085	0.9442	0.7830	0.6130	1.6130	0.2049	7.9
Al(200)	44.73	0.3805	0.1448	0.9248	0.7104	0.5047	1.5047	0.2678	5.6

The volume of unit cell, Ω of Al with fcc structure is given by the cube of the lattice parameter a. Namely, $\Omega = a^3 = (0.4049\,\text{nm})^3 = 0.06638\,\text{nm}^3$. Then, substitute these calculated values for corresponding variables in (2) for estimating R_{Al} and the results are summarized in Table 3. When considering that four atoms ($n = 4$) per unit cell being included in fcc, the following point may also be worthy of note with respect to the volume of unit cell Ω.

Table 3 Calculated results of R_{Al} for two peaks of Al

	$\frac{\sin\theta}{\lambda}$	f_{Al}	$\lvert F\rvert^2$	p	LP	e^{-2M_T}	R_{Al}
Al(111)	0.214	8.85	$16f_{\text{Al}}^2 = 1253$	8	7.9	0.936	16.82×10^6
Al(200)	0.247	8.35	$16f_{\text{Al}}^2 = 1116$	6	5.6	0.916	7.80×10^6

$$\Omega = \frac{M \times n}{\rho N_A} = \frac{26.982 \times 4}{2.70 \times 10^6 \times 0.6022 \times 10^{24}}$$
$$= 0.06638 \times 10^{-27}\text{m}^3 = 0.06638\,\text{nm}^3$$

M: Molar mass, ρ: density, N_A: Avogadro's number.

In Table 3, f is the atomic scattering factor which is taken from Appendix A.3 and the multiple factor p are referred to Table 2.2 in Chap. 2. It may also be noted for the unit of R_{Al} that the value(0.06638) of Ω is expressed in the unit of nm^3 and then we use it as it stands. For this reason, the factor of 10^{-27} is excluded.

(ii) The component of Si

Similar calculation is made for Si on the basis of fundamental information of silicon in the following. Silicon has diamond structure ($a = 0.5431\,\mathrm{nm}$), density $2.33 \times 10^6\,\mathrm{g/m^3}$, molar atomic mass 28.086 g, and Debye temperature Θ is 645 K. The results are summarized in Tables 4 and 5.

$$x = \frac{\Theta}{T} = \frac{645}{293} = 2.20 \rightarrow \phi(x) = 0.578$$

$$B_T = \frac{1.15 \times 10^4}{M} \times \frac{293}{(645)^2} \times \left(0.578 + \frac{2.20}{4}\right) = 0.325\,.$$

Table 4 Temperature factor for diffraction peak of Si

	2θ(degree)	$\frac{\sin\theta}{\lambda}$	$\left(\frac{\sin\theta}{\lambda}\right)^2$	$-2B_T\left(\frac{\sin\theta}{\lambda}\right)^2$	e^{-2M_T}
Si(111)	28.41	0.159	0.0253	−0.0164	0.984
Si(220)	47.33	0.260	0.0676	−0.0439	0.951

Table 5 Lorentz polarization factor for diffraction peaks of Si

	2θ(degree)	$\sin\theta$	$\sin^2\theta$	$\cos\theta$	$\cos 2\theta$	$\cos^2 2\theta$	$1+\cos^2 2\theta$	$2\sin^2\theta\cos\theta$	LP
Si(111)	28.41	0.2454	0.0602	0.9694	0.8796	0.7737	1.7737	0.1167	15.2
Si(220)	47.33	0.4014	0.1611	0.9159	0.6718	0.4594	1.4594	0.2951	4.9

For calculating R_{Si}, we also take into consideration that the structure factors are $32f^2$ for the (111) plane and $64f^2$ for (220) plane, respectively, (see Question 3.11). $\Omega = a^3 = (0.5431\,\mathrm{nm})^3 = 0.16019\,\mathrm{nm}^3$. The results are given in Table 6.

Table 6 Calculated results of R_{Si} for two peaks of Si

	$\frac{\sin\theta}{\lambda}$	f_{Si}	$\lvert F\rvert^2$	p	LP	e^{-2M_T}	R_{Si}
Si(111)	0.159	10.16	$32f_{\mathrm{Si}}^2 = 3303$	8	15.2	0.984	15.42×10^6
Si(220)	0.260	8.68	$64f_{\mathrm{Si}}^2 = 4822$	12	4.9	0.915	10.52×10^6

(iii) Calculation of the volume fraction c_{Al} and c_{Si}

With respect to the ratio of volume fraction of each component, (1) can be rewritten in the following form.

$$\frac{I_{\mathrm{Si}}}{I_{\mathrm{Al}}} = \frac{R_{\mathrm{Si}}c_{\mathrm{Si}}}{R_{\mathrm{Al}}c_{\mathrm{Al}}} \rightarrow \frac{c_{\mathrm{Al}}}{c_{\mathrm{Si}}} = \frac{I_{\mathrm{Al}}R_{\mathrm{Si}}}{I_{\mathrm{Si}}R_{\mathrm{Al}}}. \tag{6}$$

First, substitute the integrated intensity values of Al (111) and Si (111) planes for I_{Al} and I_{Si}, respectively and then we obtain.

$$\frac{c_{Al}}{c_{Si}} = \frac{216.2}{180.3} \times \frac{15.42 \times 10^6}{16.82 \times 10^6} = 1.099.$$

From the relationship of $c_{Al} + c_{Si} = 1$, $c_{Si} = 0.476$ is found.

Next, the similar computation is made for the integrated intensity values of A1 (200) and Si (220) planes.

$$\frac{c_{Si}}{c_{Al}} = \frac{93.1}{118.4} \times \frac{10.52 \times 10^6}{7.80 \times 10^6} = 1.061.$$

In this case, $c_{Si} = 0.485$. The average of these two results is given by $c_{Si} = 0.48$. Therefore, it can lead to the result that the volume fraction of Si in mixed sample of Al powder containing Si is 0.48 (and 0.52 for Al).

Question 4.16 Obtain basic equations required for the principle and analysis of the "internal standard method" by coupling standard substances, such as NaCl and CaF_2.

Answer 4.16 In the internal standard method, the intensity of a specific peak of substance being determined is compared with a peak from a standard substance mixed with the sample in known proportions. The problem associated with mean absorption coefficient of the object sample can be eliminated by obtaining the calibration curve, from the measurements for test samples, where a known amount of a standard substance is mixed with a known amount of original sample. For standard substances, NaCl, CaF_2, ZnO, etc. are often used, but the standard sample should be selected so as to be of good crystallinity and have diffraction peak well separated from the specific peak of substance being determined. Suppose we wish to determine the amount of $\alpha-$component in a mixture containing two or more components. (Two or three components are preferred, otherwise we face serious problems, such as the peak overlap between two components.)

The integrated intensity I_α of a diffraction peak of α – component to be determined in a mixed sample is given by the following equation if setting the volume fraction of α – component to c_α and the linear absorption coefficient of a sample to μ_m.

$$I_\alpha = \mathrm{K}_1 \frac{c_\alpha}{\mu_m}, \tag{1}$$

where K_1 is a constant.

We mix a known amount of a standard substance denoted by s to a known amount of the original sample to form a new composite sample. Let us set c_α and c'_α to the volume fractions of α – component in the original and composite samples, respectively. In addition, c_s is the volume fraction of s in the composite sample. For the convenience of discussion, two component system containing α and β are considered here. Then, (1) is rewritten as follows:

$$I_\alpha = \mathrm{K}_2 \frac{c'_\alpha}{\mu_m}. \tag{2}$$

Similarly, the intensity of a particular peak from the standard substance s is given as

$$I_s = \mathrm{K}_{3s} \frac{c_s}{\mu_m}. \tag{3}$$

When dividing (2) by (3), we obtain the following equation

$$\frac{I_\alpha}{I_s} = \frac{\mathrm{K}_2}{\mathrm{K}_{3s}} \cdot \frac{c'_\alpha}{c_s}. \tag{4}$$

The absorption coefficient of a sample μ_m is canceled out in this division process.

The volume fraction c'_α of $\alpha-$component in the new composite sample prepared by adding the standard substance is given by the following equation.

$$c'_\alpha = \frac{\frac{w'_\alpha}{\rho_\alpha}}{\frac{w'_\beta}{\rho_\beta} + \frac{w'_\beta}{\rho_\beta} + \frac{w_s}{\rho_s}}, \tag{5}$$

where w and ρ represent weight fraction and density of each component, respectively. A similar equation holds for c_s of the volume fraction of the standard substance.

$$c_s = \frac{\frac{w_s}{\rho_s}}{\frac{w'_\alpha}{\rho_\alpha} + \frac{w'_\beta}{\rho_\beta} + \frac{w_s}{\rho_s}}. \tag{6}$$

Then, the following simple equation can be derived from (5) and (6).

$$\frac{c'_\alpha}{c_s} = \frac{w'_\alpha}{w_s} \cdot \frac{\rho_s}{\rho_\alpha}. \tag{7}$$

By combining (7) with (4),

$$\frac{I_\alpha}{I_s} = \frac{\mathrm{K}_2}{\mathrm{K}_{3s}} \cdot \frac{\rho_s}{\rho_\alpha w_s} \cdot w'_\alpha, \tag{8}$$

where not only K_2 and K_{3s} but also ρ_s and ρ_α are all constants. Therefore, if the weight fraction w_s of the standard substance is kept constant in all the composite samples, (8) can be rewritten in the following simple form.

$$\frac{I_\alpha}{I_s} = \mathrm{K}_9 w'_\alpha, \tag{9}$$

where K_9 is another constant. The weight fraction of $\alpha-$component, w_α in the original sample is related to the weight fraction, w'_α in the composite sample together with the value of w_s for the standard substance in the following.

$$w'_\alpha = w_\alpha(1 - w_s). \tag{10}$$

Since w_s is a constant, (9) can be replaced by the following equation.

$$\frac{I_\alpha}{I_s} = \mathrm{K}_{11} w_\alpha. \tag{11}$$

Therefore, the intensity ratio of a peak from $\alpha -$ component to be determined to the intensity of a specific peak of the standard substance is found to change linearly with the weight fraction w_α of $\alpha -$component in the original sample. The relationship of (9) can also be explained as follows.

In a binary system, preparing the new composite sample by the addition of the standard substance z(g) to the original sample y(g), the weight fraction of $\alpha -$ component in the standard substance s in the new composite sample and in the original sample are given by $\left(\frac{z}{y+z}\right)$ and $\left(\frac{yw_\alpha}{y+z}\right)$, respectively. As a result, the integrated intensities of a peak from $\alpha -$ component to be determined and a specific peak of the standard substance can be expressed in the following equations, respectively.

$$I_\alpha = \mathrm{K}_{12} \frac{\left(\frac{yw_\alpha}{y+z}\right)}{\left(\frac{\mu_m}{\rho}\right)} \tag{12}$$

$$I_s = \mathrm{K}_{13s} \frac{\left(\frac{z}{y+z}\right)}{\left(\frac{\mu_m}{\rho}\right)}. \tag{13}$$

If the ratio of (12) and (13) is taken, the average absorption coefficient term of a sample vanishes and we obtain

$$\frac{I_\alpha}{I_s} = \frac{\mathrm{K}_\alpha}{\mathrm{K}_{13s}} \cdot \frac{y}{z} \cdot w_\alpha. \tag{14}$$

Equation (14) clearly shows that a linear relationship is well recognized between the intensity ratio of diffraction peaks and the weight fraction w_α of $\alpha -$component in the original sample, when keeping the ratio of y to z a constant.

Specifically, the composite samples are prepared by adding the amount of a reference substance in the range between 0.1 and 0.8 g to the original sample whose weight is the order of $1 \sim 3$g and the integrated intensities are measured with respect to the specific diffraction peaks attributed to both the original sample to be determined and the standard substance, to obtain the calibration line. Once the calibration curve is obtained and in case where the same X-ray diffraction equipment is employed, it can be used as it stands for quantitative analysis of α − component, only by adding the same amount of the reference substance to the desired sample.

Question 4.17 According to some preliminary measurements, an original sample of interest is found to include magnesium oxide (MgO) and calcium oxide (CaO) and the positions detected for the diffraction peaks corresponding to the (200) plane of MgO and the (111) plane of CaO are not so close together and they are relatively easy to measure their intensities as separate peaks. On the contrary, a specific peak assigned to the (200) plane of CaO is clearly detected at angles near 37 degree as easily seen from the results of Fig. A, but this peak overlaps with the diffraction peak of the (111) plane of MgO. Accordingly, we cannot use the diffraction peaks of both the (111) plane of MgO and the (200) plane of CaO for quantitative analysis of these components. Based on these results, we carried out the intensity measurements using $Cu - K\alpha_1$ radiation ($\lambda = 0.15406$ nm) for quantitatively determining the amounts of both MgO and CaO by the internal standard method with KCl as a standard substance and the results summarized in Tables A-C were obtained. Calculate the content of CaO and MgO in the sample of interest.

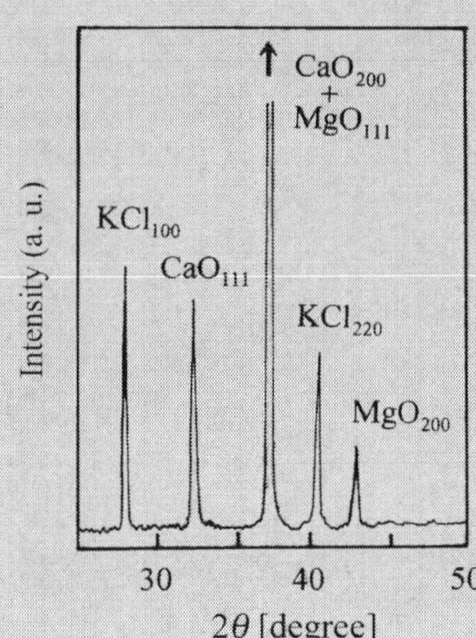

Fig. 1 Typical X ray diffraction pattern of a sample containing MgO and CaO with KCl as the internal standard substance

Some amounts of powder sample of interest are rapidly quenched so as to avoid precipitation of any crystalline phase and it is referred to as G_0. Then, the reagent (99.9% of purity) of MgO or CaO and the sample of G_0 are mixed at a rate given by G_0^* in the column of Tables A and B for obtaining the calibration curves of MgO and CaO components. Specifically, 0.2 g of the standard substance of KCl powder was added to 2 g of the mixed sample of (MgO reagent $+G_0$), and the ratio of the integrated intensities of specific peaks attributed to the MgO or KCl component were measured. The results are summarized in the right-hand side column of Table A. Similar experiments were made and the results are summarized in Table B.

Table A Diffraction data of MgO for calibration

	MgO Reagent	G_0^*	$\frac{I_{MgO(200)}}{I_{KCl(200)}}$	$\frac{I_{MgO(200)}}{I_{KCl(220)}}$
M0	0 g	100 g		
M1	3 g	97 g	0.149	0.147
M2	8 g	92 g	0.370	0.373
M3	12 g	88 g	0.552	0.566
M4	15 g	85 g	0.713	0.704

Table B Diffraction data of CaO for calibration

	CaO Reagent	G_0^*	$\frac{I_{CaO(111)}}{I_{KCl(200)}}$	$\frac{I_{CaO(111)}}{I_{KCl(220)}}$
C0	0 g	100 g		
C1	3 g	97 g	0.403	0.389
C2	8 g	92 g	0.984	0.958
C3	12 g	88 g	1.403	1.430
C4	15 g	85 g	1.802	1.785

On the contrary, the integrated intensities of diffraction peaks were measured for the composite sample prepared by mixing the original sample (it is not G_0) with 0.2 g of KCl powder. The results are given in Table C.

Table C Diffraction data of the composite sample prepared by the original sample with KCl

2θ(degree)	Integrated intensity		2θ(degree)	Integrated intensity	
28.35	245.7	KCl(200)	32.18	253.2	CaO(111)
40.51	236.0	KCl(220)	42.90	171.8	MgO(200)

Answer 4.17 At first, the calibration curves should be obtained from the experimental data of Tables A and B regarding the integrated intensities of specific peaks attributed to MgO and CaO as well as KCl as a standard substance. In these cases, the amount of MgO or CaO added to the sample of G_0 is known. In both cases, good linearity is found as shown in Fig. 1. For further convenience, we compute an analytical equation representing the calibration curve for respective component using the least-squares method, so that the relevant numerical data are summarized in Tables 1 and 2. Some essential points of the least-squares method for applying to a correlation of $y = a + bx$ are described in Appendix A.7.

Table 1 Basic data for calibration curve of MgO

	MgO			
	y	x	xy	x^2
1	0.149	3	0.447	9
2	0.370	8	2.960	64
3	0.552	12	6.624	144
4	0.713	15	10.695	225
5	0.147	3	0.441	9
6	0.373	8	2.984	64
7	0.566	12	6.792	144
8	0.704	15	10.560	225
$\sum$	3.574	76	41.503	884

Table 2 Basic data for calibration curve of CaO

	CaO			
	y	x	xy	x^2
1	0.389	3	1.167	9
2	0.958	8	7.664	64
3	1.430	12	17.160	144
4	1.785	15	26.775	225
5	0.403	3	1.209	9
6	0.984	8	7.986	64
7	1.403	12	16.836	144
8	1.802	15	27.030	225
$\sum$	9.154	76	105.737	884

The normal equations of the least-squares method for MgO and the results are as follows.

$$\left.\begin{aligned} 3.574 &= 8a + 76b \\ 41.503 &= 76a + 884b \end{aligned}\right\} \Rightarrow \left\{\begin{aligned} a &= 0.0040 \\ b &= 0.0466 \end{aligned}\right.$$

$$y = 0.0040 + 0.0466x. \tag{1}$$

The results for the CaO case are given in the following.

$$\left.\begin{aligned} 9.154 &= 8a + 76b \\ 105.737 &= 76a + 884b \end{aligned}\right\} \Rightarrow \left\{\begin{aligned} a &= 0.0433 \\ b &= 0.1159 \end{aligned}\right.$$

$$y = 0.0433 + 0.1159x. \tag{2}$$

The amount of each component can be calculated from (1) or (2), when coupling with the measured ratio of the integrated intensities of peaks corresponding to the (200) plane of MgO or the (111) plane of CaO for the composite sample prepared by mixing the original sample (it is not G_0) with the standard substance KCl.

(i) With respect to the content of MgO

$$\frac{I_{\mathrm{MgO}}(200)}{I_{\mathrm{KCl}}(200)} = \frac{171.8}{245.7} = 0.699 \rightarrow x = 14.9$$

$$\frac{I_{\mathrm{MgO}}(200)}{I_{\mathrm{KCl}}(220)} = \frac{171.8}{236.0} = 0.728 \rightarrow x = 15.5.$$

From these two results, 15.2 mass% of MgO content is obtained as a mean value.

(ii) With respect to the content of CaO

$$\frac{I_{\mathrm{CaO}}(111)}{I_{\mathrm{KCl}}(200)} = \frac{253.2}{245.7} = 1.031 \rightarrow x = 8.5$$

$$\frac{I_{CaO}(111)}{I_{KCl}(220)} = \frac{253.2}{236.0} = 1.073 \rightarrow x = 8.9.$$

These two results give 8.7 mass% of CaO content, as a mean value.

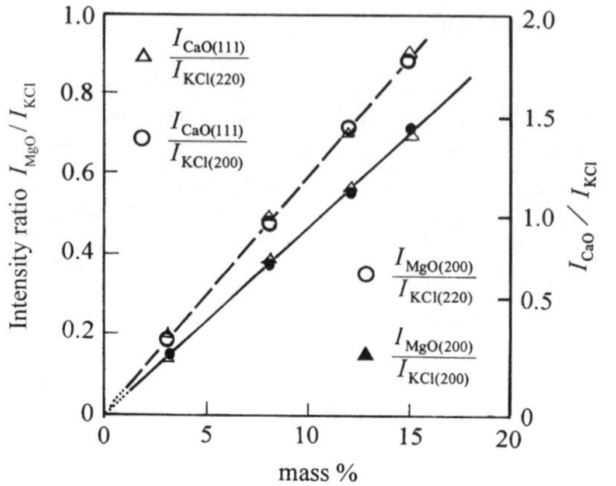

Fig. 1 Calibration curves of MgO and CaO components with KCl as the internal standard substance

Question 4.18 Magnesium oxide (MgO) powder was ground by the ball milling process to produce a fine powdered sample. With respect to this fine powdered sample, X-ray diffraction experiments using Cu-Kα radiation ($\lambda = 0.15406$ nm) with a crystal monochromator were carried out and we found the decrease in peak height as well as the peak broadening. The values of full width of half maximum intensity (FWHM) of peaks corresponding to (200), (220), (311), (222), and (400) planes are summarized in Table A. Nevertheless, similar X-ray diffraction measurements were made for magnesium oxide powder sample which were fully annealed so as to remove the lattice strain for comparison. The resultant FWHM values of five peaks are given in Table B. Compute the average size of crystallites in a sample prepared by ball milling processing.

Table A Diffraction data of ball mill processing sample

	hkl	2θ(degree)	FWHM(degree)
1	200	42.90	0.183
2	220	62.31	0.205
3	311	74.71	0.243
4	222	78.63	0.274
5	400	94.06	0.309

Table B Diffraction data of annealed sample

	hkl	2θ(degree)	FWHM(degree)
1	200	42.91	0.093
2	220	62.30	0.072
3	311	74.68	0.068
4	222	78.60	0.090
5	400	94.04	0.087

Answer 4.18 At first, the so-called instrumental broadening factor, B_i, is calculated from the sample fully annealed. Note that the FWHM values should be converted from degree to radian as shown in Table 1.

Table 1 The values of peak broadening of five peaks for the fully annealed sample

	hkl	2θ(degree)	FWHM(degree)	B_i = FWHM(radian)	B_i^2
1	200	42.91	0.093	1.62×10^{-3}	2.624×10^{-6}
2	220	62.30	0.072	1.26×10^{-3}	1.588×10^{-6}
3	311	74.68	0.068	1.19×10^{-3}	1.416×10^{-6}
4	222	78.60	0.090	1.57×10^{-3}	2.465×10^{-6}
5	400	94.04	0.087	1.52×10^{-3}	2.310×10^{-6}

(π radian = 180 degrees)

To subtract the instrumental broadening factor from the measured width value B_{obs}, we use the following relationship derived when assuming the shape of a diffraction peak is approximated by Gaussian distribution.

$$B_r^2 = B_{obs}^2 - B_i^2, \tag{1}$$

where B_r is the value of peak broadening related only to the change in size of crystallites. The relevant results are summarized in Tables 2 and 3.

Table 2 The values of peak broadening of five peaks for a sample prepared by the ball-milling process

	hkl	2θ(degree)	FWHM(degree)	B_{obs}(radian)	B_{obs}^2
1	200	42.90	0.183	3.19×10^{-3}	10.176×10^{-6}
2	220	62.31	0.205	3.58×10^{-3}	12.816×10^{-6}
3	311	74.71	0.243	4.24×10^{-3}	17.978×10^{-6}
4	222	78.63	0.274	4.78×10^{-3}	22.848×10^{-6}
5	400	94.06	0.309	5.39×10^{-3}	29.052×10^{-6}

Table 3 Some fundamental data for estimating the change in size of crystallites

	hkl	$B_r = \sqrt{B_{obs}^2 - B_i^2}$	$\cos\theta$	$B_r \cos\theta$	$\sin\theta$
1	200	2.75×10^{-3}	0.9307	2.56×10^{-3}	0.3657
2	220	3.35×10^{-3}	0.8558	2.87×10^{-3}	0.5174
3	311	4.07×10^{-3}	0.7944	3.24×10^{-3}	0.6068
4	222	4.51×10^{-3}	0.7737	3.49×10^{-3}	0.6336
5	400	5.17×10^{-3}	0.6816	3.52×10^{-3}	0.7317

The peak broadening of the sample prepared by the ball milling process is quite likely to be influenced by inhomogeneous (heterogeneous) strain, so that the Hall method is considered to be useful for the present analysis.

$$B_r \cos\theta = 2\eta \sin\theta + \frac{\lambda}{\varepsilon}, \tag{2}$$

where λ is the wavelength, ε is the average value of the size of crystallites, and 2η corresponds to the amount of inhomogeneous strain. Note that a linear relationship is obtained when plotting $B_r \cos\theta$ as a function of $\sin\theta$, the gradient of this straight line is equivalent to 2η, and the intercept Δy with the $B_r \cos\theta$ axis gives the value of $\frac{\lambda}{\varepsilon}$. Although the peak width in the Hall method is originally derived using the integral width, the FWHM values are used here. It may be safely said that this approximation is insignificant.

Correlation between $B_r \cos\theta$ and $\sin\theta$ is obtained as illustrated in Fig. 1 and the following value of ε, the size of crystallites, can be obtained using the graphical analysis together with the value (2η) corresponding to inhomogeneous strain of 2.75×10^{-3} radian.

$$\varepsilon = \frac{\lambda}{\Delta y} = \frac{0.15406 \times 10^{-9}}{1.55 \times 10^{-3}} = 99 \times 10^{-9}\mathrm{m} \fallingdotseq 100\ \mathrm{nm}.$$

However, for applying the least-squares method to draw a straight line of $y = a + bx$, the data set of Table 4 was prepared. (Refer to the Appendix A.7 for the general procedure of least-squares method).

Table 4 Data set for the least-squares method

	$y = B_r \cos\theta$	$x = \sin\theta$	xy	x^2
1	2.56×10^{-3}	0.3657	0.936×10^{-3}	0.1337
2	2.87×10^{-3}	0.5174	1.485×10^{-3}	0.2677
3	3.24×10^{-3}	0.6068	1.966×10^{-3}	0.3682
4	3.49×10^{-3}	0.6336	2.211×10^{-3}	0.4014
5	3.52×10^{-3}	0.7317	2.596×10^{-3}	0.5354
$\sum$	15.68×10^{-3}	2.8552	9.174×10^{-3}	1.7064

$$\left.\begin{aligned} \sum y = \sum a + b\sum x \rightarrow 15.62 \times 10^{-3} &= 5a + 2.8552b \\ \sum xy = a\sum x + b\sum s^2 \rightarrow 9.174 \times 10^{-3} &= 2.8552a + 1.7064b \end{aligned}\right\}. \tag{3}$$

Finding a and b satisfying simultaneously two relational expressions given by (3)

$$a = 1.48 \times 10^{-3} \rightarrow \frac{\lambda}{\varepsilon}$$

$$b = 2.90 \times 10^{-3} \rightarrow 2\eta.$$

Since the value of a is equivalent to the intercept Δy,

$$\varepsilon = \frac{\lambda}{a} = \frac{0.15406 \times 10^{-9}}{1.48 \times 10^{-3}} = 104 \times 10^{-9}\mathrm{m} \fallingdotseq 100\,\mathrm{nm}.$$

Although the agreement between the two results presently obtained is acceptable, it may be suggested here that analysis by the least-squares method is preferable for minimizing random error.

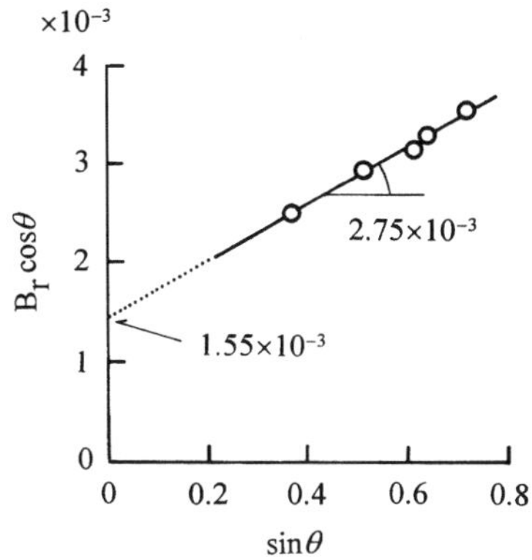

Fig. 1 Correlation between $B_r \cos\theta$ and $\sin\theta$

Chapter 5
Reciprocal Lattice and Integrated Intensities of Crystals

The Bragg law enables us to explain all the diffraction phenomena of X-rays by a crystal described in the previous chapters. However, there are some diffraction phenomena that may not be explained by the Bragg law. The diffuse scattering at non-Bragg angles is a particular example. For this purpose, we need a more generalized theory of diffraction using the vector representation. Particularly, the concept "reciprocal lattice" is extremely effective for handling all the diffraction phenomena. In other words, the reciprocal-lattice theory of diffraction, being general, is applicable to all diffraction phenomena of X-rays by a crystal from the simplest one to the most complex case. It may be added that the usual set of three-dimensional atomic coordinates is called the crystal lattice or real-space lattice, as opposed to the reciprocal lattice.

5.1 Mathematical Definition of Reciprocal Lattice

At first, let us consider the crystal lattice having a unit cell defined by the vectors $\boldsymbol{a}_1$, $\boldsymbol{a}_2$, and $\boldsymbol{a}_3$ (sometime called the primitive vectors of the crystal lattice or the primitive crystal-lattice vectors), and we obtain the corresponding reciprocal lattice having a unit cell defined by the vectors $\boldsymbol{b}_1$, $\boldsymbol{b}_2$ and $\boldsymbol{b}_3$ in the following form.

$$\left.\begin{array}{l} \boldsymbol{b}_1 = \dfrac{\boldsymbol{a}_2 \times \boldsymbol{a}_3}{V}, \quad \boldsymbol{b}_2 = \dfrac{\boldsymbol{a}_3 \times \boldsymbol{a}_1}{V}, \quad \boldsymbol{b}_3 = \dfrac{\boldsymbol{a}_1 \times \boldsymbol{a}_2}{V} \\ V = \boldsymbol{a}_1 \cdot (\boldsymbol{a}_2 \times \boldsymbol{a}_3) \end{array}\right\} \tag{5.1}$$

where, V is the volume of the unit cell. In solid-state physics, the factor of 2π is usually included in the form of (5.1), such as $\boldsymbol{b}_1 = (2\pi)(\boldsymbol{a}_2 \times \boldsymbol{a}_3)/V$, but it is commonly omitted in crystallography. It is also worth mentioning that the reverse of the reciprocal-lattice vector is the crystal-lattice vector and that the reciprocal of the volume of the unit cell, which is described by the triple-scalar product of the three vectors as $\boldsymbol{a}_1 \cdot (\boldsymbol{a}_2 \times \boldsymbol{a}_3)$, corresponds to the volume $V^*$of the unit cell of the reciprocal lattice. They are represented as follows.

$$\left.\begin{aligned} \boldsymbol{a}_1 &= \frac{\boldsymbol{b}_2 \times \boldsymbol{b}_3}{V^*} = (\boldsymbol{b}_2 \times \boldsymbol{b}_3) \cdot V \\ \boldsymbol{a}_2 &= \frac{\boldsymbol{b}_3 \times \boldsymbol{b}_1}{V^*} = (\boldsymbol{b}_3 \times \boldsymbol{b}_1) \cdot V \\ \boldsymbol{a}_3 &= \frac{\boldsymbol{b}_1 \times \boldsymbol{b}_2}{V^*} = (\boldsymbol{b}_1 \times \boldsymbol{b}_2) \cdot V \\ V^* &= \boldsymbol{b}_1 \cdot (\boldsymbol{b}_2 \times \boldsymbol{b}_3) = \frac{\boldsymbol{a}_2 \times \boldsymbol{a}_3}{V} \cdot \frac{\boldsymbol{a}_1}{V} = \frac{1}{V} \end{aligned}\right\} \tag{5.2}$$

For example, it is shown in (5.1) that the reciprocal-lattice axis $\boldsymbol{b}_1$ is found to be perpendicular to the plane of the crystal-lattice vectors $\boldsymbol{a}_2$ and $\boldsymbol{a}_3$, and its length is equal to the reciprocal of spacing of the (100) plane. The same relationship is confirmed with respect to other reciprocal-lattice vectors $\boldsymbol{b}_2$ and $\boldsymbol{b}_3$. Namely, the reciprocal-lattice axes $\boldsymbol{b}_2$ and $\boldsymbol{b}_3$ are normal to the (010) and (001) planes of the crystal lattice, respectively, and their lengths are equal to the reciprocals of the spacing of these planes. In other words, the point at the end of the $\boldsymbol{b}_1$ vector is labeled (100), that at the end of the $\boldsymbol{b}_2$ vector is labeled (010), and that at the end of the $\boldsymbol{b}_3$ vector is labeled (001), and then an array of points each of which is labeled with its coordinates by the basic vectors. Similar relationships are confirmed for all the planes with an arbitrary Miller indices $(h\,k\,l)$ of the crystal lattice. Of course, $h\,k\,l$ are three integers.

Considering these results, a vector $\boldsymbol{H}_{hkl}$ perpendicular to the $(h\,k\,l)$ plane is given by the following equation with the reciprocal-lattice vectors, $\boldsymbol{b}_1$, $\boldsymbol{b}_2$, and $\boldsymbol{b}_3$.

$$\boldsymbol{H}_{hkl} = h\boldsymbol{b}_1 + k\boldsymbol{b}_2 + l\boldsymbol{b}_3 \tag{5.3}$$

The length of this reciprocal-lattice vector $\boldsymbol{H}_{hkl}$ is equal to the reciprocal of a spacing d of the $(h\,k\,l)$ plane, as follows.

$$|\boldsymbol{H}_{hkl}| = \frac{1}{d_{hkl}} \tag{5.4}$$

Every crystal structure has two lattices, the real-space lattice in the dimensions of (length), and the reciprocal lattice in the dimensions of (1/length) and a diffraction pattern produced by a crystal is a map of the reciprocal lattice of the crystal. It may also be noted that the reciprocal lattice is a lattice in the Fourier space associated with the crystal.

Each point of reciprocal lattice can represent the spacing as well as the direction of a related crystal plane. That is, the wave diffracted by the crystal planes with a certain periodicity of atoms in the real-space lattice appears as a diffracted spot in reciprocal space, as seen in the Laue photographs. Such spots also produce a certain regular sequence to form the reciprocal lattice. Fourier transform that is widely utilized in crystallography corresponds to an exchange operation from the real space to the inverse space or from the inverse space to the real space.

According to (5.1), we have following relationships between a reciprocal-lattice vector $\boldsymbol{b}_j$ and the real-space-lattice vector $\boldsymbol{a}_k$.

$$\boldsymbol{b}_j \cdot \boldsymbol{a}_k = \delta_{jk}\ ;\ \delta_{jk} = 0(j \neq k),\ \delta_{jk} = 1(j = k) \tag{5.5}$$

The complete set of relationships is as follows.

$$\boldsymbol{b}_1 \cdot \boldsymbol{a}_2 = \boldsymbol{b}_1 \cdot \boldsymbol{a}_3 = \boldsymbol{b}_2 \cdot \boldsymbol{a}_1 = \boldsymbol{b}_2 \cdot \boldsymbol{a}_3 = \boldsymbol{b}_3 \cdot \boldsymbol{a}_1 = \boldsymbol{b}_3 \cdot \boldsymbol{a}_2 = 0$$
$$\boldsymbol{b}_1 \cdot \boldsymbol{a}_1 = \boldsymbol{b}_2 \cdot \boldsymbol{a}_2 = \boldsymbol{b}_3 \cdot \boldsymbol{a}_3 = 1$$

The reciprocal-lattice vector $\boldsymbol{K}$ may be expressed by $\boldsymbol{b}_j$ and integers k_j as follows.

$$\boldsymbol{K} = k_1\boldsymbol{b}_1 + k_2\boldsymbol{b}_2 + k_3\boldsymbol{b}_3 \tag{5.6}$$

Similarly, the real-lattice vector $\boldsymbol{R}$ may be given by $\boldsymbol{a}_j$ and integers $\boldsymbol{n}_j$ in the following form.

$$\boldsymbol{R} = n_1\boldsymbol{a}_1 + n_2\boldsymbol{a}_2 + n_3\boldsymbol{a}_3 \tag{5.7}$$

Since the scalar product of the reciprocal-lattice vector and the real-space-lattice vector can be expressed in the following equation, the product of $\boldsymbol{K}$ with $\boldsymbol{R}$ is found to be integers.

$$\boldsymbol{K} \cdot \boldsymbol{R} = k_1 n_1 + k_2 n_2 + k_3 n_3 \tag{5.8}$$

As the scalar product of $\boldsymbol{K} \cdot \boldsymbol{R}$ is an integer, the relationship of $\mathrm{e}^{\mathrm{i}2\pi\boldsymbol{K}\cdot\boldsymbol{R}} = 1$ is established. Therefore, it may safely be said that the reciprocal lattice, $\boldsymbol{K}$, corresponds to one set of the wave vector $2\pi\boldsymbol{K}$ that satisfies the relationship of $\mathrm{e}^{\mathrm{i}2\pi\boldsymbol{K}\cdot\boldsymbol{R}} = 1$ with respect to all the real-space vectors $\boldsymbol{R}$.

When the real-space lattice is primitive (simple), the reciprocal lattice becomes primitive, and similarly, if the real-space lattice is classified into the complex (nonprimitive) lattice, the reciprocal lattice is also a complex one. Such general properties that take on a very simple relationship for any crystal whose unit cell is based on mutually perpendicular vectors (are checked by (5.1)). Some examples for explaining the relationship between the real-space lattice and the reciprocal lattice are illustrated in Fig. 5.1 using two cases of cubic and hexagonal crystals. $\boldsymbol{H}_{hkl}$ is found to be normal to the (hkl) plane, and its length is equal to the reciprocal of the spacing d_{hkl}. We recall that if the volume of Bravais lattices is set to V, then the volume of the reciprocal lattice is $1/V$.

5.2 Intensity from Scattering by Electrons and Atoms

X-rays are characterized by the wave–particle duality. An X-ray beam is known to be an electromagnetic wave characterized by an electric field that exerts its force on an electron. For this reason, an electron is continuously accelerated and decelerated by the field of X-rays. In other words, an electron as set into oscillation by an electric field of X-rays is continuously accelerating and decelerating during its motion so that the corresponding electron emits a new electromagnetic wave. In this case, an electron is said to scatter X-rays. The scattered beam has the exactly

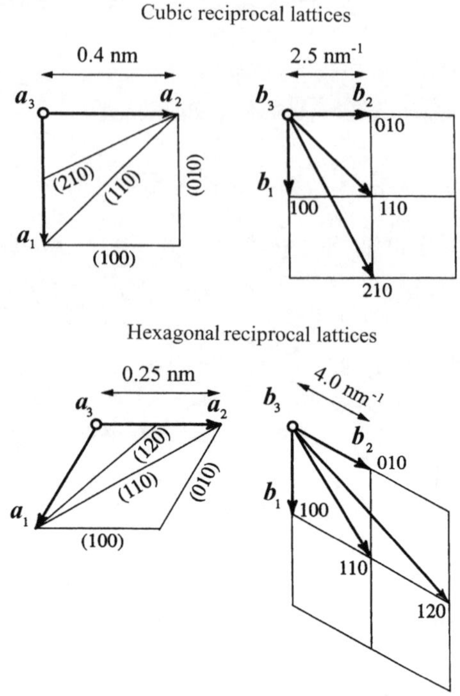

Fig. 5.1 The relationships between the reciprocal lattice and the real crystal lattice in two cases. A cubic crystal with lattice parameter $a_1 = 0.4\,\text{nm}$ (4 Å) and a hexagonal crystal with lattice parameter $a_1 = 0.25\,\text{nm}$ (2.5 Å). The axes of $\boldsymbol{a}_3$ and $\boldsymbol{b}_3$ in both cases are normal to the drawing

same wavelength and frequency as the incident X-ray beam. This is called coherent scattering or unmodified scattering.

On the contrary, the interaction of the X-ray beam with electrons may also involve the exchange of energy and momentum. This is frequently represented by an elastic collision like that of two billiard balls. In this process, a loosely bound electron is knocked out by an incident X-ray photon (or quantum), and some of the energy of the incident X-ray photon is consumed by giving kinetic energy to the electron. Such scattering is called incoherent scattering, modified scattering, or Compton scattering.

In addition, there is another way in which the X-ray beam interacts with an electron. When the incident X-ray photon has sufficient energy to knock out an inner shell electron to produce a photo electron that is emitted, the atom will be left in the excited state with a hole in the inner electron shell. This is called photoelectron effect by X-rays. Usually, the resultant hole is quickly filled by an electron located in the outer electron shell, and we obtain an X-ray photon with an energy equal to the difference in the relevant electron energy levels. This corresponds to the production process for characteristic radiation.

Taking these interactions between X-rays and electron into account, some fundamental points for the scattering intensity from an atom and a crystal are summarized below.

When an unpolarized X-ray beam, such as that irradiated from an X-ray tube, encounters a single electron located at the origin, the scattering intensity I_e at the distance R from the origin is given by the following equation, which is frequently called Thomson equation.

$$I_e = I_0 \frac{e^4}{m^2 c^4 R^2} \left(\frac{1 + \cos^2 2\theta}{2} \right) \tag{5.9}$$

where, I_0 is intensity of the incident X-ray beam, e is an elementary charge, m is a rest mass of electron, c is the speed of light in vacuum, and 2θ is the scattering angle. The last term within parentheses in (5.9) is called polarization factor, and an alternative expression of this factor is needed when using a crystal monochromator. The constant term $(e^2/mc^2)^2$ of (5.9) is equivalent to the square of the classical electron radius r_e given as $(2.8179 \times 10^{-15})^2\ \mathrm{m}^2$ in SI units.

The amplitude, f, of the coherent scattering per atom including more than one electron is given by the sum of the amplitude f_{ej} per one electron (refer to Chap. 3), and it is as follows.

$$f = \sum_j f_{ej} = \sum_j \int_0^\infty 4\pi r^2 \rho_j(r) \frac{\sin Qr}{Qr} \mathrm{d}r \tag{5.10}$$

The atomic scattering factor f is a function of Q or $(\sin\theta/\lambda)$ and is equivalent to the efficiency of scattering of a given atom in a given electron. Since the number of electrons in the atom is equal to the atomic number Z, we obtain $\sum_j \int_0^\infty 4\pi r^2 \rho_j(r)\mathrm{d}r = Z$. Then, it is clear that f is close to the atomic number Z for any atom if the value of Q or $(\sin\theta/\lambda)$ becomes very small (scattering in the forward direction).

The atomic scattering factors mentioned here satisfy the following two assumptions:

(1) The distribution of electrons around the nucleus in atom can be well approximated by spherical symmetry.
(2) The wavelength of the incident X-ray beam is much shorter than, or far from, that of an absorption edge of the scattering atom.

There are not many cases, but apparent deviation from the perfect spherical symmetry is found in the electron distribution around the nucleus in a carbon atom with diamond structure, and this is one example where the assumption of (1) is not satisfied. The wavelength of the incident X-ray beam is close to the absorption edge of the scattering atom, the assumption of (2) is no longer accepted, and a correction of the atomic scattering factor in the following complex form is required.

$$f = f_0 + f' + \mathrm{i}f'' \tag{5.11}$$

where f' and f'' are the real and imaginary components of the anomalous dispersion, and both components depend on the incident X-ray energy (wavelength). f_0 in

(5.11) corresponds to the normal atomic scattering factor whose numerical values are given in Table (see Appendix A.3). On the contrary, f' and f'' are related only to inner-shell electrons, such as K or L. Since the spatial distribution of inner electrons is considerably smaller than the magnitude of the X-ray wavelength, the dipole approximation [$\exp(-\mathrm{i}\boldsymbol{Q}\cdot\boldsymbol{r}) \fallingdotseq 1$] is well accepted. Thus, the Q-dependence of the anomalous dispersion components can usually be ignored. Normal X-ray diffraction measurements are carried out using wavelengths away from the absorption edge region without correction of the anomalous dispersion effect.

The main mechanism of X-ray diffraction by crystals is coherent scattering, but incoherent scattering also occurs. The incoherent scattering intensity $i(M)$ per atom can be described by the simple sum of the incoherent scattering intensity of respective electrons as follows (refer to Chap. 3.2):

$$i(M) = \sum_j i_{ej} = Z - \sum_{j=1}^{Z} f_{ej}^2 \tag{5.12}$$

5.3 Intensity from Scattering by a Small Crystal

When setting the origin of a certain unit cell in a crystal to a position O, a position vector of another unit cell from the origin of O may be expressed in the form $m_1\boldsymbol{a}_1 + m_2\boldsymbol{a}_2 + m_3\boldsymbol{a}_3$. Here, $\boldsymbol{a}_1$, $\boldsymbol{a}_2$, and $\boldsymbol{a}_3$ are the basis vectors of the unit cell, whereas m_1, m_2, and m_3 are integers and the coordinates. When setting the position vector of j-atom with respect to the origin in a unit cell to the vector $\boldsymbol{r}_j$, the positions of j-atoms in the crystal can be described by $\boldsymbol{R}_{mj} = m_1\boldsymbol{a}_1 + m_2\boldsymbol{a}_2 + m_3\boldsymbol{a}_3 + \boldsymbol{r}_j$.

Let us consider the case where the monochromatic X-ray beam with the wavelength λ and its intensity denoted by I_0 encounters a very small single crystal. The incident and diffracted X-rays are usually represented by the wave vectors $\boldsymbol{s}_0$ and $\boldsymbol{s}$, respectively, and $\boldsymbol{R}_{mj}$, $\boldsymbol{s}_0$, and $\boldsymbol{s}$ are, in general, not coplanar ($|\boldsymbol{s}_0|$ and $|\boldsymbol{s}| = 1/\lambda$ where λ is wavelength). This crystal is assumed to be so small, relative to the distance between the X-ray source and the corresponding crystal, and therefore, the incident X-ray beam can be treated by the plane wave approximation. Suppose the plane waves are passing through the origin O and are scattered by j-atoms located at the position vectors $\boldsymbol{R}_{mj}$ and we wish to detect the scattered intensity I from a j-atom at P at a distance R from the origin O, I is given as follows.

$$\begin{aligned} I = I_e FF^* &\frac{\sin^2\left\{\left(\frac{\pi}{\lambda}\right)(\boldsymbol{s}-\boldsymbol{s}_0)\cdot N_1\boldsymbol{a}_1\right\}}{\sin^2\left\{\left(\frac{\pi}{\lambda}\right)(\boldsymbol{s}-\boldsymbol{s}_0)\cdot \boldsymbol{a}_1\right\}} \cdot \\ &\frac{\sin^2\left\{\left(\frac{\pi}{\lambda}\right)(\boldsymbol{s}-\boldsymbol{s}_0)\cdot N_2\boldsymbol{a}_2\right\}}{\sin^2\left\{\left(\frac{\pi}{\lambda}\right)(\boldsymbol{s}-\boldsymbol{s}_0)\cdot \boldsymbol{a}_2\right\}} \cdot \frac{\sin^2\left\{\left(\frac{\pi}{\lambda}\right)(\boldsymbol{s}-\boldsymbol{s}_0)\cdot N_3\boldsymbol{a}_3\right\}}{\sin^2\left\{\left(\frac{\pi}{\lambda}\right)(\boldsymbol{s}-\boldsymbol{s}_0)\cdot \boldsymbol{a}_3\right\}} \end{aligned} \tag{5.13}$$

$$I_e = I_0\left(\frac{e^2}{mc^2R}\right)^2 \tag{5.14}$$

Here, F is a structure factor. The scattering intensity by a single electron (Thomson's scattering equation (5.9)) is also used here. In addition, since (5.13) is obtained under the condition that the incident X-ray beam is polarized vertically to the plane of the drawing, the polarization factor becomes unity. When the incident X-ray beam is not polarized at all, it is necessary to use the polarization factor described by $(1 + \cos^2 2\theta)/2$.

It is also noted that a function of $(\sin^2(Nk)/\sin^2 k)$ that appears in (5.13) is called "Laue function." When k is integer multiple (n) of π, both denominator and numerator of the Laue function are zero. This condition of $k = n\pi$ (n: integer) is equivalent to the so-called "diffraction by a crystal" whose lattice is characterized by the atomic array in a line at the regular interval, and then the phase difference of waves generated from such structure is given by integer multiples of the wavelength. It may also be added that if $k \to n\pi$, the value of the limit of the Laue function is given by N^2. Accordingly, when the scattering intensity is represented by the Laue function normalized by N^2, a peak at the position of $k = n\pi$ is found to be sharper with increasing n. Since n is considerably large in a crystal lattice, a peak is usually approximated by the delta function at the position of $k = n\pi$. On the contrary, let us consider that the value of a changes in the form of $k = (s-s_0)\cdot a$, at constant n. If the absolute value, $|a|$ becomes triple, the frequency of Laue function shows the reverse, as 1/3. That is, the length of the real-space lattice can be described as 1/(length) in the reciprocal space lattice, and their relationships are *vice versa* (see Question 5.8).

5.4 Integrated Intensity from Small Single Crystals

Since the incident X-ray beam is usually not completely parallel and strictly a monochromatic radiation, this produces variation in its direction vector s_0. For example, the usual monochromatic beam is simply said to be one strong Kα component, but it has a width of about 0.0004 nm. For this reason, we have to check the effect on diffraction phenomena by a crystal by considering some deviation from the ideal case. As already shown in Chap. 4.7 (2), the so-called destructive interference is not perfect so that we will detect a certain diffracted intensity in the close vicinity of the Bragg angle region near $2\theta_B$. This is particularly true when the number of planes completely satisfying the Bragg condition is not sufficient, for example, by reducing the particle size of a crystalline powder sample, even one consisting of perfect single crystals. Accordingly a measured diffracted peak has some width in 2θ. In other words, if the crystal is perfect, we have to consider the small size alone a crystal imperfection.

A single crystal is often found to have a periodic array of defects such as dislocation if its structure is examined in detail. Such a single crystal is said to have a kind of crystal imperfection that affects, more or less, the diffraction phenomena. This is a kind of substructure into which a single crystal is broken up as shown in the schematic diagram in Fig. 5.2, and this is called "mosaic structure." In other words, a crystal with mosaic structure does not have its atomic arrangement with a

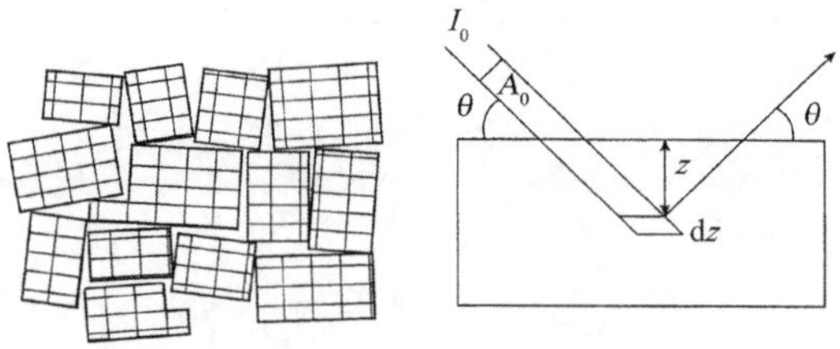

Fig. 5.2 Schematic diagram of mosaic structure of a real crystal and geometric arrangement for calculating the integrated intensity from a crystal plate with mosaic structure

completely regular crystal lattice extending from one side of the crystal to the other side. The crystal lattice is rather broken up into a number of tiny crystal blocks in the order of 100 nm in length, and each block is slightly disorientated from one another disturbing the coherency in their crystal planes. These tiny blocks are quite likely to be identical with subgrains, and the region between the blocks are considered the dislocation wall. In addition, the maximum angle of disorientation between tiny crystal blocks changes from a very small value to as much as 1°. In this case, the incident X-ray beam changes its direction, little by little, at each mosaic crystal structure so as to form a relatively sharp diffraction peak with a certain range in width. As a result, the integrated intensity of the diffracted beam from a crystal with mosaic structure increases, relative to that computed theoretically for an ideally perfect single crystal. As shown in the schematic diagram in Fig. 5.3, a large number of atoms arranged with a perfectly three-dimensional periodic array to form a crystal scatter X-rays in relatively few directions because the structural periodicity causes the so-called destructive interference of the scattered X-rays in all directions except only those predicted by the Bragg law in which constructive interference takes place. It may be added that a single atom scatters the incident X-ray beam in all directions in space.

Although the scattering intensity from a small single crystal is expressed by (5.13), we have to use the integrated intensity for comparing with the measured intensity data. Here, let us consider the scattering intensity in the angular region between a slightly smaller angle near $2\theta_B$ and a slightly larger angle when rotating the crystal with its angular speed of ω. Here, $2\theta_B$ is the Bragg angle. The integrated intensity P from the $(h\,k\,l)$ plane may be given by the following equation.

$$P = \frac{I_0}{\omega}\left(\frac{e^4}{m^2c^4}\right)\frac{\lambda N_{\mathrm{uc}} F_{hkl}^2}{v_a}\left(\frac{1+\cos^2 2\theta}{2\sin 2\theta}\right) \tag{5.15}$$

where v_a is the volume of a unit cell and N_{uc} is the number of unit cell included in a crystal. The factor given in the second parenthesis to the right-hand side of (5.15) is called the Lorentz polarization factor with respect to the measurement for a single crystal using an unpolarized incident X-ray beam. Equation (5.15) clearly suggests that the experimental results can provide the structure factor F_{hkl}, from which the unknown crystal structure is able to be determined if the integrated intensity P

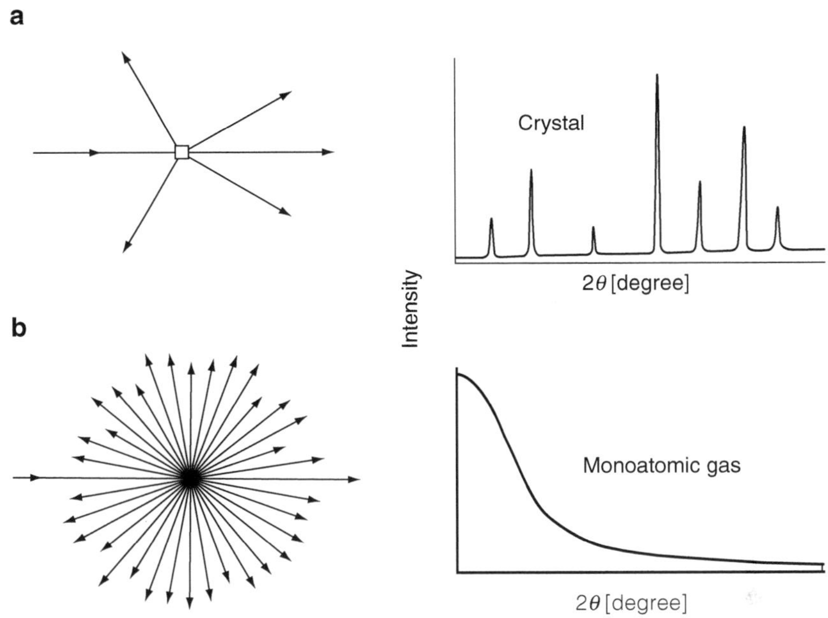

Fig. 5.3 Schematic diagrams for scattering by a crystal (**a**) and by a single atom (**b**) together with their scattering patterns for comparison. The vertical scales are not equal

from a (hkl) plane and the intensity I_0 of the incident X-ray beam can be quantitatively obtained. If the F_{hkl} value for a pure substance with simple structure can be estimated from the measured intensity data with sufficient reliability, the atomic scattering factor f may also be computed.

5.5 Integrated Intensity from Mosaic Crystals and Polycrystalline Samples

The integrated intensity of a single crystal with mosaic structure can be represented as follows. Each mosaic block in the single crystal is considered a perfect crystal, but a slight difference in direction of adjoining mosaic blocks is recognized so as not to interfere with each other. This is called "ideal mosaic single crystal." A real single crystal is usually described by the intermediate state between perfect single crystal and ideal single mosaic crystal. If a desired crystal is regarded as an ideal mosaic single crystal, the integrated intensity can be computed by assuming that (5.15) is allowed to be used in each mosaic block.

Let us consider the intensity of the incident X-ray beam I_0 and its cross-section A_0 and that the average volume of mosaic blocks is set to δv (see Fig. 5.2). If the intensity I_0 of the incident X-ray beam reaches the mosaic block located in a portion between z and $z + \mathrm{d}z$ deep from the crystal surface, its intensity will be given by the form of $I_0 \exp(-2z\mu/\sin\theta)$, where μ is the linear absorption coefficient for a substance of mosaic block. Considering the intensity $P_0 = I_0 A_0$ of the incident X-ray beam, the integrated intensity from an ideal mosaic single crystal can be given by the following equation.

$$P = \frac{P_0}{\omega}\left(\frac{e^4}{m^2c^4}\right)\frac{\lambda^3 F_{hkl}^2}{2\mu v_a^2}\left(\frac{1+\cos^2 2\theta}{2\sin 2\theta}\right) \tag{5.16}$$

A crystalline powder sample is approximated by a very small poly-crystalline aggregate, and each grain is found to have a crystallographic orientation different from that of its neighbors. When considered as a whole, the orientations of all the grains are randomly distributed. Let us consider that a monochromatic X-ray beam with the wavelength λ hits a powder crystalline sample where a tiny single crystal in a small part of this sample shows its direction so as to satisfy the Bragg law with respect to the $(h\,k\,l)$ plane. In this case, the plane-normal vector $\boldsymbol{H}_{hkl}$ of the $(h\,k\,l)$ plane of a tiny single crystal coincides with the direction of the scattering vector $(\boldsymbol{s} - \boldsymbol{s}_0)$, which is defined by the vector $\boldsymbol{s}_0$ of incident X-ray beam and the vector $\boldsymbol{s}$ of diffracted X-ray beam. It is also noteworthy that the concepts of the reciprocal lattice and the Ewald sphere as well as the limiting sphere are very convenient for understanding the relationships between the incident and diffracted beam vectors and crystal plane. The details of such information are available in Question 5.7.

When the total number of small single crystals in a powder sample is set to N_{sc} and multiplicity factor p_{hkl}, which indicates the number of the equivalent $(h\,k\,l)$ plane with the same spacing and structure factor, the total number of the plane-normal vector $\boldsymbol{H}_{hkl}$ in the powder sample is given by $N_{\mathrm{sc}} p_{hkl}$. Here, we take into account the distribution of the scattering vector from a powder crystalline sample satisfying the Bragg law, if the incident angle is in the region between $(\theta + \alpha)$ and $(\theta + \alpha + \mathrm{d}\alpha)$. Then, the diffracted intensity from the $(h\,k\,l)$ plane can be obtained by integrating all directions and the cross-sectional area followed by multiplying the scattered intensity I of one tiny single crystal given by (5.13), the number of crystals $\mathrm{d}N_{\mathrm{sc}}$ in $\mathrm{d}\alpha$ of a sample and the area element $\mathrm{d}A$ in the cross-section.

The intensity P computed in this procedure is equivalent to the total diffraction intensity uniformly distributed on the Debye ring so as to form a cone of half apex angle 2θ, as is seen in Fig. 5.4. Accordingly, the intensity that can be measured using a diffractometer corresponds to the intensity per unit length of this Debye ring. That is, it is equivalent to the intensity P', which is divided P by the circumference of the Debye ring $(2\pi R \sin 2\theta)$. Therefore, when the unpolarized X-ray beam with its intensity of I_0 is employed, the integrated intensity measured at a distance R from the crystalline powder sample is given by the following equation:

$$P' = \frac{I_0}{16\pi R}\left(\frac{e^4}{m^2c^4}\right)\frac{V\lambda^3 m_{hkl} F_{hkl}^2}{v_a^2}\left(\frac{1+\cos^2 2\theta}{\sin\theta \sin 2\theta}\right) \tag{5.17}$$

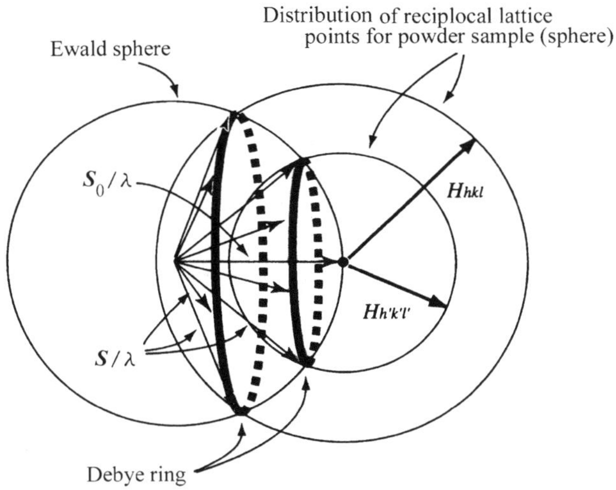

Fig. 5.4 The relationships between the reciprocal lattice and the Ewald sphere (formation of a cone of diffracted X-rays) in the Debye–Scherrer method

where V is the irradiated volume. It is worth mentioning that the last term to the right-hand side of (5.17) is the Lorentz polarization factor for a crystalline powder sample.

Equation (5.17) shows that the integrated intensity for the $(h\,k\,l)$ plane of the crystalline powder sample is proportional to three factors: multiplicity factor p_{hkl}, structure factor F_{hkl}, and Lorentz polarization factor. The most common X-ray diffraction measurement for crystalline powder samples is known to be the method using a diffractometer. In this case, the conditions are adjusted so as to satisfy the sample being a flat plate with infinite thickness. In this experimental condition, the irradiated volume of (5.17) may be replaced by $V = A_0/2\mu$.

5.6 Solved Problems (18 Examples)

Question 5.1 The reciprocal lattice corresponding to a unit cell described by the primitive crystal-lattice vectors $\boldsymbol{a}_1$, $\boldsymbol{a}_2$, and $\boldsymbol{a}_3$ has a unit cell defined by the vectors $\boldsymbol{b}_1$, $\boldsymbol{b}_2$, and $\boldsymbol{b}_3$ given by the following equations, when the volume of the crystal unit cell is set to V.

$$\boldsymbol{b}_1 = \frac{\boldsymbol{a}_2 \times \boldsymbol{a}_3}{V}, \quad \boldsymbol{b}_2 = \frac{\boldsymbol{a}_3 \times \boldsymbol{a}_1}{V}, \quad \boldsymbol{b}_3 = \frac{\boldsymbol{a}_1 \times \boldsymbol{a}_2}{V}$$

This corresponds to the definition of the reciprocal lattice as a function of the crystal lattice. Show the crystal lattice as a function of the reciprocal lattice.

Answer 5.1 Recall a few fundamentals of vector algebra.

(1) Product of two vectors A and B (it is called cross product or vector product)

$$(\boldsymbol{A} \times \boldsymbol{B}) = -(\boldsymbol{B} \times \boldsymbol{A})$$

(2) Product of three vectors A, B, and C (it is called triple-scalar product)

$$\begin{aligned}\boldsymbol{A} \cdot (\boldsymbol{B} \times \boldsymbol{C}) &= \boldsymbol{B} \cdot (\boldsymbol{C} \times \boldsymbol{A}) = \boldsymbol{C} \cdot (\boldsymbol{A} \times \boldsymbol{B}) \\ &= -\boldsymbol{A} \cdot (\boldsymbol{C} \times \boldsymbol{B}) = -\boldsymbol{B} \cdot (\boldsymbol{A} \times \boldsymbol{C}) = -\boldsymbol{C} \cdot (\boldsymbol{B} \times \boldsymbol{A})\end{aligned}$$

When three vectors $\boldsymbol{A}$, $\boldsymbol{B}$, and $\boldsymbol{C}$ correspond to the sides of parallelepiped as shown in Fig. 1, $(\boldsymbol{B} \times \boldsymbol{C})$ is an equivalent vector to the area of parallelogram that constitutes the base of parallelepiped. Note that the vector product of two vectors $\boldsymbol{A}$ and $\boldsymbol{B}$ written by $\boldsymbol{A} \times \boldsymbol{B}$ is corresponding to a vector $\boldsymbol{C}$ normal to the plane of $\boldsymbol{A}$ and $\boldsymbol{B}$. The magnitude of $\boldsymbol{C}$ is equal to the area of the parallelogram constituted of $\boldsymbol{A}$ and $\boldsymbol{B}$, whereas the direction of $\boldsymbol{C}$ is along the direction of movement forward of the right-hand screw if rotated in such a way so as to bring $\boldsymbol{A}$ to $\boldsymbol{B}$. Therefore, the direction is found to be perpendicular to the plane made by $\boldsymbol{B}$ and $\boldsymbol{C}$. Consequently, the triple-scalar product of three vectors, $\boldsymbol{A} \cdot (\boldsymbol{B} \times \boldsymbol{C})$, may be described by the projection of the vector $\boldsymbol{A}$ onto $(\boldsymbol{B} \times \boldsymbol{C})$ This is equivalent to the volume V of parallelepiped.

Reference: The scalar product of two vectors $\boldsymbol{A}$ and $\boldsymbol{B}$ or the inner product (or dot product) is expressed as $\boldsymbol{A} \cdot \boldsymbol{B}$. It is also noted that $\boldsymbol{A} \cdot \boldsymbol{B} = \boldsymbol{B} \cdot \boldsymbol{A}$ due to a scalar quantity.

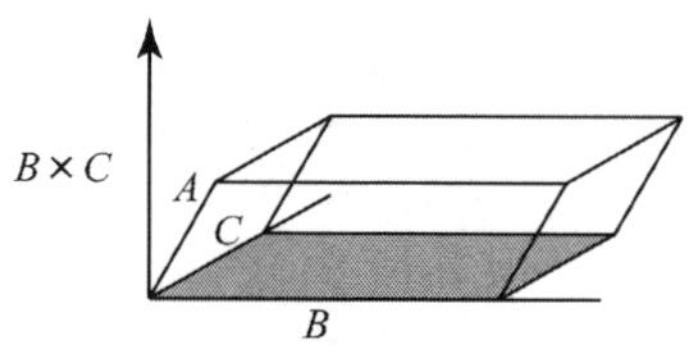

Fig. 1 Graphical representation of $\boldsymbol{A} \cdot (\boldsymbol{B} \times \boldsymbol{C})$

Next, obtain the cross product of the reciprocal-lattice vectors $\boldsymbol{b}_2$ and $\boldsymbol{b}_3$, i.e., $(\boldsymbol{b}_2 \times \boldsymbol{b}_3)$

$$\boldsymbol{b}_2 \times \boldsymbol{b}_3 = \frac{\boldsymbol{a}_3 \times \boldsymbol{a}_1}{V} \times \frac{\boldsymbol{a}_1 \times \boldsymbol{a}_2}{V} = \frac{1}{V^2}\{(\boldsymbol{a}_3 \times \boldsymbol{a}_1) \times (\boldsymbol{a}_1 \times \boldsymbol{a}_2)\} \tag{1}$$

When $\boldsymbol{u} = (\boldsymbol{a}_3 \times \boldsymbol{a}_1)$ is set and the known formula $\boldsymbol{u} \times (\boldsymbol{v} \times \boldsymbol{w}) = \boldsymbol{v}(\boldsymbol{u} \cdot \boldsymbol{w}) - \boldsymbol{w}(\boldsymbol{u} \cdot \boldsymbol{v})$ for vector products is simultaneously employed, the following equations may be readily obtained.

$$\begin{aligned}\boldsymbol{u} \times (\boldsymbol{a}_1 \times \boldsymbol{a}_2) &= \boldsymbol{a}_1(\boldsymbol{u} \cdot \boldsymbol{a}_2) - \boldsymbol{a}_2(\boldsymbol{u} \cdot \boldsymbol{a}_1) \\ &= \boldsymbol{a}_1\{(\boldsymbol{a}_3 \times \boldsymbol{a}_1) \cdot \boldsymbol{a}_2\} - \boldsymbol{a}_2\{(\boldsymbol{a}_3 \times \boldsymbol{a}_1) \cdot \boldsymbol{a}_1\}\end{aligned} \tag{2}$$

The second term of (2) contains the same vector $\boldsymbol{a}_1$ twice so that it vanishes as one of the characteristic features of vectors includes $\boldsymbol{u} \cdot (\boldsymbol{u} \times \boldsymbol{w}) = 0$. In addition, if the relationship of $(\boldsymbol{a}_3 \times \boldsymbol{a}_1) \cdot \boldsymbol{a}_2 = V$ is taken into account, (1) may be rearranged in a simplest form as follows.

$$\boldsymbol{b}_2 \times \boldsymbol{b}_3 = \frac{\boldsymbol{a}_1}{V} \tag{3}$$

On the contrary, the volume V^* of reciprocal lattices is given by the following equation.

$$V^* = \boldsymbol{b}_1 \cdot (\boldsymbol{b}_2 \times \boldsymbol{b}_3) = \frac{(\boldsymbol{a}_2 \times \boldsymbol{a}_3) \cdot \boldsymbol{a}_1}{V^2} = \frac{1}{V} \tag{4}$$

Therefore, the following relationship can be obtained by combining (3) and (4).

$$\boldsymbol{a}_1 = \frac{\boldsymbol{b}_2 \times \boldsymbol{b}_3}{V^*}, \quad \boldsymbol{a}_2 = \frac{\boldsymbol{b}_3 \times \boldsymbol{b}_1}{V^*}, \quad \boldsymbol{b}_3 = \frac{\boldsymbol{b}_1 \times \boldsymbol{b}_2}{V^*} \tag{5}$$

When comparing the given equation in question with (5), it turns out that similarity between two formulas is well recognized, and only right-hand side and left-hand side are changed. Namely, the reverse of the reciprocal-lattice vectors is equal to the crystal-lattice vectors, and the reciprocal of the volume of the crystal lattice is equivalent to the volume of the reciprocal lattice.

Question 5.2 Demonstrate that the reciprocal-lattice vector $\boldsymbol{H}_{hkl} = h\boldsymbol{b}_1 + k\boldsymbol{b}_2 + l\boldsymbol{b}_3$ is perpendicular to the (hkl) plane, regardless of crystal systems, and that the magnitude of this reciprocal-lattice vector is equal to the reciprocal of the spacing (d_{hkl}).

Answer 5.2 The relationship between the crystal-lattice vectors $\boldsymbol{a}_1$, $\boldsymbol{a}_2$, and $\boldsymbol{a}_3$ with respect to the (*hkl*) plane and the reciprocal lattice $\boldsymbol{H}_{hkl}$ is shown in Fig. 1 from which the target is given in the following. If the scalar product (inner product) of two vectors included in the (*hkl*) plane and the reciprocal-lattice vector is obtained, and its value is found to be zero and then the two kinds of vectors are said to show the perpendicular relationship. In Fig. 1, two non parallel vectors in the (*hkl*) plane are found, $\left(\frac{\boldsymbol{a}_1}{h} - \frac{\boldsymbol{a}_2}{k}\right)$ and $\left(\frac{\boldsymbol{a}_2}{k} - \frac{\boldsymbol{a}_3}{l}\right)$.

$$\begin{aligned} \boldsymbol{H}_{hkl} \cdot \left(\frac{\boldsymbol{a}_1}{h} - \frac{\boldsymbol{a}_2}{k}\right) &= (h\boldsymbol{b}_1 + k\boldsymbol{b}_2 + l\boldsymbol{b}_3) \cdot \left(\frac{\boldsymbol{a}_1}{h} - \frac{\boldsymbol{a}_2}{k}\right) \\ &= \left(h\boldsymbol{b}_1 \cdot \frac{\boldsymbol{a}_1}{h} + 0 + 0\right) - \left(0 + k\boldsymbol{b}_2 \cdot \frac{\boldsymbol{a}_2}{k} + 0\right) = 1 - 1 = 0 \end{aligned} \tag{1}$$

Here, we use the characteristic properties of real and reciprocal-lattice vectors, such as $\boldsymbol{b}_j \cdot \boldsymbol{a}_k = 1$ if $j = k$ and $\boldsymbol{b}_j \cdot \boldsymbol{a}_k = 0$ if $j \neq k$. In other words, this is based on the useful relationship that since $\boldsymbol{b}_3$, for example, is perpendicular to both $\boldsymbol{a}_1$ and $\boldsymbol{a}_2$, the scalar product with either one of the crystal and reciprocal vectors is zero. The

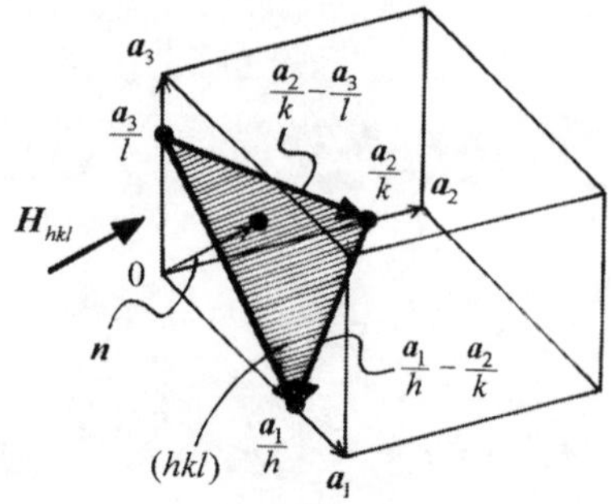

Fig. 1 The relation of vector and unit vector n of normal direction that is in $(h\ k\ l)$ plane

same result is readily confirmed in the combination of $\boldsymbol{H}_{hkl} \cdot \left(\frac{a_1}{h} - \frac{a_2}{k}\right)$. Therefore, $\boldsymbol{H}_{hkl}$ is said to be perpendicular to the $(h\,k\,l)$ plane.

If the unit vector on the normal that goes to the $(h\,k\,l)$ plane is set as $\boldsymbol{n}$, the following equation is obtained.

$$\boldsymbol{n} = \frac{\boldsymbol{H}_{hkl}}{|\boldsymbol{H}_{hkl}|} \tag{2}$$

The spacing d_{hkl} corresponds to the projection to the direction of normal vector $\boldsymbol{n}$ that goes to the target plane from the origin 0 in Fig. 1. Therefore, the following relationship is proved with respect to the vector $\dfrac{\boldsymbol{a}_1}{h}$.

$$\begin{aligned} d_{hkl} &= \frac{\boldsymbol{a}_1}{h} \cdot \boldsymbol{n} = \frac{\boldsymbol{a}_1}{h} \cdot \frac{\boldsymbol{H}_{hkl}}{|\boldsymbol{H}_{hkl}|} = \frac{1}{|\boldsymbol{H}_{hkl}|} \frac{\boldsymbol{a}_1}{h} \cdot (h\boldsymbol{b}_1 + k\boldsymbol{b}_2 + l\boldsymbol{b}_3) \\ &= \frac{1}{|\boldsymbol{H}_{hkl}|} \frac{h}{h} (\boldsymbol{a}_1 \cdot \boldsymbol{b}_1 + 0 + 0) = \frac{1}{|\boldsymbol{H}_{hkl}|} \end{aligned} \tag{3}$$

Reference: Let us confirm some fundamental points for vector, area, volume, and the relevant matrices and determinants.

Using any orthogonal set of unit vectors, $\boldsymbol{e}_1$, $\boldsymbol{e}_2$, and $\boldsymbol{e}_3$, whose direction is the same as a coordinate axis, three vectors $\boldsymbol{A}$, $\boldsymbol{B}$, and $\boldsymbol{C}$ situated on a plane are expressed by the following equation.

$$\left. \begin{aligned} \boldsymbol{A} &= a_1\boldsymbol{e}_1 + a_2\boldsymbol{e}_2 + a_3\boldsymbol{e}_3 \\ \boldsymbol{B} &= b_1\boldsymbol{e}_1 + b_2\boldsymbol{e}_2 + b_3\boldsymbol{e}_3 \\ \boldsymbol{C} &= c_1\boldsymbol{e}_1 + c_2\boldsymbol{e}_2 + c_3\boldsymbol{e}_3 \end{aligned} \right\} \tag{4}$$

In this case, the vector product of $\boldsymbol{A}$ and $\boldsymbol{B}$ is given as follows.

$$\boldsymbol{A} \times \boldsymbol{B} = (a_2b_3 - a_3b_2)\boldsymbol{e}_1 + (a_3b_1 - a_1b_3)\boldsymbol{e}_2 + (a_1b_2 - a_2b_1)\boldsymbol{e}_3 \tag{5}$$

The area (S) spanned by two vectors $\boldsymbol{A}$ and $\boldsymbol{B}$ situated on a plane corresponds to the case where the coefficients a_3 and b_3 for $\boldsymbol{e}_3$ of a unit vector perpendicular to the plane of the drawing are zero. Therefore, the following relationship may be obtained from (5).

$$\boldsymbol{A} \times \boldsymbol{B} = (a_1 b_2 - a_2 b_1)\boldsymbol{e}_3 = S\boldsymbol{e}_3 \tag{6}$$

Equation (6) shows that the area of the parallelogram formed by two vectors $\boldsymbol{A}$ and $\boldsymbol{B}$ on a plane is expressed by $(a_1 b_2 - a_2 b_1)$, and if the direction of $\boldsymbol{A} \to \boldsymbol{B}$ corresponds to that of $\boldsymbol{e}_1 \to \boldsymbol{e}_2$, it becomes positive sign, and if reverse, it becomes negative sign. It is also helpful to use a determinant for expressing these relationships. For example, in the case of (6), one obtains as follows:

$$\begin{vmatrix} a_1 & a_2 \\ b_1 & b_2 \end{vmatrix} = a_1 b_2 - a_2 b_1 \tag{7}$$

On the contrary, the volume (V) may be expressed in the following relation if three vectors $\boldsymbol{A}$, $\boldsymbol{B}$, and $\boldsymbol{C}$ are in the space.

$$\boldsymbol{B} \times \boldsymbol{C} = (b_2 c_3 - b_3 c_2)\boldsymbol{e}_1 + (b_3 c_1 - b_1 c_3)\boldsymbol{e}_2 + (b_1 c_2 - b_2 c_1)\boldsymbol{e}_3 \tag{8}$$

$$\boldsymbol{A} \cdot (\boldsymbol{B} \times \boldsymbol{C}) = a_1(b_2 c_3 - b_3 c_2) + a_2(b_3 c_1 - b_1 c_3) + a_3(b_1 c_2 - b_2 c_1) = V \tag{9}$$

Here, the relationships of $\boldsymbol{e}_1 \cdot \boldsymbol{e}_1 = \boldsymbol{e}_2 \cdot \boldsymbol{e}_2 = \boldsymbol{e}_3 \cdot \boldsymbol{e}_3 = 1$ and $\boldsymbol{e}_1 \cdot \boldsymbol{e}_2 = \boldsymbol{e}_2 \cdot \boldsymbol{e}_3 = \boldsymbol{e}_3 \cdot \boldsymbol{e}_1 = 0$ for unit vectors are utilized.

$$\begin{vmatrix} a_1 & a_2 & a_3 \\ b_1 & b_2 & b_3 \\ c_1 & c_2 & c_3 \end{vmatrix} = a_1(b_2 c_3 - b_3 c_2) - a_2(b_1 c_3 - b_3 c_1) + a_3(b_1 c_2 - b_2 c_1) \tag{10}$$

$$= a_1 b_2 c_3 + a_2 b_3 c_1 + a_3 b_1 c_2 - a_1 b_3 c_2 - a_2 b_1 c_3 - a_3 b_2 c_1 \tag{11}$$

Equation (9) shows that the direction of vector $\boldsymbol{A}$, on the basis of a plane made by $\boldsymbol{A}$ and $\boldsymbol{B}$, has positive sign if the vector $\boldsymbol{A}$ is situated at the same side of a plane made by vectors $(\boldsymbol{B} \times \boldsymbol{C})$. If it is on the opposite side, the sign becomes negative. As mentioned previously, $\boldsymbol{A} \cdot (\boldsymbol{B} \times \boldsymbol{C})$ corresponds to a volume, and it is expressed by a determinant given by (10).

Question 5.3 The direction of the nodal line of two planes, (hkl) and $(h'k'l')$, which are not parallel to each other, is called a zone axis and is usually expressed as $[uvw]$.

(1) Obtain a relationship with uvw, hkl, and $h'k'l'$.
(2) Check the Weiss zone law, $hu + kv + lw = 0$, using the fact that a reciprocal-lattice vector is perpendicular to the corresponding crystal plane.

Answer 5.3

(1) If the reciprocal-lattice vectors of the (hkl) and $(h'k'l')$ planes are $\boldsymbol{H}_{hkl}$ and $\boldsymbol{H}_{h'k'l'}$, respectively, a zone axis $[uvw]$ may be expressed by the vector $\boldsymbol{Z}_{uvw}$ whose direction is parallel to the vector product of two reciprocal-lattice vectors $\boldsymbol{H}_{hkl}$ and $\boldsymbol{H}_{h'k'l'}$. Such relationships are easily seen in Fig. 1.

$$\boldsymbol{H}_{hkl} \times \boldsymbol{H}_{h'k'l'} = (h\boldsymbol{b}_1 + k\boldsymbol{b}_2 + l\boldsymbol{b}_3) \times (h'\boldsymbol{b}_1 + k'\boldsymbol{b}_2 + l'\boldsymbol{b}_3) \quad (1)$$
$$= (hk' - h'k)\boldsymbol{b}_1 \times \boldsymbol{b}_2 + (kl' - k'l)\boldsymbol{b}_2 \times \boldsymbol{b}_3 + (lh' - l'h)\boldsymbol{b}_3 \times \boldsymbol{b}_1 \quad (2)$$

Here, we use the relationship that the vector product of the same vector (example: $\boldsymbol{b}_1 \times \boldsymbol{b}_1$) is zero. In addition, when using the relationships $\boldsymbol{a}_1 = \frac{\boldsymbol{b}_2 \times \boldsymbol{b}_3}{V^*}$, $\boldsymbol{a}_2 = \frac{\boldsymbol{b}_3 \times \boldsymbol{b}_1}{V^*}$, $\boldsymbol{a}_3 = \frac{\boldsymbol{b}_1 \times \boldsymbol{b}_2}{V^*}$, and $V^* = \frac{1}{V}$, (2) can be rewritten as follows.

$$\boldsymbol{H}_{hkl} \times \boldsymbol{H}_{h'k'l'} = (hk' - h'k)\frac{\boldsymbol{a}_3}{V} + (kl' - k'l)\frac{\boldsymbol{a}_1}{V} + (lh' - l'h)\frac{\boldsymbol{a}_2}{V} \quad (3)$$
$$= \frac{1}{V}\begin{vmatrix} \boldsymbol{a}_1 & \boldsymbol{a}_2 & \boldsymbol{a}_3 \\ h & k & l \\ h' & k' & l' \end{vmatrix} \quad (4)$$

The vector $\boldsymbol{Z}_{uvw}$ may be expressed in the following form, when the unit cell is defined by $\boldsymbol{a}_1, \boldsymbol{a}_2$, and $\boldsymbol{a}_3$, called the primitive crystal-lattice vectors.

$$\boldsymbol{Z}_{uvw} = u\boldsymbol{a}_1 + v\boldsymbol{a}_2 + w\boldsymbol{a}_3 \quad (5)$$

Therefore, the following relationship is obtained from (3) and (5).

$$\begin{vmatrix} u \\ v \\ w \end{vmatrix} = \begin{pmatrix} kl' - k'l \\ lh' - l'h \\ hk' - h'k \end{pmatrix}$$

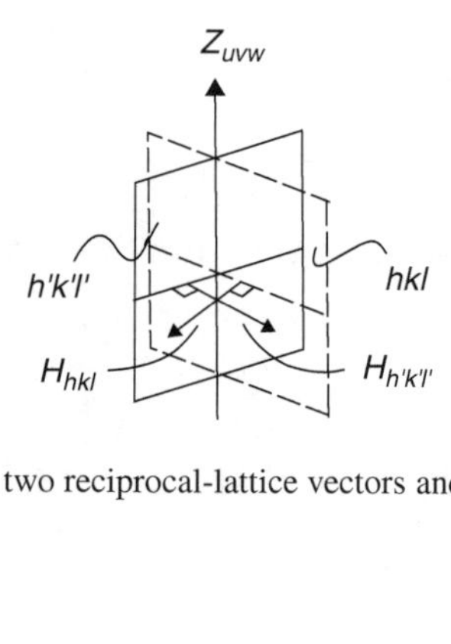

Fig. 1 The correlation between two reciprocal-lattice vectors and a zone axis

(2) On the contrary, since a normal vector of a plane is perpendicular to any vector included in that plane and if the (hkl) plane belongs to the zone axis $[uvw]$, the scalar product of the reciprocal-lattice vector $\boldsymbol{H}_{hkl}$, corresponding to the plane, and the vector $\boldsymbol{Z}_{uvw}$ showing the direction of the zone axis $[uvw]$ should be zero i.e., $\boldsymbol{H}_{hkl} \cdot \boldsymbol{Z}_{uvw} = 0$.

$$\boldsymbol{H}_{hkl} \cdot \boldsymbol{Z}_{uvw} = (h\boldsymbol{b}_1 + k\boldsymbol{b}_2 + l\boldsymbol{b}_3) \cdot (u\boldsymbol{a}_1 + v\boldsymbol{a}_2 + w\boldsymbol{a}_3) \tag{6}$$

$$\begin{aligned} &= hu\boldsymbol{b}_1 \cdot \boldsymbol{a}_1 + hv\boldsymbol{b}_1 \cdot \boldsymbol{a}_2 + hw\boldsymbol{b}_1 \cdot \boldsymbol{a}_3 \\ &\quad + ku\boldsymbol{b}_2 \cdot \boldsymbol{a}_1 + kv\boldsymbol{b}_2 \cdot \boldsymbol{a}_2 + kw\boldsymbol{b}_2 \cdot \boldsymbol{a}_3 \\ &\quad + lu\boldsymbol{b}_3 \cdot \boldsymbol{a}_1 + lv\boldsymbol{b}_3 \cdot \boldsymbol{a}_2 + lw\boldsymbol{b}_3 \cdot \boldsymbol{a}_3 \end{aligned} \tag{7}$$

From the definition of the reciprocal-lattice vectors $\boldsymbol{b}_1$, $\boldsymbol{b}_2$, and $\boldsymbol{b}_3$, the primitive crystal-lattice vectors $\boldsymbol{a}_1$, $\boldsymbol{a}_2$, and $\boldsymbol{a}_3$, the following useful relationships are obtained, since $\boldsymbol{b}_1$, for example, is known to be perpendicular to both the vectors $\boldsymbol{a}_2$ and $\boldsymbol{a}_3$.

$$\left.\begin{aligned} \boldsymbol{b}_j \cdot \boldsymbol{a}_k &= 1 \quad (j = k) \\ \boldsymbol{b}_j \cdot \boldsymbol{a}_k &= 0 \quad (j \neq k) \end{aligned}\right\} \tag{8}$$

By applying these relationships to $\boldsymbol{H}_{hkl} \cdot \boldsymbol{Z}_{uvw}$, the Weiss zone law may be readily obtained.

$$hu + kv + lw = 0 \tag{9}$$

Question 5.4 When two crystallographic directions are expressed by two vectors, $\boldsymbol{A}_u = u_1\boldsymbol{a}_1 + u_2\boldsymbol{a}_2 + u_3\boldsymbol{a}_3$ and $\boldsymbol{A}_v = v_1\boldsymbol{a}_1 + v_2\boldsymbol{a}_2 + v_3\boldsymbol{a}_3$, the angle ϕ formed by two vectors $\boldsymbol{A}_u$ and $\boldsymbol{A}_v$ may be given by $\cos\phi = \frac{\boldsymbol{A}_u \cdot \boldsymbol{A}_v}{|\boldsymbol{A}_u| \cdot |\boldsymbol{A}_v|}$.

Obtain an equation for providing $\cos\phi$ in two cases, (1) orthorhombic system and (2) cubic system.

Answer 5.4

(1) Orthorhombic system $|\boldsymbol{a}_1| \neq |\boldsymbol{a}_2| \neq |\boldsymbol{a}_3| \qquad \boldsymbol{a}_1 \perp \boldsymbol{a}_2 \perp \boldsymbol{a}_3$

$$|\boldsymbol{A}_u| = \sqrt{\boldsymbol{A}_u \cdot \boldsymbol{A}_u} = \sqrt{|u_1\boldsymbol{a}_1|^2 + |u_2\boldsymbol{a}_2|^2 + |u_3\boldsymbol{a}_3|^2}$$

$$|\boldsymbol{A}_v| = \sqrt{\boldsymbol{A}_v \cdot \boldsymbol{A}_v} = \sqrt{|v_1\boldsymbol{a}_1|^2 + |v_2\boldsymbol{a}_2|^2 + |v_3\boldsymbol{a}_3|^2}$$

$$\boldsymbol{A}_u \cdot \boldsymbol{A}_v = u_1v_1|\boldsymbol{a}_1|^2 + u_2v_2|\boldsymbol{a}_2|^2 + u_3v_3|\boldsymbol{a}_3|^2$$

$$\cos\phi = \frac{u_1v_1|\boldsymbol{a}_1|^2 + u_2v_2|\boldsymbol{a}_2|^2 + u_3v_3|\boldsymbol{a}_3|^2}{\sqrt{|u_1\boldsymbol{a}_1|^2 + |u_2\boldsymbol{a}_2|^2 + |u_3\boldsymbol{a}_3|^2}\sqrt{|v_1\boldsymbol{a}_1|^2 + |v_2\boldsymbol{a}_2|^2 + |v_3\boldsymbol{a}_3|^2}}$$

(2) Cubic systems $|\boldsymbol{a}_1| = |\boldsymbol{a}_2| = |\boldsymbol{a}_3| = a \qquad \boldsymbol{a}_1 \perp \boldsymbol{a}_2 \perp \boldsymbol{a}_3$

$$|\boldsymbol{A}_u| = \sqrt{u_1^2 + u_2^2 + u_3^2} \times a \qquad |\boldsymbol{A}_v| = \sqrt{v_1^2 + v_2^2 + v_3^2} \times a$$

$$\boldsymbol{A}_u \cdot \boldsymbol{A}_v = (u_1 v_1 + u_2 v_2 + u_3 v_3) a^2$$
$$\cos\phi = \frac{u_1 v_1 + u_2 v_2 + u_3 v_3}{\sqrt{u_1^2 + u_2^2 + u_3^2}\sqrt{v_1^2 + v_2^2 + v_3^2}}$$

Question 5.5 If the orthogonal vectors of unit length are set to $\boldsymbol{e}_x$, $\boldsymbol{e}_y$ and $\boldsymbol{e}_z$, the primitive translation vectors of the hexagonal close-packed (hcp) structure may be given as follows.

$$\boldsymbol{a}_1 = \frac{\sqrt{3}}{2} a\boldsymbol{e}_x + \frac{a}{2}\boldsymbol{e}_y, \quad \boldsymbol{a}_2 = -\frac{\sqrt{3}}{2} a\boldsymbol{e}_x + \frac{a}{2}\boldsymbol{e}_y, \quad \boldsymbol{a}_3 = c\boldsymbol{e}_z$$

(1) Obtain the unit vectors of reciprocal lattices $\boldsymbol{b}_1$, $\boldsymbol{b}_2$, and $\boldsymbol{b}_3$.
(2) Show the first Brillouin zone of the hcp lattice.

Answer 5.5

(1) For example, the definition of reciprocal-lattice vector $\boldsymbol{b}_1$ is as follows.

$$\boldsymbol{b}_1 = \frac{\boldsymbol{a}_1 \times \boldsymbol{a}_3}{V} = \frac{\boldsymbol{a}_2 \times \boldsymbol{a}_3}{\boldsymbol{a}_1 \cdot (\boldsymbol{a}_2 \times \boldsymbol{a}_3)}$$

Therefore, when calculating $\boldsymbol{a}_1 \cdot (\boldsymbol{a}_2 \times \boldsymbol{a}_3)$ called the triple-scalar product using matrices, we obtain the following results.

$$\begin{aligned}
(\boldsymbol{a}_2 \times \boldsymbol{a}_3) &= \begin{vmatrix} \boldsymbol{e}_x & \boldsymbol{e}_y & \boldsymbol{e}_z \\ -\frac{\sqrt{3}}{2}a & \frac{a}{2} & 0 \\ 0 & 0 & c \end{vmatrix} \\
&= \boldsymbol{e}_x \begin{vmatrix} \frac{a}{2} & 0 \\ 0 & c \end{vmatrix} - \boldsymbol{e}_y \begin{vmatrix} -\frac{\sqrt{3}}{2}a & 0 \\ 0 & c \end{vmatrix} + \boldsymbol{e}_z \begin{vmatrix} -\frac{\sqrt{3}}{2}a & \frac{a}{2} \\ 0 & 0 \end{vmatrix} \\
&= \frac{ac}{2}\boldsymbol{e}_x + \frac{\sqrt{3}}{2}ac\boldsymbol{e}_y \qquad (1)
\end{aligned}$$

$$\begin{aligned}
\boldsymbol{a}_1 \cdot (\boldsymbol{a}_2 \times \boldsymbol{a}_3) &= \left(\frac{\sqrt{3}}{2}a\boldsymbol{e}_x + \frac{a}{2}\boldsymbol{e}_y\right) \cdot \left(\frac{ac}{2}\boldsymbol{e}_x + \frac{\sqrt{3}}{2}ac\boldsymbol{e}_y\right) \\
&= \frac{\sqrt{3}}{4}a^2 c\boldsymbol{e}_x \cdot \boldsymbol{e}_x + \frac{\sqrt{3}}{4}a^2 c\boldsymbol{e}_y \cdot \boldsymbol{e}_y \qquad (\because \boldsymbol{e}_x \cdot \boldsymbol{e}_y \text{ is zero}) \\
&= \frac{\sqrt{3}}{2}a^2 c \qquad (\because \boldsymbol{e}_x \cdot \boldsymbol{e}_x = \boldsymbol{e}_y \cdot \boldsymbol{e}_y = 1) \qquad (2)
\end{aligned}$$

The vector products of $(\boldsymbol{a}_3 \times \boldsymbol{a}_1)$ and $(\boldsymbol{a}_1 \times \boldsymbol{a}_2)$ are also calculated as follows.

$$(\boldsymbol{a}_3 \times \boldsymbol{a}_1) = -\frac{ac}{2}\boldsymbol{e}_x + \frac{\sqrt{3}}{2}ac\boldsymbol{e}_y \tag{3}$$

$$(\boldsymbol{a}_1 \times \boldsymbol{a}_2) = \frac{\sqrt{3}a^2}{2}\boldsymbol{e}_z \tag{4}$$

From (1) to (4), the unit vectors of the reciprocal lattice $\boldsymbol{b}_1$, $\boldsymbol{b}_2$, and $\boldsymbol{b}_3$ can be obtained as follows.

$$\left.\begin{aligned} \boldsymbol{b}_1 &= \frac{\boldsymbol{a}_2 \times \boldsymbol{a}_3}{\boldsymbol{a}_1 \cdot (\boldsymbol{a}_2 \times \boldsymbol{a}_3)} = \frac{\frac{ac}{2}\boldsymbol{e}_x + \frac{\sqrt{3}}{2}ac\boldsymbol{e}_y}{\frac{\sqrt{3}}{2}a^2c} = \frac{1}{a}\left(\frac{1}{\sqrt{3}}\boldsymbol{e}_x + \boldsymbol{e}_y\right) \\ \boldsymbol{b}_2 &= \frac{\boldsymbol{a}_3 \times \boldsymbol{a}_1}{\boldsymbol{a}_1 \cdot (\boldsymbol{a}_2 \times \boldsymbol{a}_3)} = \frac{1}{a}\left(-\frac{1}{\sqrt{3}}\boldsymbol{e}_x + \boldsymbol{e}_y\right) \\ \boldsymbol{b}_3 &= \frac{\boldsymbol{a}_1 \times \boldsymbol{a}_2}{\boldsymbol{a}_1 \cdot (\boldsymbol{a}_2 \times \boldsymbol{a}_3)} = \frac{1}{c}\boldsymbol{e}_z \end{aligned}\right\} \tag{5}$$

(2) In solid-state physics, the polyhedron called Wigner–Seitz cell is widely used. We take the central (Winger–Seitz) cell of the reciprocal lattice as the first Brillouin zone. Each such cell involves one reciprocal lattice point at the center of the cell. It may be suggested that the construction of the Wigner–Seitz type unit cell in crystal lattice is the same as that of the first Brillouin zone in reciprocal lattice.

We may recall some fundamental points about the first Brillouin zone. In this unit cell, one reciprocal lattice point is set to the origin and connects with all points that are adjacent to the origin by a straight line. Next, normal planes perpendicular to the lines are constructed at their middle points, and we choose a polyhedron that has a minimum volume from the set of polyhedra enclosed by these planes. It is known that if the first Brillouin zone of the body-centered cubic lattice is bounded by the planes normal to the twelve vectors at their midpoints, then the resultant zone becomes a regular rhombic dodecahedron. With respect to the face-centered cubic lattice, the boundaries of the central cell are determined for the most part by eight planes normal to these vectors at their midpoints to form an octahedron. However, the corners of this octahedron are cut by planes that are the perpendicular bisectors of six other reciprocal-lattice vectors.

The first Brillouin zone of the hcp lattice can be obtained by the following procedure. From (5), the arbitrary reciprocal-lattice vector $\boldsymbol{H}_{pqr}$ may be provided in the following form.

$$\boldsymbol{H}_{pqr} = (p\boldsymbol{b}_1 + q\boldsymbol{b}_2 + r\boldsymbol{b}_3) \tag{6}$$

$$= \frac{1}{\sqrt{3}a}(p-q)\boldsymbol{e}_x + \frac{1}{a}(p+q)\boldsymbol{e}_y + \frac{r}{c}\boldsymbol{e}_z \tag{7}$$

The shortest nonzero vectors may be given when the values of prq are any of the combinations of 100 or 010. All these results are summarized as follows.

$$\left.\begin{aligned} H_{100} &= \frac{1}{a}\left(\frac{1}{\sqrt{3}}\boldsymbol{e}_x + \boldsymbol{e}_y\right), \quad & H_{\bar{1}00} &= \frac{1}{a}\left(\frac{-1}{\sqrt{3}}\boldsymbol{e}_x + \boldsymbol{e}_y\right) \\ H_{010} &= \frac{1}{a}\left(\frac{-1}{\sqrt{3}}\boldsymbol{e}_x + \boldsymbol{e}_y\right), \quad & H_{0\bar{1}0} &= \frac{1}{a}\left(\frac{1}{\sqrt{3}}\boldsymbol{e}_x - \boldsymbol{e}_y\right) \\ H_{1\bar{1}0} &= \frac{1}{a}\left(\frac{2}{\sqrt{3}}\boldsymbol{e}_x\right), \quad & H_{\bar{1}10} &= \frac{1}{a}\left(\frac{-2}{\sqrt{3}}\boldsymbol{e}_x\right) \end{aligned}\right\} \tag{8}$$

$$H_{001} = \frac{1}{c}\boldsymbol{e}_z, \qquad H_{00\bar{1}} = \frac{-1}{c}\boldsymbol{e}_z \tag{9}$$

The first Brillouin zone of the hcp lattice can be obtained from the perpendicular bisector planes of these eight reciprocal-lattice vectors given by (8) and (9). The results are shown in Fig. 1a. Namely, the first six reciprocal-lattice vectors make six side planes of the regular hexagonal prism and the two remaining reciprocal-lattice vectors provide both top and bottom planes. As shown in Fig. 1b, the first Brillouin zone of the hcp lattice is described, in another way, by the hexagonal prism with its height given by $|\boldsymbol{b}_3| = 1/c$. Here, the hexagonal base is formed by lines that are obtained by drawing six vectors (solid lines) in the reciprocal-lattice space and further drawing the perpendicular bisector (broken lines) of each vector.

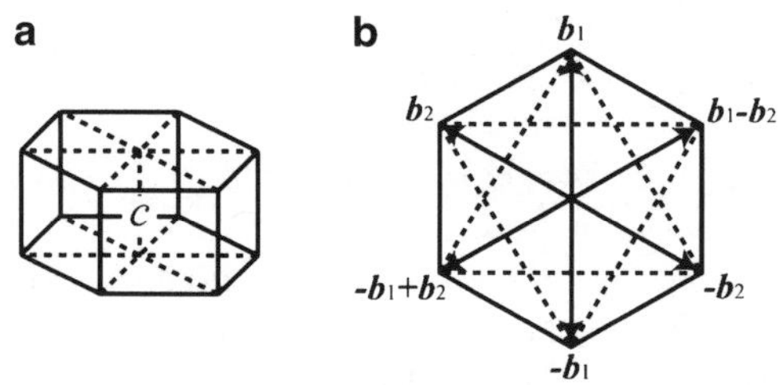

Fig. 1 The first Brillouin zone of the hcp lattice

Note: In the real hcp crystals, since the position of $\left(\frac{2}{3}\,\frac{1}{3}\,\frac{1}{2}\right)$ is occupied by an atom with the c/a value deviated from the ideal case ($c/a = \sqrt{8/3} = 1.633$), the Brillouin zone shows a more complicated form. Such information is available in other monographs on Solid-State Physics, see for example "Introduction to the Electron Theory of Metals" by Uichiro Mizutani, Cambridge University Press, (2001) Chap. 5.

Question 5.6 Explain that the constructive interference is observed when the scattering vector corresponds to a reciprocal-lattice vector.

Answer 5.6 Let us consider the case where the incident X-ray beam with a wave vector $\boldsymbol{s}_0$ encounters the scatterer located at point B, which is defined by vector $\boldsymbol{r}$ in the real space, and produces the coherent scattering (see Fig. 1) by coupling the scatterer located at point A. It is also assumed here that the scattered waves in the direction of $\boldsymbol{s}$ are measured at the position P with a sufficiently long distance R from points A and B.

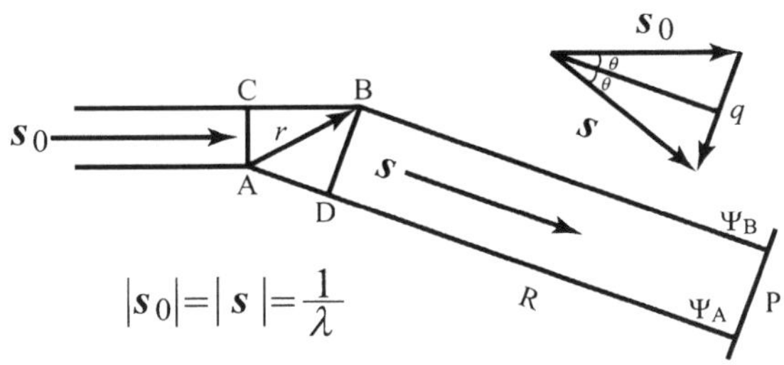

Fig. 1 The waves scattered from two points A and B and the scattering vector

The scattering vector $\boldsymbol{q}$ is given by $\boldsymbol{q} = \boldsymbol{s} - \boldsymbol{s}_0$, and the optical path difference Δ produced from the scatterers at points A and B is $\Delta = \boldsymbol{r} \cdot \boldsymbol{q}$. Therefore, the scattered wave detected at the position of P is represented as superposition of the waves produced at points A and B, and it may be expressed in the following form.

$$\Psi = \Psi_A + \Psi_B = e^{2\pi i(qR - \nu t)} + e^{2\pi i\{(qR + \Delta) - \nu t\}} \tag{1}$$

$$= e^{2\pi i(qR - \nu t)}[1 + e^{2\pi i \boldsymbol{q}\cdot\boldsymbol{r}}] \tag{2}$$

The first term to the right-hand side of (2) is equivalent to a common wave phase factor for all waves, and the second term corresponds to the amplitude of the scattered wave. Therefore, the interference effect proves only if the exponent portion of the second term is integer multiples of $2\pi i$. Namely, when the optical path difference Δ is integer multiples of the wavelength, the scattered wave becomes in phase.

In order to extend the result for two points A and B to a general case containing many scatterers, the scattering power of n-th scatterer is set to f_n. Then, a generalized form of the amplitude of the scattered waves may be expressed as follows.

$$\Psi = \sum_n \Psi_n = e^{2\pi i(qR - \nu t)} G(\boldsymbol{q}) \tag{3}$$

$$G(\boldsymbol{q}) = \sum_n f_n e^{2\pi i \boldsymbol{q}\cdot\boldsymbol{r}_n} \tag{4}$$

With respect to the amplitude of the scattered waves, (4) shows that the maximum value is obtained when all the exponent portions are integer multiples of $2\pi\mathrm{i}$. This is because the scattered waves by all the scatterers located at $\boldsymbol{r}_n$ are in phase and reinforce one another (so-called constructive interference) to form a diffracted beam in the direction of $\boldsymbol{q}$. Note that in all other directions of space, the scattered waves are out of phase and cancel one another. That is, $\boldsymbol{r}_n$ is characterized by a certain periodicity (with a regular interval) so that $\boldsymbol{q} \cdot \boldsymbol{r}_n$ is expressed by integers.

It may be helpful to recall the relationships between the crystal-lattice vector $\boldsymbol{R}_{pqr}$ showing arbitrary lattice points and the reciprocal-lattice vector $\boldsymbol{H}_{hkl}$.

$$\boldsymbol{R}_{pqr} = p\boldsymbol{a}_1 + q\boldsymbol{a}_2 + r\boldsymbol{a}_3 \tag{5}$$

$$\boldsymbol{H}_{hkl} = h\boldsymbol{b}_1 + k\boldsymbol{b}_2 + l\boldsymbol{b}_3 \tag{6}$$

Here, $\boldsymbol{a}_1$, $\boldsymbol{a}_2$, and $\boldsymbol{a}_3$ are the primitive vectors of the crystal lattice, and $\boldsymbol{b}_1$, $\boldsymbol{b}_2$, and $\boldsymbol{b}_3$ are those of the reciprocal lattice, pqr and hkl are integers. If the scalar product between these two vectors is calculated, the following relationship may be confirmed.

$$\boldsymbol{R}_{pqr} \cdot \boldsymbol{H}_{hkl} = (p\boldsymbol{a}_1 + q\boldsymbol{a}_2 + r\boldsymbol{a}_3)(h\boldsymbol{b}_1 + k\boldsymbol{b}_2 + l\boldsymbol{b}_3) \tag{7}$$

$$= ph + qk + rl = \text{integer} \tag{8}$$

Here, we use the vector property that $\boldsymbol{b}_j \cdot \boldsymbol{a}_k$ is unity when $j = k$ and that it is zero in the $j \neq k$ case. Equation (8) is one of the important relationships that always holds in a crystal. Accordingly, since the crystal-lattice vector $\boldsymbol{R}_{pqr}$ can be set to $\boldsymbol{r}_n$, the condition, for which all exponents of the exponential functions in (4) are integer multiples of $2\pi\mathrm{i}$, may be given by the following equation.

$$\boldsymbol{q} = \boldsymbol{H}_{hkl} \tag{9}$$

This means that the so-called constructive interference can be obtained only when the scattering vector corresponds to a reciprocal-lattice vector. Schematic diagram for this relationship is given in Fig. 2.

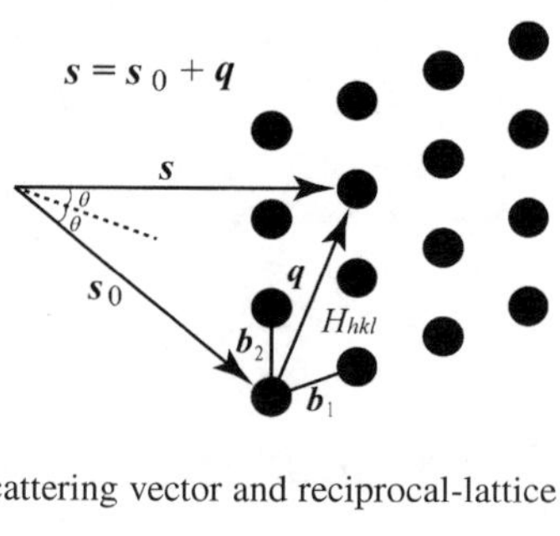

Fig. 2 Correlation between scattering vector and reciprocal-lattice vector

Note: This is equivalent to the Bragg law. Further information is available in other monographs, see for example "Structure Determination by X-ray Crystallography" by Mark Ladd and Rex Palmer, Fourth Edition, Kluwer Academic/Plenum Publishers, London, (2003).

Question 5.7 With respect to X-ray diffraction from a crystal, explain the relationships between the incident beam, the scattered beam, and the direction of a crystal for detecting the scattering intensity using the sphere of reflection (Ewald sphere) or the limiting sphere.

Answer 5.7 The present requirement is to consider the following point under the $\boldsymbol{q} = \boldsymbol{H}_{hkl}$ condition. When the starting point of the scattering vector is set to the origin of the reciprocal-lattice vector, the wave vector $\boldsymbol{s}_0$ of the incident beam corresponds to the vector of $(1/\lambda)$, which points in the direction of $\boldsymbol{q}$(= equivalent to the origin of the reciprocal-lattice vector). Here, λ is the wavelength of the incident beam. The scattered beam wave vector $\boldsymbol{s}$ turns into the vector orientated from the starting point of $\boldsymbol{s}_0$ to the terminal point of $\boldsymbol{q}$, and the length of $\boldsymbol{s}_0$ and $\boldsymbol{s}$ is given by $(1/\lambda)$, and it may be simplified by $\boldsymbol{q} = \boldsymbol{H}_{hkl} = (\boldsymbol{s} - \boldsymbol{s}_0)/\lambda$. These relationships may be expressed as shown in Fig. 1, explaining the conditions for diffraction graphically. The sphere of radius $(1/\lambda)$ in this figure is the Ewald sphere (it is also called Ewald reflection sphere). Constructive interference will be obtained if the condition of $\boldsymbol{q} = \boldsymbol{H}_{hkl}$ is satisfied. This corresponds to the case where a reciprocal-lattice point hkl touches the surface of the Ewald sphere drawn around the origin. Accordingly, we can detect the diffracted intensity, when a detector is set to the terminal point of $\boldsymbol{q}$. When any reciprocal lattice point of hkl does not touch the surface of the Ewald sphere, it is difficult to measure the diffraction intensity related to the hkl plane. Figure 1 shows the case where a crystal (real lattice) sample slightly rotates and also shows its relevance to the reciprocal-lattice points. However, it should also be kept in mind that even when a crystal sample is rotated, not all reciprocal-lattice points necessarily touch the surface of the Ewald sphere. This is also supported by the fact that only the reciprocal-lattice points, which are located at inside the diameter $(2/\lambda)$ of the Ewald sphere drawn around a point of (000), satisfy the condition of $\boldsymbol{q} = \boldsymbol{H}_{hkl}$ as easily seen in Fig. 2 so that the detection of diffraction intensity is not always possible. In other words, the diffraction intensities for the reciprocal-lattice points can be observed when satisfying the following condition (see also Fig. 3) using the incident beam with the wavelength of λ. This sphere is called limiting sphere.

$$\frac{1}{d_{hkl}} \leq \frac{2}{\lambda} \quad \text{or} \quad d_{hkl} \geq \frac{\lambda}{2} \tag{1}$$

When there are many small crystal grains and the direction of their crystal planes is distributed at random, the corresponding reciprocal-lattice points will be smeared out onto spheres around the point (000). As a result, with respect to the reciprocal-lattice points touching the surface of the Ewald sphere drawn by the radius of

$(1/d_{hkl})$ around the origin, the higher the probability we obtain, the more frequently we will be able to detect the diffraction intensity (see Fig. 3). This is a fundamental principle for measuring the diffraction intensity using a goniometer for crystalline powder samples.

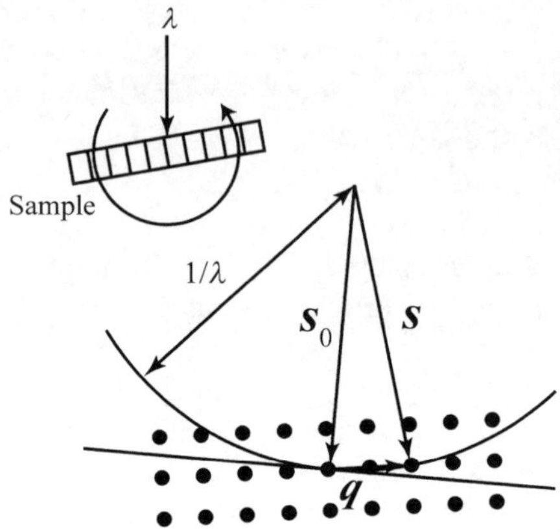

Fig. 1 Rotation of a crystal sample and its relevance to the reciprocal-lattice points

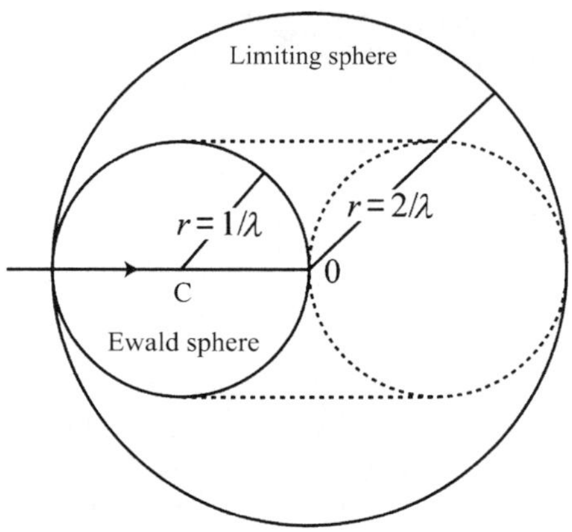

Fig. 2 Graphical representation for the relationship between the rotation of a crystal sample and the limiting sphere

Note: Considering the terminal point of the incident beam vector being the initial point of the reciprocal-lattice vector, Fig. 4 shows the relationships between the wavelength of the incident X-ray beam and the limiting sphere using the case of λ_1 and λ_2. In this case, at the reciprocal-lattice points overlapped by the Ewald sphere with radius of $1/\lambda_1$ or $1/\lambda_2$, the diffraction intensity will be detectable in the direction oriented from the center of the Ewald sphere to the respective reciprocal-lattice points. If the wavelength of the incident beam continuously changes from

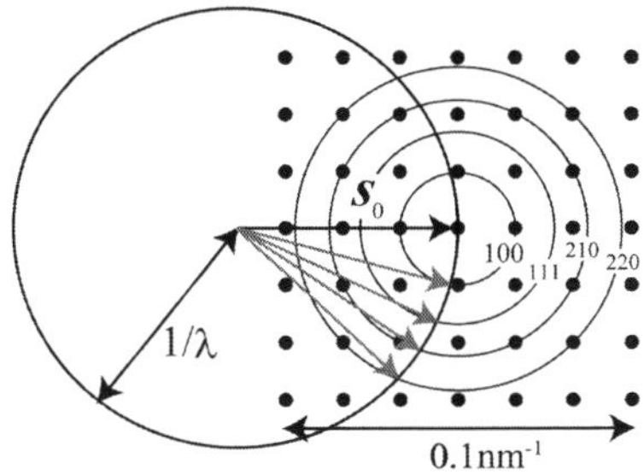

Fig. 3 Relationships between the wavelength of the incident beam and the limiting sphere for a polycrystalline powder sample

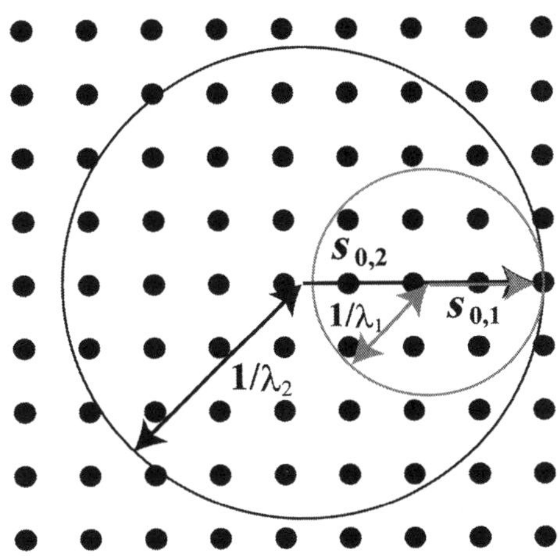

Fig. 4 Relationships between the terminal point of the incident beam and the initial point of the reciprocal-lattice vector when two beams with wavelengths of λ_1 and λ_2 are simultaneously incident on a sample

λ_{min} to λ_{max}, we will be able to detect the diffraction intensity with sufficiently high probability, since there are many Ewald spheres with various radii (Laue method).

Question 5.8 Explain Huygens principle for a typical optical phenomenon where light passes through a small hole (or slit) and the relevant Kirchhoff theory of diffraction that handles it mathematically using wave equation.

Answer 5.8 As shown in Fig. 1a, when a plane wave perpendicularly encounters a plate with a small hole (or slit), a spherical wave is produced from the hole as a central point and propagates. Two or more spherical waves are produced from each hole if the plate has several small holes so that the interference of waves is observed at a position apart from the plate (see Fig. 2b). Such interference behavior is known to depend on the periodicity of small holes. It may be suggested that Huygens principle is the generalized method for covering various wave phenomena including a relatively big hole. In such a case, a big hole is assumed to be the adjoined and connected small holes.

Let us consider the optical phenomena with respect to a small hole located at a distance $\boldsymbol{l}_0$ from a light source $\mathbf{P}_0$. This small hole is assumed to be a small lattice with $\mathrm{d}x$ and $\mathrm{d}y$, and produces the spherical wave and its behavior is observed at a position P located at a distance $\boldsymbol{l}$ from the hole. For convenience of discussion by the Kirchhoff theory of diffraction, we set up the condition as shown in Fig. 2. Here, it is supposed that the plate is in the $x - y$ plane and that a small hole is located at the position characterized by the vector $\boldsymbol{r}$ from the origin.

The incident wave hits a plate from the negative side of z-axis of Fig. 2, and only the wave encounters a small area characterized $\mathrm{d}\boldsymbol{r} = \mathrm{d}x \cdot \mathrm{d}y$ will receive modulation. Here, $\mathrm{d}\boldsymbol{r}$ is defined by the position with $g(\boldsymbol{r}) = g(x, y, z = 0)$ as a function of $\boldsymbol{r}$. For example, considering that if a hole is open, it may be described as $g(\boldsymbol{r}) = 1$ and conversely, $g(\boldsymbol{r}) = 0$ for a closed hole. Then, we may express the case where a spherical wave is produced from a small hole $\mathrm{d}\boldsymbol{r}$ by satisfying the condition $g(\boldsymbol{r}) = 1$. Next, let us consider that such a spherical wave is observed at the point P which is denoted by the vector $\boldsymbol{l}$ from $\mathrm{d}\boldsymbol{r}$, as well as the distance $|\boldsymbol{l}'|$ from the origin.

Point P is characterized by the position using axes x, y, and z and angles α, β, and γ, as shown in Fig. 2. According to the Kirchhoff theory of diffraction, if the wave $\Psi_0(\boldsymbol{l}_0)$ from the light source is incident on a small hole $\mathrm{d}\boldsymbol{r}$ and passes through it, and further reaches the point P, the contribution $\mathrm{d}\Psi(\mathrm{P})$ corresponding to that wave may be given by the following equations.

$$\mathrm{d}\Psi(\mathrm{P}) = \mathrm{i}s_0\Phi \cdot \frac{\mathrm{e}^{-2\pi \mathrm{i} s_0 |\boldsymbol{l}|}}{|\boldsymbol{l}|} g(\boldsymbol{r}) \cdot \Psi_0(\boldsymbol{l}_0)\mathrm{d}\boldsymbol{r} \tag{1}$$

$$\Phi = \frac{\cos(n, \boldsymbol{l}_0) - \cos(n, \boldsymbol{l})}{2} \tag{2}$$

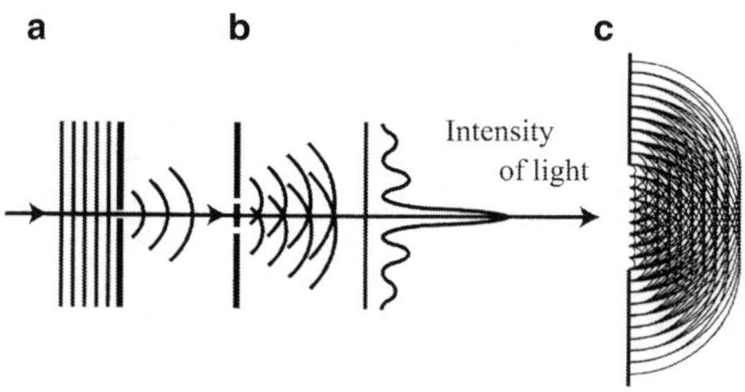

Fig. 1 (**a**) The incident (*plane*) wave encounters one small hole and produces a spherical wave, (**b**) the interference effect found in two small holes and the related intensity distribution and (**c**) image of the interference effect found in a big hole

As shown in Fig. 3, $\boldsymbol{n}$ in the small area defined by $(\mathrm{d}x \cdot \mathrm{d}y)$ is a normal vector whose direction is oriented to the observation point P, and $(n, \boldsymbol{l}_0)$ and $(n, \boldsymbol{l})$ are the angles formed by the corresponding two vectors. It may be added, as readily seen in Fig. 3, that $\Phi = 1$ is well accepted when the light source P_0, the hole $\mathrm{d}\boldsymbol{r}$, and the observation point are located near a straight line.

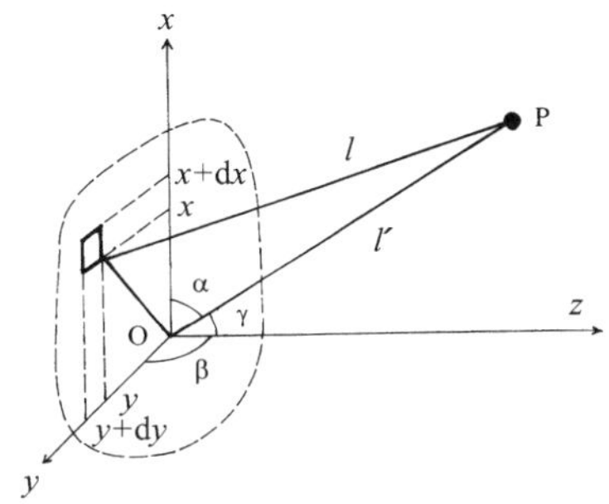

Fig. 2 Spatial relationship between the small area dr in the $x-y$ plane and an observation point **P**

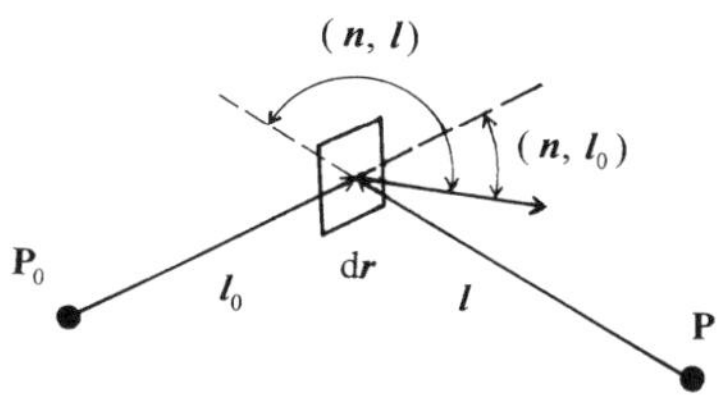

Fig. 3 Positional relationships between the normal vector and a light source, a hole and an observation point

Since the phase of the incident wave $\Psi_0(\boldsymbol{l}_0)$ emitted from the light source is quite likely to be equal in a small area $(\mathrm{d}x \cdot \mathrm{d}y)$ of the scatterer set in the $x-y$ plane, it is referred to as $\Psi_0(\boldsymbol{l}_0) = 1$. In addition, considering the relation of $\boldsymbol{l} = \boldsymbol{r} - \boldsymbol{l}'$ (see Fig. 2), the integration of (1), i.e., the amplitude of the wave observed at point P is given by the following equation.

$$\Psi(\mathrm{P}) = is_0 \int \frac{\mathrm{e}^{-2\pi i s_0 |\boldsymbol{l}|}}{|\boldsymbol{r} - \boldsymbol{l}'|} \cdot \Phi g(\boldsymbol{r}) \mathrm{d}\boldsymbol{r} \tag{3}$$

$|\boldsymbol{l}| = |\boldsymbol{r} - \boldsymbol{l}'|$ corresponds to a change in the location of the scatterer (plate) in which a small area $(\mathrm{d}x \cdot \mathrm{d}y)$ is set up, and it varies significantly in comparison with the wavelength λ. Therefore, the term given by an exponential function in (3) is expected to show considerably large variation, whereas it is thought that $1/|\boldsymbol{r}-\boldsymbol{l}'|$, a denominator part of (3), shows monotonic change only. In addition, Φ is considered to be constant under the given condition of the scatterers. If these points are taken into account, (3) may be simplified in the following form:

$$\Psi(\mathrm{P}) = \frac{\mathrm{i}s_0 \Phi}{|\boldsymbol{l}|} \int \boldsymbol{e}^{-2\pi \mathrm{i} s_0 |\boldsymbol{l}|} g(\boldsymbol{r}) \mathrm{d}\boldsymbol{r} \tag{4}$$

The coordinates of the observation point P are set to $(x_\mathrm{p}, y_\mathrm{p}, z_\mathrm{p})$. As previously mentioned, the coordinates of the small area $\mathrm{d}\boldsymbol{r}$, i.e., a scattered object, can be expressed by $(x, y, 0)$ so that we obtain the following equations (see to Fig. 2).

$$\left.\begin{aligned} |\boldsymbol{l}|^2 &= (x_\mathrm{p} - x)^2 + (y_\mathrm{p} - y)^2 + z_\mathrm{p}^2 \\ |\boldsymbol{l}'|^2 &= x_\mathrm{p}^2 + y_\mathrm{p}^2 + z_\mathrm{p}^2 \end{aligned}\right\} \tag{5}$$

$$\left.\begin{aligned} |\boldsymbol{l}|^2 &= |\boldsymbol{l}'|^2 - 2(x_\mathrm{p} x + y y_\mathrm{p}) + x^2 + y^2 \\ &= |\boldsymbol{l}'|^2 \left\{ 1 - 2\frac{x_\mathrm{p} x + y y_\mathrm{p}}{|\boldsymbol{l}'|^2} + \frac{x^2 + y^2}{|\boldsymbol{l}'|^2} \right\} \end{aligned}\right\} \tag{6}$$

$$|\boldsymbol{l}| = |\boldsymbol{l}'| \cdot \sqrt{1 - 2\frac{x_\mathrm{p} x + y y_\mathrm{p}}{|\boldsymbol{l}'|^2} + \frac{x^2 + y^2}{|\boldsymbol{l}'|^2}} \tag{7}$$

$|\boldsymbol{l}'|$ is sufficiently large in comparison with the absolute values of x and y. That is, since the relationships of $|\boldsymbol{l}'| \gg x$ and $|\boldsymbol{l}'| \gg y$ hold, the following approximation is obtained.

$$\left.\begin{aligned} |\boldsymbol{l}| &= |\boldsymbol{l}'| - x\frac{x_\mathrm{p}}{|\boldsymbol{l}'|} - y\frac{y_\mathrm{p}}{|\boldsymbol{l}'|} \\ &= |\boldsymbol{l}'| - x \cdot \cos\alpha - y \cdot \cos\beta \end{aligned}\right\} \tag{8}$$

Considering that $g(\boldsymbol{r})\mathrm{d}\boldsymbol{r}$ is expressed by $g(x, y)\mathrm{d}x \cdot \mathrm{d}y$, the relationships of (8) can be rearranged by coupling (4) in the following form.

$$\Psi(\mathrm{P}) = \frac{\mathrm{i}s_0 \Phi}{|\boldsymbol{l}'|} \mathrm{e}^{-2\pi \mathrm{i} s_0 |\boldsymbol{l}'|} \int g(x, y) \mathrm{e}^{2\pi \mathrm{i} s_0 (x \cos\alpha + y \cos\beta)} \mathrm{d}x \cdot \mathrm{d}y \tag{9}$$

The term, which can be excluded from the integration of (9), is set to a constant C, and the components of x-axis and y-axis are described in the following equations when using the unit vector $\boldsymbol{s}_0$ of the incident wave.

$$\left.\begin{aligned} s_x &= s_0 \cos\alpha = \frac{\cos\alpha}{\lambda} \\ s_y &= s_0 \cos\beta = \frac{\cos\beta}{\lambda} \end{aligned}\right\} \tag{10}$$

Therefore, (9) is given in the following form.

$$\Psi(s_x, s_y) = \mathrm{C} \int g(x, y) \mathrm{e}^{2\pi \mathrm{i}(x \cdot s_x + y \cdot s_y)} \mathrm{d}x \cdot \mathrm{d}y \tag{11}$$

Here, $\Psi(s_x, s_y)$ corresponds to the two-dimensional Fourier transform of $g(x, y)$. Equation (11) may also be expressed as follows.

$$\Psi(s_x, s_y) = \mathrm{C} \int g(x, y) \mathrm{e}^{2\pi \mathrm{i}(x \cdot \frac{\cos\alpha}{\lambda} + y \cdot \frac{\cos\beta}{\lambda})} \mathrm{d}x \cdot \mathrm{d}y \tag{12}$$

Question 5.9 Calculate the diffraction intensity produced from the case where m slits with aperture width L are aligned at an interval d on a line (one-dimensional array).

Answer 5.9 A schematic diagram of the given condition is illustrated in Fig. 1. Considering that if a hole is open, it may be described as $g(\boldsymbol{r}) = 1$ and in reverse $g(\boldsymbol{r}) = 0$ for a closed hole. Similar to this idea, a mathematical representation of slits is given by the following equations.

$$g(x) = \left\{ \begin{matrix} 1 & (j-1)d \leq x \leq (j-1)d + L \\ 0 & (j-1)d + L < x < jd \\ & (j = 1, 2, \dots m) \end{matrix} \right\} \tag{1}$$

where m slits are aligned at an interval d along the x-axis. We also use the scattering amplitude Ψ given by the Kirchhoff theory of diffraction.

$$\Psi(\mathbf{P}) = C \int g(x, y) \mathrm{e}^{2\pi \mathrm{i} s_0 (x \cdot \cos\alpha + y \cdot \cos\beta)} \mathrm{d}x \cdot \mathrm{d}y \tag{2}$$

In the present case, the constant C in (2) is omitted, and because of the one-dimensional arrangement, $g(x, y)$ is replaced by $g(x)$ as well as $\alpha + \beta = \frac{\pi}{2}$. Then, one obtains the following equations.

$$\Psi(\mathrm{P}) = C \int_{-\infty}^{\infty} g(x) \mathrm{e}^{2\pi \mathrm{i} s_0 x \cdot \cos\alpha} \mathrm{d}x \tag{3}$$

$$\begin{aligned} &= \int_0^L 1 \cdot \mathrm{e}^{2\pi \mathrm{i} s_0 x \cdot \cos\alpha} \mathrm{d}x + \int_d^{d+L} 1 \cdot \mathrm{e}^{2\pi \mathrm{i} s_0 x \cdot \cos\alpha} \mathrm{d}x \\ &+ \int_{2d}^{2d+L} 1 \cdot \mathrm{e}^{2\pi \mathrm{i} s_0 x \cdot \cos\alpha} \mathrm{d}x + \cdots + \int_{(m-1)d}^{(m-1)d+L} 1 \cdot \mathrm{e}^{2\pi \mathrm{i} s_0 x \cdot \cos\alpha} \mathrm{d}x \end{aligned} \tag{4}$$

Equation (4) can be rewritten as follows.

$$\Psi(\mathrm{P}) = \sum_{j=1}^{m} \int_{(j-1)d}^{(j-1)d+L} \mathrm{e}^{2\pi \mathrm{i} x \cdot \cos\alpha} \mathrm{d}x \tag{5}$$

Here, we use again $\alpha + \beta = \frac{\pi}{2}$ and $\boldsymbol{s}_0 \cos\alpha = \frac{\cos\alpha}{\lambda} = \frac{\sin\gamma}{\lambda}$ (see Question 5.7 and Fig. 2).

$$\Psi(\mathrm{P}) = \sum_{j=1}^{m} \int_{(j-1)d}^{(j-1)d+L} \mathrm{e}^{2\pi \mathrm{i} x \frac{\sin\gamma}{\lambda}} \mathrm{d}x \tag{6}$$

Using that a wave vector of $\boldsymbol{s}_0$ is equal to $1/\lambda$ and setting as $\boldsymbol{t} = 2\pi s_0 \sin\gamma$ simultaneously, (6) can be arranged in the following form.

$$\Psi(\mathrm{P}) = \sum_{j=1}^{m} \int_{(j-1)d}^{(j-1)d+L} \mathrm{e}^{\mathrm{i}tx} \mathrm{d}x \tag{7}$$

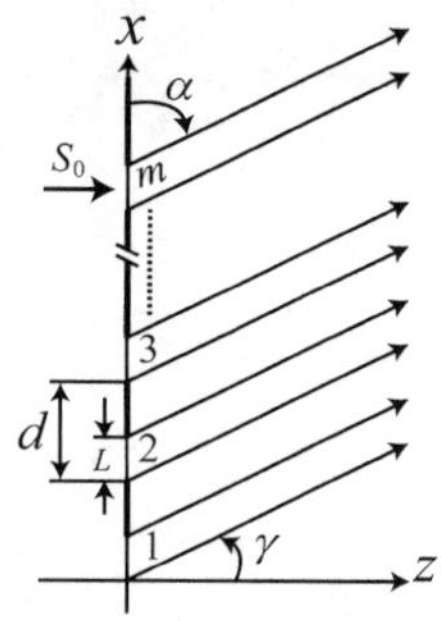

Fig. 1 Schematic diagram of diffraction in one-dimensional slit system

This problem is related to the integration using $\int \mathrm{e}^{kx}\mathrm{d}x = \dfrac{\mathrm{e}^{kx}}{k}$, but the sum of the definite integral of (7) can be estimated by the following method.

$$T = \sum_{j=1}^{m} \left(\int_{(j-1)d}^{(j-1)d+L} \mathrm{e}^{\mathrm{i}tx}\mathrm{d}x \right) = \left[\frac{\mathrm{e}^{\mathrm{i}tx}}{\mathrm{i}t} \right]_0^L + \left[\frac{\mathrm{e}^{\mathrm{i}tx}}{\mathrm{i}t} \right]_d^{d+L}$$
$$+ \left[\frac{\mathrm{e}^{\mathrm{i}tx}}{\mathrm{i}t} \right]_{2d}^{2d+L} + \cdots + \left[\frac{\mathrm{e}^{\mathrm{i}tx}}{\mathrm{i}t} \right]_{(j-1)d}^{(j-1)d+L} \tag{8}$$

$$\mathrm{i}tT_1 = \mathrm{e}^{\mathrm{i}tL} + \mathrm{e}^{\mathrm{i}t(d+L)} + \mathrm{e}^{\mathrm{i}t(2d+L)} + \cdots + \mathrm{e}^{\mathrm{i}t\{(j-1)d+L\}} \tag{9}$$
$$= \mathrm{e}^{\mathrm{i}tL}\{1 + \mathrm{e}^{\mathrm{i}td} + \mathrm{e}^{2\mathrm{i}td} + \mathrm{e}^{3\mathrm{i}td} + \cdots + \mathrm{e}^{(j-1)\mathrm{i}td}\} \tag{10}$$
$$\mathrm{i}tT_2 = 1 + \mathrm{e}^{\mathrm{i}td} + \mathrm{e}^{2\mathrm{i}td} + \mathrm{e}^{3\mathrm{i}td} + \cdots \mathrm{e}^{(j-1)\mathrm{i}td} \tag{11}$$

We find that the right-hand side of { } of (10) and (11) corresponds to geometric progression expressed by the first term = 1 and the common ratio = $\mathrm{e}^{\mathrm{i}td}$. In the geometric progression $\{\boldsymbol{a}_m\} = \boldsymbol{a}_0 + \boldsymbol{a}_0\boldsymbol{r} + \boldsymbol{a}_0\boldsymbol{r}^2 + \cdots \boldsymbol{a}\boldsymbol{r}^{m-1}$, the partial sum $\boldsymbol{S}_n$ up to the nth term is given by the following equation so that we also obtain the sum of (11).

$$S_n = \frac{a_0(1-r^m)}{1-r} = \mathrm{i}tT_2 = \frac{1\cdot(1-\mathrm{e}^{\mathrm{i}tdm})}{1-\mathrm{e}^{\mathrm{i}td}} \tag{12}$$

The integrated value is given by $T = \frac{1}{\mathrm{i}t}(T_1 - T_2)$, and then the following equation is proved.

$$T = \frac{1}{\mathrm{i}t}(\mathrm{e}^{\mathrm{i}tL}T_2 - T_2) = \frac{1}{\mathrm{i}t}T_2(\mathrm{e}^{\mathrm{i}tL} - 1) \tag{13}$$

Substituting (13) for (12), the integrated value becomes as follows.

$$T = \left(\frac{\mathrm{e}^{\mathrm{i}tL}-1}{\mathrm{i}t}\right)\left(\frac{1-\mathrm{e}^{\mathrm{i}tdm}}{1-\mathrm{e}^{\mathrm{i}td}}\right) \tag{14}$$

We use $\mathrm{e}^{\mathrm{i}x} - \mathrm{e}^{-\mathrm{i}x} = 2\mathrm{i}\sin x$, which is one of the relationships between exponential functions and trigonometric functions.

$$\mathrm{e}^{\mathrm{i}x} - 1 = 2\left(\frac{\mathrm{e}^{\mathrm{i}\frac{x}{2}} - \mathrm{e}^{-\mathrm{i}\frac{x}{2}}}{2}\right)\mathrm{e}^{\mathrm{i}\frac{x}{2}} = 2\mathrm{i}\sin\left(\frac{x}{2}\right)\cdot\mathrm{e}^{\mathrm{i}\frac{x}{2}} \tag{15}$$

$$\frac{tdm}{2} = m\pi d\cdot s_0\sin\gamma,\quad \frac{td}{2} = \pi d\cdot s_0\sin\gamma,\quad \frac{tL}{2} = \pi L\cdot s_0\sin\gamma \tag{16}$$

Rearranging (14) using (15) and (16), we will get the solution of (7) in the following equations.

$$\Psi(\mathrm{P}) = \frac{2\mathrm{i}\sin(\pi L\cdot s_0\sin\gamma)}{\mathrm{i}2\pi s_0\sin\gamma}\mathrm{e}^{\mathrm{i}\frac{tL}{2}} \times \frac{-2\mathrm{i}\sin(m\pi d\cdot s_0\sin\gamma)\cdot\mathrm{e}^{\mathrm{i}\frac{tdm}{2}}}{-2\mathrm{i}\sin(\pi d\cdot s_0\sin\gamma)\cdot\mathrm{e}^{\mathrm{i}\frac{td}{2}}} \tag{17}$$

$$= \frac{L\sin(\pi L\cdot s_0\sin\gamma)}{\pi L s_0\sin\gamma}\cdot\frac{\sin(m\pi d\cdot s_0\sin\gamma)}{\sin(\pi d\cdot s_0\sin\gamma)}\cdot\mathrm{e}^{\mathrm{i}\frac{tL}{2}}\cdot\mathrm{e}^{\mathrm{i}(m-1)\frac{td}{2}} \tag{18}$$

The 1st term of (18) is equivalent to the contribution of one slit, and the 2nd term corresponds to the contribution from the arrangement m slits being aligned at an interval d. Note that L in the 1st term is introduced into both denominator and numerator for getting unity of description with that of a sine function. Diffraction intensity is known to be the square of amplitude, and it can be obtained by multiplying the expression given for the amplitude by its complex conjugate. Then, it will be given as follows.

$$I = |\Psi(\mathrm{P})|^2 = \left|\frac{L\sin(\pi L\cdot s_0\sin\gamma)}{\pi L s_0\sin\gamma}\right|^2\cdot\left|\frac{\sin(m\pi d\cdot s_0\sin\gamma)}{\sin(\pi d\cdot s_0\sin\gamma)}\right|^2 \tag{19}$$

The 2nd term of (19) is called the Laue function, and in the present case, this term shows the effect of m slits aligned at an interval d on the diffraction intensity. Therefore, it also leads to the case where atoms form a three-dimensional array with regular periodicity, that is, crystal. For example, if setting as $k = \pi d \cdot s_0 \sin\gamma$ and setting the 2nd term of (19) to $F_L(k, m)$ again, we obtain the following simple equation as a function of k and m.

$$|F_L(k,m)|^2 = \frac{\sin^2(mk)}{\sin^2 k} \tag{20}$$

When k is an integer multiple (m) of π, both denominator and numerator turn out to be zero. That is, the condition $k = m\pi$ (m: integer) turns out to be $d \cdot \sin\gamma = m \cdot \lambda$, recalling $\pi d \cdot s_0 \sin\gamma = m\pi$ and $s_0 = 1/\lambda$. When the phase difference of waves generated from slits aligned on the line at the regular interval is integer multiples of the wavelength, the so-called constructive interference forms a diffracted beam. This is equivalent to the diffraction condition in a crystal. By applying mathematical procedures, the limit value is found to become m^2 if $k \to m\pi$ in (20). Using this relationship and normalizing the Laue function by m^2 to represent the diffraction intensity, we obtain the results as shown in Fig. 2.

When increasing the number of slits m, the peak at $k = m\pi$ will be sharp. On the contrary, if the value of d changes under the fixed m value, a different behavior is found. For example, when doubling the value of d, we find the frequency of the Laue function reduced to one half (1/2). (Suggestion: Try to check such a behavior by yourself. For example, for fixed $m = 10$, the value of d is varied as 0.1, 0.2, and 0.4 mm).

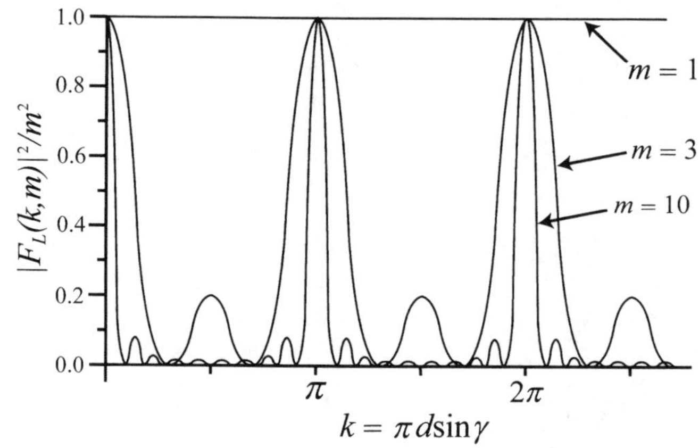

Fig. 2 Examples of the normalized Laue function

Question 5.10 X-rays are known to be electromagnetic waves, so that their periodic oscillations of both the electric and magnetic fields are polarized. Explain the polarization of an electromagnetic wave.

Answer 5.10 In order to facilitate the understanding of the nature of electromagnetic waves, a schematic diagram is given in Fig. 1, which represents the oscillatory components vibrating in the $x - y$ plane. The x-axis is the direction of propagation of the wave. It is also noted that electromagnetic waves including X-rays are assumed to be plane-polarized so that we can draw the electric field vector (E) always in the y-plane only and the magnetic field vector (H) perpendicular to E as shown in Fig. 1. That is, the electric field and magnetic field vectors perpendicularly intersect to form the so-called orthogonal set.

The electric field intensity ϕ of the plane-polarized waves shows cyclic variations with time t, and it is represented in the following form as a function of the distance x in the direction of the x-axis.

$$\phi = A\cos 2\pi(\nu t + \delta) \tag{1}$$

Where A is amplitude, ν is frequency, and δ is phase. Wavelength (λ) and frequency (ν) are correlated in the form of $\lambda = c/\nu$ using the speed of light c. The wavelength (λ) is equivalent to one cycle unit in variation of the electric field E against x with the fixed time t. This periodicity is a key point for handling waves.

Using angular frequency (or angular vibration), the variation of the electric field intensity $\phi(y, t)$ as a function of time t is also represented in the following form.

$$\phi_1(x, t) = e_y A_y \cdot \mathrm{e}^{\mathrm{i}(\omega t - kx + \delta_y)} \tag{2}$$

where e_y is a unit vector in the direction of y-axis, ω is angular frequency given by $\omega = 2\pi\nu$, and k is wave number given by $k = (2\pi)/\lambda$. Note that $s = 1/\lambda$ is frequently used as wave number so that careful treatment is required for discussing equations of waves.

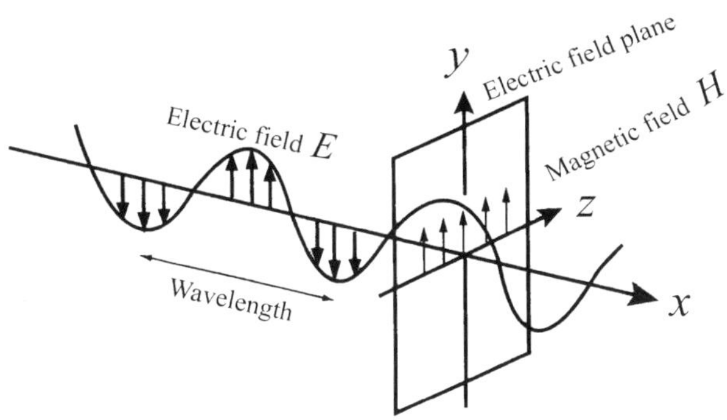

Fig. 1 Schematic diagram for propagation of X-rays

Equation (2) describes the case where the direction of wave oscillation is restricted to the $x - y$ plane only. For discussing a more generalized case, it is necessary to handle the state where a wave with the common angular frequency

ω and wave vector k propagates along the x-direction and vibrates in both $x-y$ and $y-z$ planes. These two planes perpendicularly intersect to form the so-called orthogonal set. Therefore, the z-axis component should be included as is expressed by the following equation.

$$\phi_2(x,y) = e_z A_z \mathrm{e}^{\mathrm{i}(\omega - kx + \delta z)} \tag{3}$$

δ_y and δ_z of (2) and (3) are the initial phases of the y- and z-components of each wave. When two waves are synthesized, the variation in the electric field intensity $\phi(y,t)$ of the resultant wave may be given as follows.

$$\phi(x,t) = (e_y A_y \mathrm{e}^{\mathrm{i}\delta y} + e_z A_z \mathrm{e}^{\mathrm{i}\delta_z})\mathrm{e}^{\mathrm{i}(\omega t - kx)} \tag{4}$$

Setting phase to $\delta = \delta_z - \delta_y$, the wave of ϕ_2 is shifted to the positive side by a value of δ, when compared with the wave of ϕ_1. Since the variation in the electric field intensity is expressed by a real part of the complex number representation, each component may be given by the following equations.

$$\left.\begin{aligned} \phi_y &= A_y \cos(\omega t - kx + \delta_y) \\ \phi_z &= A_z \cos(\omega t - kx + \delta_z) \end{aligned}\right\} \tag{5}$$

It is noted that a wave front is a surface perpendicular to the direction of wave propagation. Let us consider that the wave propagates with vibration in both $x-y$ and $y-z$ planes under the condition of fixed time t. The locus of a wave front of the electric field vector for this wave may be illustrated as shown in Fig. 2. The locus turns out to be a spiral of a right-hand screw in the direction of propagation of the wave in the region of $0 < \delta < \pi$, and its cycle is expressed by λ. Looking at this locus with its relevance to the $y-z$ plane, it will be observed as an oval (ellipse). The electromagnetic wave exhibiting this behavior is called "clockwise elliptic polarization of light." This oval nature is connected with the components given by (5) in the following equation.

$$\left(\frac{\phi_y}{A_y}\right)^2 + \left(\frac{\phi_z}{A_z}\right)^2 - 2\left(\frac{\phi_y}{A_y}\right)\left(\frac{\phi_z}{A_z}\right)\cos\delta = \sin^2\delta \tag{6}$$

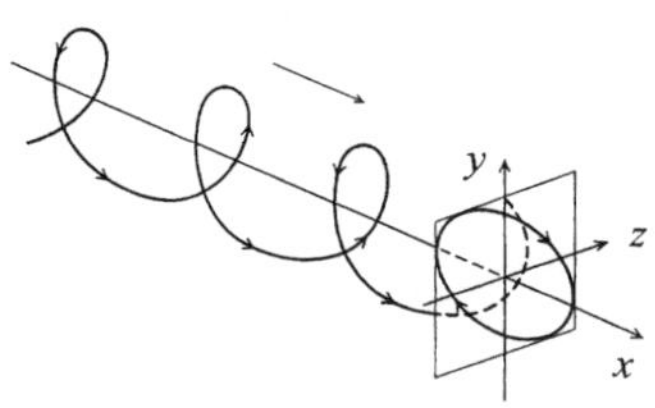

Fig. 2 Propagation of the elliptic polarized wave

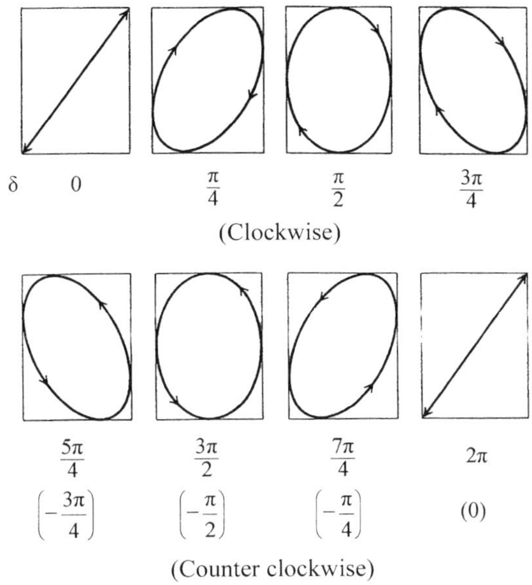

Fig. 3 Variation in polarization: counterclockwise (a left-handed screw), which is looked from an observer

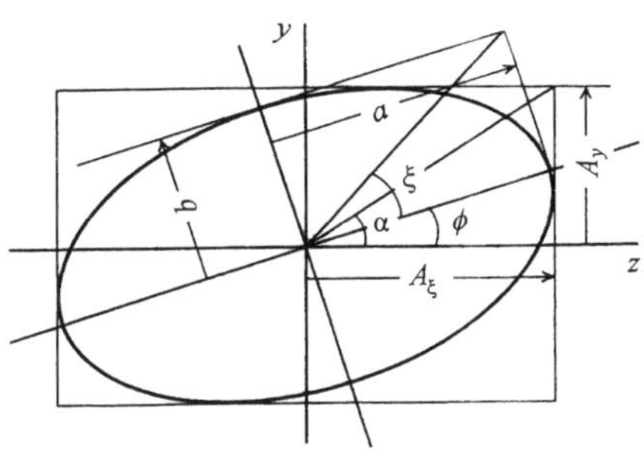

Fig. 4 Description for the elliptic polarization

If phase δ is set to the region $\pi < \delta < 2\pi$, the electromagnetic wave becomes "counterclockwise elliptic polarization of light". The shape of the ellipse depends on $\tan\alpha = A_y/A_z$ and δ. For convenience, Fig. 3 shows the form of the ellipse polarization when δ changes from 0 to 2π under the condition of $\tan\alpha = 1.3$. (The details are given in other monographs, for example, Masao Tsuruta : Applied Optics, Baifukan, Tokyo, (1990)).

Reference: The relationships describing elliptic polarization of light are given below, for the case where the angle ϕ formed by a long axis and the z-axis shows the variation of $0 \leq \phi \leq \pi$. a and b are the long radius and the short radius of the ellipse, respectively (see also Fig. 4).

$$\left.\begin{aligned} a^2 + b^2 &= A_y^2 + A_z^2 \\ \tan\alpha &= \frac{A_y}{A_z} \quad \left(0 \le \alpha \le \frac{\pi}{2}\right) \end{aligned}\right\} \tag{7}$$

$$\left.\begin{aligned} \tan 2\phi &= \tan 2\alpha \cdot \cos\delta \quad (\delta = \delta_z - \delta_y) \\ \sin 2\xi &= \sin 2\alpha \cdot \sin\delta \quad (\delta = \delta_z - \delta_y) \end{aligned}\right\} \tag{8}$$

$$\tan\xi = \pm\frac{b}{a} \quad \left(-\frac{\pi}{4} \le \xi \le \frac{\pi}{4}\right) \tag{9}$$

Note: clockwise rotation: positive sign, counterclockwise rotation : negative sign, and $\tan\xi$ is called ellipticity.

Question 5.11 Derive an equation for providing the diffraction intensity of a diatomic molecular gas.

Answer 5.11 At first, let us consider a monatomic gas where each atom is moving at random without any correlation. In such a case, since X-rays scattered from each atom do not mutually interfere, the resultant diffraction intensity I can be computed as the simple sum of scattering power $f(\boldsymbol{q})$ of each atom.

$$I = \sum_{l}^{N} |f(\boldsymbol{q})|^2 = N|f(\boldsymbol{q})|^2 \tag{1}$$

Where N is the total number of atoms in the system.

On the contrary, let $\boldsymbol{r}_{mn}$ be the coordinate of n-th atom located at a distance from the position of m-th atom, and then if all the sum of N number of atoms is taken, we obtain the following equation.

$$I = \sum_{n}\sum_{m} |f(\boldsymbol{q})|^2 \mathrm{e}^{-2\pi \mathrm{i}\boldsymbol{q}\cdot\boldsymbol{r}_{mn}} \tag{2}$$

$$= N|f(\boldsymbol{q})|^2 \left(1 + \sum_{n\neq m} \mathrm{e}^{-2\pi \mathrm{i}\boldsymbol{q}\cdot\boldsymbol{r}_{mn}}\right) \tag{3}$$

In (3), if there is no correlation between atoms, the exponential term is zero, and (3) is readily found to be equal to (1).

Supposing the case where diatomic molecules are arranged along a straight line with a distance d and molecules do not move at all. In addition, we find the relationship displayed in the central part of Fig. 1. Here, $\boldsymbol{d}$ is a vector representing the positional interrelation (=spatial correlation) of atoms and $\boldsymbol{R}$ is equivalent to a vector directed to the center of molecule from the origin. Let $\boldsymbol{q}$ be the scattering vector. In this case, it can be treated along the way similar to the scattering amplitude of diffraction by the slit aperture of d. (see Question 5.9) That is, the scattering amplitude $F(\boldsymbol{q})$ is the Fourier transform of the distribution function given by the delta function in the region between $-d/2$ and $d/2$.

$$F(\boldsymbol{q}) = \int_{-\infty}^{\infty} \mathrm{e}^{-2\pi \mathrm{i} qx} \left\{ \delta\left(x - \frac{d}{2}\right) + \delta\left(x + \frac{d}{2}\right) \right\} dx \tag{4}$$

$$= \mathrm{e}^{\mathrm{i}\pi qd} + \mathrm{e}^{-\mathrm{i}\pi qd} = 2\cos(\pi qd) \tag{5}$$

Therefore, the scattering amplitude $G(\boldsymbol{q})$ of diatomic molecules with its scattering power $f(\boldsymbol{q})$ is given by the following equation.

$$G(\boldsymbol{q}) = f(\boldsymbol{q}) \cdot 2\cos(\pi \boldsymbol{q} \cdot \boldsymbol{d}) \tag{6}$$

Note that if the distribution for the direction of vector $\boldsymbol{q}$ is calculated with respect to the scattering amplitude, the term of $\mathrm{e}^{-2\pi \mathrm{i}\boldsymbol{q}\cdot\boldsymbol{R}}$, equivalent to the phase, will appear. However, it appears to be unrelated to molecular orientation, so we ignore this term.

The direction of vectors d and q affects the value $G(\boldsymbol{q})$. For example, the value of $G(\boldsymbol{q})$ depends on whether the direction of $\boldsymbol{d}$ is parallel or perpendicular to q. However, molecules are considered to be very actively moving in the gaseous state, so the actual diffraction intensity may be obtained by averaging $|G(\boldsymbol{q})|^2$ over time. If the time-averaging operation is expressed by $\langle\rangle$, we obtain the following equation.

$$I(\boldsymbol{q}) = N\langle |G(\boldsymbol{q})|^2 \rangle = N|f(q)|^2 \cdot 4\langle \cos^2(\pi \boldsymbol{q} \cdot \boldsymbol{d}) \rangle \tag{7}$$

It is supposed that the direction of the molecules is completely random, and every direction is distributed at equal probability for the scattering vector called isotropic distribution. This corresponds to the condition that the terminal point of vector $\boldsymbol{d}$ expressing the mutual spatial correlation of two atoms exists on the sphere surface of radius $\boldsymbol{d}_\mathrm{S}$ at equal probability (see Fig. 2). Therefore, the time-averaging (7) can be expressed by the following equations.

$$\langle \cos^2(\pi \boldsymbol{q} \cdot \boldsymbol{d}) \rangle = \frac{1}{4\pi \boldsymbol{d}_\mathrm{S}^2} \int_0^{2\pi} \boldsymbol{d}_\mathrm{S} \mathrm{d}\phi \int_0^{\pi} (\pi \boldsymbol{q} \cdot \boldsymbol{d}) \boldsymbol{d}_\mathrm{S} \sin\theta \mathrm{d}\theta \tag{8}$$

$$= \frac{2\pi \boldsymbol{d}_\mathrm{S} \times \boldsymbol{d}_\mathrm{S}}{4\pi \boldsymbol{d}_\mathrm{S}^2} \int_0^{\pi} \cos^2(\pi qd \cos\theta) \sin\theta \mathrm{d}\theta \tag{9}$$

$$= \frac{1}{2} \left\{ 1 + \frac{\sin(2\pi qd)}{2\pi qd} \right\} \tag{10}$$

With respect to the scalar product of two vectors in Fig. 2, we use $\boldsymbol{q} \cdot \boldsymbol{d} = qd\cos\theta$ by using polar coordinates. Applying (10)–(7), we obtain a formula for calculating the diffraction intensity of a diatomic molecular gas as follows.

$$I(q) = N|f(q)|^2 \cdot 2 \left\{ 1 + \frac{\sin(2\pi qd)}{2\pi qd} \right\} \tag{11}$$

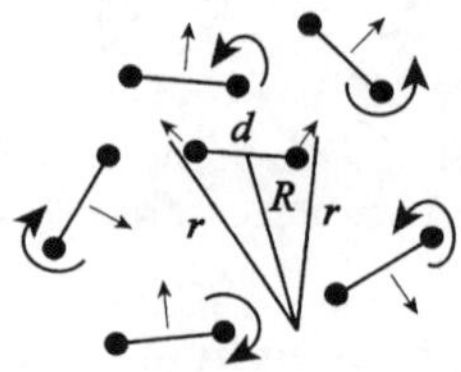

Fig. 1 Model of diatomic molecular gas

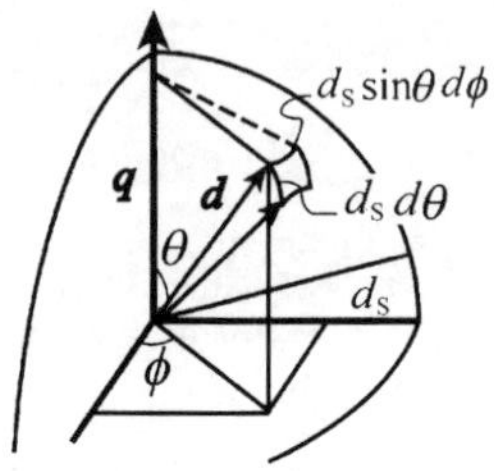

Fig. 2 Relationship between $\boldsymbol{d}$ and $\boldsymbol{q}\cdot\boldsymbol{d}$ in the isotropic distribution

Note:

$$\cos\theta = x \rightarrow -\sin\theta \mathrm{d}\theta = \mathrm{d}x$$

$$\cos(\pi) = -1, \quad \cos(0) = 1$$

$$\int_a^b f(x)\mathrm{d}x = -\int_b^a f(x)\mathrm{d}x$$

$$\cos^2\frac{\alpha}{2} = \frac{1}{2}(1+\cos\alpha)$$

$$\int_0^\pi \cos^2(\pi q \mathrm{d}\cos\theta)\sin\theta \mathrm{d}\theta = -\int_1^{-1}\cos^2(\pi q \mathrm{d}x)\mathrm{d}x = \int_{-1}^1 \left(\frac{2\pi q \mathrm{d}x}{2}\right)\mathrm{d}x$$

$$= 2\int_0^1 \frac{1}{2}\{1+\cos(2\pi q \mathrm{d}x)\}\mathrm{d}x = \left[x + \frac{\sin(2\pi q \mathrm{d}x)}{2\pi q \mathrm{d}x}\right]_0^1 = 1 + \frac{\sin 2\pi q \mathrm{d}}{2\pi q \mathrm{d}}$$

Question 5.12 Derive the Debye formula in the following form providing the diffraction intensity for a polyatomic molecular gas.

$$I(\boldsymbol{q}) = \sum_m \left[f_m^2 + \sum_{n\neq m} f_m f_n \frac{\sin(2\pi \boldsymbol{q}\cdot\boldsymbol{r}_{mn})}{2\pi \boldsymbol{q}\cdot\boldsymbol{r}_{mn}} \right]$$

Answer 5.12 As found in computation for the diffraction intensity of a diatomic molecular gas, the diffraction intensity from a set of scattering objects consisting of some atoms may be given by the square of the sum of scattering amplitude of each scatterer. Representing the vector showing the position of n-th atom as $\boldsymbol{r}_n$ and the scattering amplitude as $f_n(\boldsymbol{q})$, we may obtain a generalized form of the diffraction intensity $I(\boldsymbol{q})$ in the following equations.

$$
\begin{aligned}
I(\boldsymbol{q}) &= G(\boldsymbol{q})^* G(\boldsymbol{q}) && (1)\\
&= \sum_m \mathrm{e}^{2\pi \mathrm{i}\boldsymbol{q}\cdot\boldsymbol{r}_{mn}} f_m^*(\boldsymbol{q}) \cdot \sum_n \mathrm{e}^{-2\pi \mathrm{i}\boldsymbol{q}\cdot\boldsymbol{r}_{mn}} f_n(\boldsymbol{q}) && (2)\\
&= \sum_m \sum_n f_m^*(\boldsymbol{q}) f_n(\boldsymbol{q}) \mathrm{e}^{-2\pi \mathrm{i}\boldsymbol{q}\cdot(\boldsymbol{r}_n - \boldsymbol{r}_m)} && (3)\\
&= \sum_m \sum_n f_m^*(\boldsymbol{q}) f_n(\boldsymbol{q}) \mathrm{e}^{-2\pi \mathrm{i}\boldsymbol{q}\cdot\boldsymbol{r}_{mn}} \quad (\because \boldsymbol{r}_{mn} = \boldsymbol{r}_n - \boldsymbol{r}_m) && (4)
\end{aligned}
$$

The exponential term of (4) including the vector $\boldsymbol{r}_{mn}$ and the scattering vector $\boldsymbol{q}$ shows correlations of the scattering object formed by some atoms. The time-averaging process is applied along the way similar to the case of a diatomic molecular gas. The results are as follows.

$$
\begin{aligned}
\langle \mathrm{e}^{-2\pi \mathrm{i}\boldsymbol{q}\cdot\boldsymbol{r}_{mn}} \rangle &= \frac{1}{4\pi} \int_0^{2\pi} \mathrm{d}\phi \int_0^{\pi} \mathrm{e}^{-2\pi \mathrm{i}\boldsymbol{q}\cdot\boldsymbol{r}_{mn}\cdot\cos\theta} \sin\theta \mathrm{d}\theta && (5)\\
&= \frac{1}{2} \int_{-1}^{1} \mathrm{e}^{-2\pi \mathrm{i}\boldsymbol{q}\cdot\boldsymbol{r}_{mn} x} \mathrm{d}x && (6)\\
&= \frac{\sin(2\pi q r_{mn})}{2\pi q r_{mn}} && (7)
\end{aligned}
$$

where setting $t = 2\pi q r_{mn}$ and the following relationship is used.

$$\int_{-1}^{1} \mathrm{e}^{-\mathrm{i}tx} \mathrm{d}x = \left[\frac{\mathrm{e}^{-\mathrm{i}tx}}{-\mathrm{i}t} \right]_{-1}^{1} = \frac{\mathrm{e}^{\mathrm{i}t} - \mathrm{e}^{-\mathrm{i}t}}{\mathrm{i}t} = \frac{2\mathrm{i}\sin t}{\mathrm{i}t} = \frac{2\sin t}{t}$$

By coupling (4) with (7) and doing the sum of $m \times n = N^2$, the diffraction intensity can be obtained in the following equation.

$$I(\boldsymbol{q}) = \sum_m^N \sum_n^N f_m f_n \frac{\sin(2\pi q r_{mn})}{2\pi q r_{mn}} \tag{8}$$

Equation (8) is called Debye formula. Since (8) includes the term reflecting intra-molecular correlations, the following formula is also widely used.

$$I(\boldsymbol{q}) = \sum_m \left[f_m^2 + \sum_{n \neq m} f_m f_n \frac{\sin(2\pi \boldsymbol{q}\cdot\boldsymbol{r}_{mn})}{2\pi \boldsymbol{q}\cdot\boldsymbol{r}_{mn}} \right] \tag{9}$$

It is noted that the parenthesis of (9) is called the interference function of molecules. For example, suppose a diatomic molecular gas consists of one kind of atom and the distance of two atoms is characterized by d only, the diffraction intensity is given as follows when setting the total number of atoms to N (namely, number of molecules is $N/2$).

$$I(\boldsymbol{q}) = Nf^2 \left\{ 1 + (N-1)\frac{\sin(2\pi q d)}{2\pi q d} \right\} \tag{10}$$

Question 5.13 In an X-ray diffraction measurement, one has the option of placing a crystal monochromator in the incident beam generated from an X-ray tube. In this experimental condition, the polarization factor may be given in the following equation.

$$P = \frac{1 + \cos^2 2\theta_{\mathrm{M}} \cos^2 2\theta}{1 + \cos^2 2\theta_{\mathrm{M}}}$$

where $2\theta_{\mathrm{M}}$ is the diffraction angle of a monochromator. Derive this equation.

Answer 5.13 Let us consider the measurement using a diffractometer with a crystal monochromator placed in the incident beam as illustrated in Fig. 1. In this experimental condition, the characteristic X-ray beam generated from an X-ray tube is first diffracted at a monochromator and then further diffracted by a sample. Let 2θ and $2\theta_{\mathrm{M}}$ be the diffraction angle of a sample and that of a monochromator, respectively. Note that the X-ray source, usually having the line focal spot on the target, is perpendicular to the plane of the drawing and it is also parallel to the diffractometer axis.

Let us set the y-direction to the direction contained in a plane perpendicular to the direction of propagation of the X-ray beam and parallel to the diffractometer axis, whereas the x-direction is perpendicular to that plane (see Fig. 1).

Using two components of the amplitude, Eox and Eoy, for the variations in electric field intensity of the incident X-ray beam, two components of the amplitudes Ex' and Ey' of X-rays diffracted by the crystal monochromator may be expressed in the following equations.

$$\left.\begin{aligned} Ex' &= Eox \\ Ey' &= Eoy \cos 2\theta_{\mathrm{M}} \\ \langle E'^2 \rangle &= \langle Ex'^2 \rangle + \langle Ey'^2 \rangle = \langle Eox^2 \rangle + \langle Eoy^2 \rangle \cos^2 2\theta_{\mathrm{M}} \end{aligned}\right\} \tag{1}$$

Since the characteristic X-ray beam generated from an X-ray tube is not polarized, the following relationship holds.

$$\frac{1}{2}\langle Eo^2 \rangle = \langle Eox^2 \rangle = \langle Eoy^2 \rangle \tag{2}$$

By taking account of the relationship of (2), $\langle Eox^2\rangle$ and $\langle Eoy^2\rangle$ terms are expressed by $\langle Eo^2\rangle$, and the diffraction intensity is known to be proportional to the square of the amplitude for the variations in the electric field intensity. Then, we obtain the intensity formula as follows.

$$I' = KI_0 \frac{(1+\cos^2 2\theta_M)}{2} \tag{3}$$

where K is a constant and I_0 is the intensity of the incident X-ray beam.

Similarly, the amplitudes of X-rays diffracted by the crystalline sample are in the following equations.

$$\left.\begin{aligned} Ex'' &= Ex' = Eox \\ Ey'' &= Ey'\cos 2\theta = Eoy\cos 2\theta \cos 2\theta_M \\ \langle E''^2\rangle &= \langle Ex''^2\rangle + \langle Ey''^2\rangle \\ &= \langle Eox^2\rangle + \langle Eoy^2\rangle \cos^2 2\theta \cos^2 2\theta_M \end{aligned}\right\} \tag{4}$$

In this case, we obtain the intensity formula as follows.

$$I'' = KI_0 \left(\frac{1+\cos^2 2\theta \cos^2 2\theta_M}{2}\right) \tag{5}$$

If I_0 is expressed by I' based on (3) and substituted into (5), the following equation is obtained.

$$I'' = I' \left(\frac{1+\cos^2 2\theta_M \cos^2 2\theta}{1+\cos^2 2\theta_M}\right) \tag{6}$$

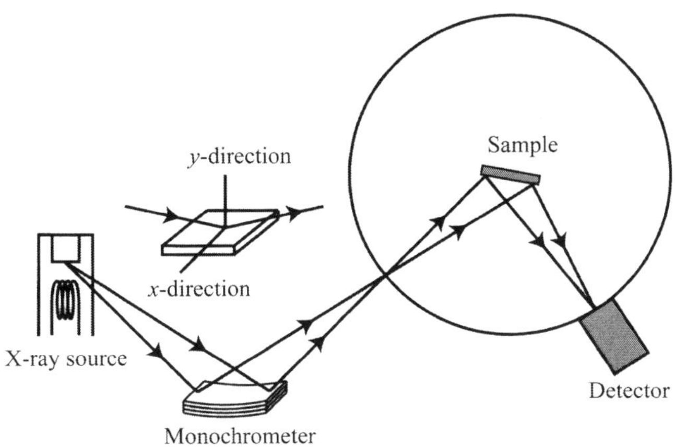

Fig. 1 Schematic diagram of geometry for a diffractometer when placing a crystal monochromator in the incident beam

Question 5.14 Answer the following questions concerning the effect of an anomalous dispersion term on the structure factor of zinc blende (ZnS). Let set X ray atomic scattering factors of Zn and S be f_{Zn} and f_{S}, and the real part and imaginary part of anomalous dispersion terms are f'_{Zn}, f''_{Zn} and f'_{S}, f''_{S}, respectively. The positions of Zn in zinc blende are (0 0 0), (0 1/2 1/2), (1/2 0 1/2), (1/2 1/2 0), and those of S are (1/4 1/4 1/4), (1/4 3/4 3/4), (3/4 1/4 3/4), (3/4 3/4 1/4).

(1) Obtain the structure factor F_{hkl} and $|F_{hkl}|^2$ including the anomalous dispersion terms.
(2) Obtain the structure factors for two diffraction peaks 111 and $\bar{1}\bar{1}\bar{1}$.
(3) The values of the anomalous dispersion terms for Cu-Kα radiation are known to be $f'_{\mathrm{Zn}} = -1.6$, $f''_{\mathrm{Zn}} = 0.68$, $f'_{\mathrm{S}} = 0.32$, and $f''_{\mathrm{S}} = 0.56$. Compute the $|F_{hkl}|^2$ values for two cases, $hkl = 111$ and $\bar{1}\bar{1}\bar{1}$.
(4) Compute the possible % difference detected in the diffraction intensity between 111 and $\bar{1}\bar{1}\bar{1}$.

Answer 5.14

(1) The structure factor of zinc blende is as follows (see Question 3.13).

$$\begin{aligned} F_{hkl} &= \left\{ f_{\mathrm{Zn}} + f'_{\mathrm{Zn}} + \mathrm{i} f''_{\mathrm{Zn}} + (f_{\mathrm{S}} + f'_{\mathrm{S}} + \mathrm{i} f''_{\mathrm{S}})\mathrm{e}^{\mathrm{i}\pi\left(\frac{h+k+l}{2}\right)} \right\} \\ &\quad (1 + \mathrm{e}^{\mathrm{i}\pi(h+k)} + \mathrm{e}^{\mathrm{i}\pi(k+l)} + \mathrm{e}^{\mathrm{i}\pi(l+h)}) \\ |F_{hkl}|^2 &= \left\{ f_{\mathrm{Zn}} + f'_{\mathrm{Zn}} + \mathrm{i} f''_{\mathrm{Zn}} + (f_{\mathrm{S}} + f'_{\mathrm{S}} - \mathrm{i} f''_{\mathrm{S}})\mathrm{e}^{\mathrm{i}\pi\left(\frac{h+k+l}{2}\right)} \right\} \\ &\quad (1 + \mathrm{e}^{\mathrm{i}\pi(h+k)} + \mathrm{e}^{\mathrm{i}\pi(k+l)} + \mathrm{e}^{\mathrm{i}\pi(l+h)}) \\ &\quad \times \left\{ f_{\mathrm{Zn}} + f'_{\mathrm{Zn}} - \mathrm{i} f''_{\mathrm{Zn}} + (f_{\mathrm{S}} + f'_{\mathrm{S}} - \mathrm{i} f''_{\mathrm{S}})\mathrm{e}^{-\mathrm{i}\pi\left(\frac{h+k+l}{2}\right)} \right\} \\ &\quad (1 + \mathrm{e}^{-\mathrm{i}\pi(h+k)} + \mathrm{e}^{-\mathrm{i}\pi(k+l)} + \mathrm{e}^{-\mathrm{i}\pi(l+h)}) \end{aligned}$$

(2) The structure factors of 111 and $\bar{1}\bar{1}\bar{1}$ are described in the following equations. In the case of $hkl = 111$,

$$F_{111} = 4\{(f_{\mathrm{Zn}} + f'_{\mathrm{Zn}} + f''_{\mathrm{S}}) - \mathrm{i}(f_{\mathrm{S}} + f'_{\mathrm{S}} - f''_{\mathrm{Zn}})\}$$

$$|F_{111}|^2 = 16\{(f_{\mathrm{Zn}} + f'_{\mathrm{Zn}} + f''_{\mathrm{S}})^2 + (f_{\mathrm{S}} + f'_{\mathrm{S}} - f''_{\mathrm{Zn}})^2\}$$

In the case of $hkl = \bar{1}\bar{1}\bar{1}$,

$$F_{\bar{1}\bar{1}\bar{1}} = 4\{(f_{\mathrm{Zn}} + f'_{\mathrm{Zn}} - f''_{\mathrm{S}}) + i(f_{\mathrm{S}} + f'_{\mathrm{S}} + f''_{\mathrm{Zn}})\}$$

$$|F_{\bar{1}\bar{1}\bar{1}}|^2 = 16\{(f_{\mathrm{Zn}} + f'_{\mathrm{Zn}} - f''_{\mathrm{S}})^2 + (f_{\mathrm{S}} + f'_{\mathrm{S}} + f''_{\mathrm{Zn}})^2\}$$

(3) We can find $\left(\frac{\sin\theta}{\lambda}\right) = \frac{\sqrt{3}}{2a}$ in both cases of 111 and $\bar{1}\bar{1}\bar{1}$ by using the relationship of $\left(\frac{\sin\theta}{\lambda}\right)^2 = \frac{h^2+k^2+l^2}{4a^2}$. Then, it turns out to be $\left(\frac{\sin\theta}{\lambda}\right) = \frac{\sqrt{3}}{2\times 5.4109} = 0.16\text{Å}^{-1}$. We also obtain $f_{\text{Zn}} = 24.16$ and $f_{\text{S}} = 11.86$ for the atomic scattering factors based on the numerical data compiled in Appendix A.3. Substitute these numerical values for the corresponding terms in the equations acquired in question (2). Including the given values for the anomalous dispersion terms, we obtain the following results.

$$|F_{111}|^2 = 16\{(24.16 - 1.6 + 0.56)^2 + (11.86 + 0.22 - 0.68)^2\} = 10669$$

$$|F_{\bar{1}\bar{1}\bar{1}}|^2 = 16\{(24.16 - 1.6 - 0.56)^2 + (11.86 + 0.32 + 0.68)^2\} = 10390$$

(4)
$$\frac{|F_{111}|^2 - |F_{\bar{1}\bar{1}\bar{1}}|^2}{|F_{111}|^2} = \frac{279}{10699} = 0.026$$

Therefore, this result implies that 2.6% of difference can be detected. It is also noteworthy that such difference in the intensities of two planes is observed because of the noncentrosymmetric nature of the zinc blende structure.

Question 5.15 Cu_3Au is known to have a cubic lattice, and below the critical temperature 663 K, the Cu and Au atoms in Cu_3Au are arranged to form a perfectly ordered phase in which each unit cell contains one Au atom and three Cu atoms. Their positions are characterized as follows. The position of Au is 000, and the positions of Cu are $\frac{1}{2}\frac{1}{2}0$, $\frac{1}{2}0\frac{1}{2}$, and $0\frac{1}{2}\frac{1}{2}$. On the contrary, in disordered phase, Cu and Au atoms are arranged at random on the atomic sites so that the probability that a particular site is occupied by a specific element is simply proportional to the atomic fraction. That is, each site is occupied by a statistical average of $\frac{1}{4}$Au and $\frac{3}{4}$Cu:

(1) Derive the expression of F for the ordered phase.
(2) Derive the expression of F for the disordered phase.
(3) For what reflections will the F value be the same in both the ordered and disordered phases? Obtain also the reflections for not-equal case.
(4) Derive the expression of F^2 for the ordered phase by introducing the anomalous dispersion terms f' and f'' of both Cu and Au atoms.

Answer 5.15

(1) In the ordered phase, each unit cell contains the following particular atomic arrangement (see Fig. 1a).
Au : One site 000
Cu : Three sites $\frac{1}{2}\frac{1}{2}0$, $\frac{1}{2}0\frac{1}{2}$, $0\frac{1}{2}\frac{1}{2}$

$$F = f_{\text{Au}} + f_{\text{Cu}}[\text{e}^{\pi\text{i}(h+k)} + \text{e}^{\pi\text{i}(k+l)} + \text{e}^{\pi\text{i}(l+h)}]$$

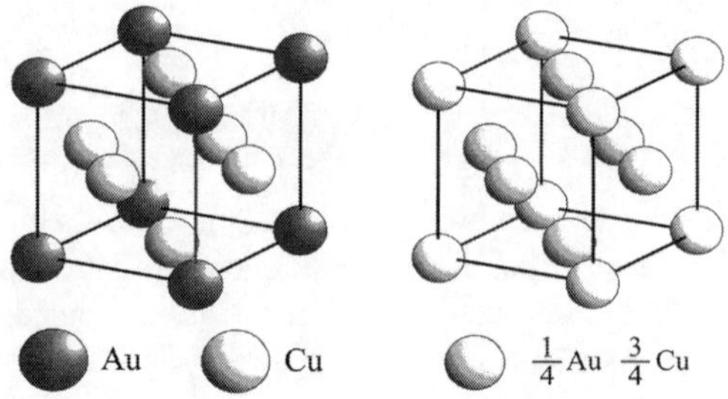

Fig. 1 Unit cells of Cu_3Au. (**a**) ordered phase and (**b**) disordered phase

(i) $h,\ k,\ l$: even/odd unmixed

$$\cos \to +1, \quad \sin \to 0 \quad (\text{when}\, h+k, k+l, l+h : \text{All are even})$$

$$F = f_{\text{Au}} + 3f_{\text{Cu}}$$

(ii) $h,\ k,\ l$: even/odd mixed

$$\cos \to -1, \quad \sin \to 0 \quad (h+k, k+l, l+h : \text{Any of two is odd})$$

$$F = f_{\text{Au}} - f_{\text{Cu}}$$

(2) In the disordered phase (see Fig. 1b), the atomic scattering factor of the average Cu-Au atom is as follows.

$$f_{av} = \frac{1}{4} f_{\text{Au}} + \frac{3}{4} f_{\text{Cu}}$$

$$\therefore F = f_{av}\left[1 + e^{2\pi i\left(\frac{h+k}{2}\right)} + e^{2\pi i\left(\frac{k+l}{2}\right)} + e^{2\pi i\left(\frac{l+h}{2}\right)}\right]$$

(i) $h,\ k,\ l$: even/odd unmixed

$$\cos \to 1, \quad \sin \to 0 \quad (h+k, k+l, l+h : \text{All are even})$$

$$F = f_{\text{Au}} + 3f_{\text{Cu}} = 4f_{av}$$

(ii) $h,\ k,\ l$: even/odd mixed

$$\cos \to -1, \quad \sin \to 0 \quad (h+k, k+l, l+h : \text{Any of two is odd}).$$

$$F = 0$$

Note that no reflections of mixed indices are observed.

(3) If $h,\ k,\ l$ are not even/odd unmixed, the F value of the ordered phase is found to be equal to that of the disordered phase. On the contrary, the difference is

found when h, k, l are even/odd number mixed. For this reason, the Bravais lattice of the ordered phase is simple cubic and that of the disordered phase is face-centered cubic.

(4) F^2 of the ordered phase
(i) even/odd unmixed

$$F^2 = (f^0_{\text{Au}} + f'_{\text{Au}})^2 + (f''_{\text{Au}})^2 + 6[(f^0_{\text{Au}} + f'_{\text{Au}}) \cdot (f^0_{\text{Cu}} + f'_{\text{Cu}}) + f''_{\text{Au}} \cdot f''_{\text{Cu}}] \\ +9[(f^0_{\text{Cu}} + f'_{\text{Cu}})^2 + (f''_{\text{Cu}})^2]$$

(ii) even/odd mixed

$$F^2 = (f^0_{\text{Au}} + f'_{\text{Au}})^2 + (f''_{\text{Au}})^2 - 2[(f^0_{\text{Au}} + f'_{\text{Au}}) \cdot (f^0_{\text{Cu}} + f'_{\text{Cu}}) + f''_{\text{Au}} \cdot f''_{\text{Cu}}] \\ +(f^0_{\text{Cu}} + f'_{\text{Cu}})^2 + (f''_{\text{Cu}})^2$$

Question 5.16 Answer the following questions concerning a Laue function in the form of $\frac{\sin^2(\pi NQd)}{\sin^2(\pi Qd)}$, using the case of $d = 0.3$ nm as an example.

(1) Explain the conditions where the Laue function gives a peak, its maximum value, and the reason why some small peaks appear in the region around the main peak.
(2) Compute the height of the second peak in close vicinity of the main peak using the case of $N = 20$ as an example.

Answer 5.16

(1) In the Laue function, N is the number of unit cells, d is a lattice spacing, and Q is a wave vector. If we set $x = \pi Qd$, both the denominator and the numerator of the Laue function will be zero for $x = n\pi$ (n is integer). The value of the Laue function for this limit can be found as follows.

$$\lim_{x \to n\pi} \left\{ \frac{\sin^2(Nx)}{\sin^2 x} \right\} = \lim_{x \to n\pi} \left\{ \frac{2N \sin(Nx) \cdot \cos(Nx)}{2 \sin x \cdot \cos x} \right\} \quad (1)$$

$$= \lim_{x \to n\pi} \frac{2N \times \frac{1}{2}\{\sin(2Nx)\}}{2 \times \frac{1}{2}\{\sin(2x)\}} \quad (2)$$

$$= \lim_{x \to n\pi} \frac{N \sin(N \cdot 2x)}{\sin 2x} = N^2 \quad (3)$$

where the following relationships are used.

$$\sin(at) \cdot \cos(bt) = \frac{1}{2}\{\sin(a+b)t + \sin(a-b)t\}$$

$$\lim_{t \to 0} \frac{\sin t}{t} = \lim_{t \to 0} \frac{t}{\sin t} = 1$$

$$\lim_{t \to 0} \frac{\sin bt}{\sin at} = \lim_{t \to 0} \frac{\sin bt}{bt} \cdot \frac{at}{\sin at} \cdot \frac{b}{a} = 1 \cdot 1 \cdot \frac{b}{a} = \frac{b}{a}$$

Therefore, the Laue function has the limit of N^2, corresponding to its maximum value, if x is zero, π, 2π, and 3π etc. For this reason, the Laue function is frequently represented in the normalized form by N^2.

The value of N itself stipulates the full peak width, i.e., it is deeply correlated with the peak sharpness in the vicinity of a reciprocal-lattice point. The larger the value of N is, the sharper the full peak width becomes. In other words, the peak width is inversely proportional to the number of unit cell found in the corresponding direction. Suppose $d = 0.3$ nm, peaks will be observed at the interval of $Q = 3.33\,\text{nm}^{-1}$ in the reciprocal lattice. Since Q and d correspond to variables in reciprocal space and real space, respectively, they exhibit mutually reciprocal relationships. For example, for peaks at a relatively small interval of Q, the d-value is large and, inversely, we find peaks at a relatively large interval of Q if the d-value becomes small.

Concerning the reason why a small peak appears in the region around the main peak, let us consider an extreme case where a sample consists only of three unit cells ($N = 3$). As shown in Fig. 1, a small peak observed in the middle of the two peaks may be called the partial interference result, which is characterized by the phase difference of π between the wave generated from a central unit cell and those of two unit cells of both ends.

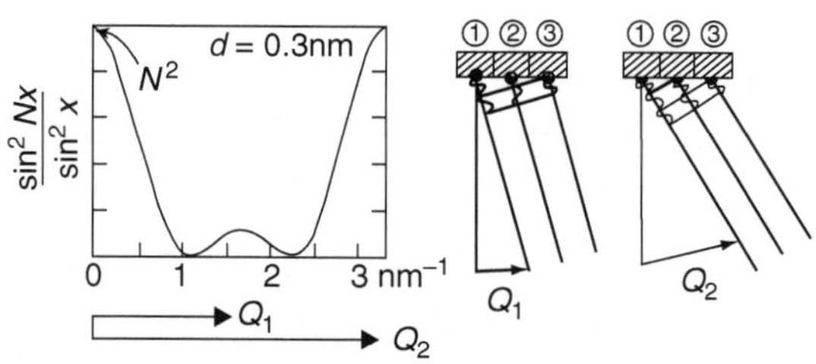

Fig. 1 Peak structure and its relevant factor appeared in the Laue function ($N = 3$)

(2) If the Laue function is calculated for the case of $N = 20$, we get the results illustrated in Fig. 2. The global maximum is found to be $N^2 = 400$. As a result, the local maximum of the 2nd peak is 18.7, that is about 4.7% of the global maximum.

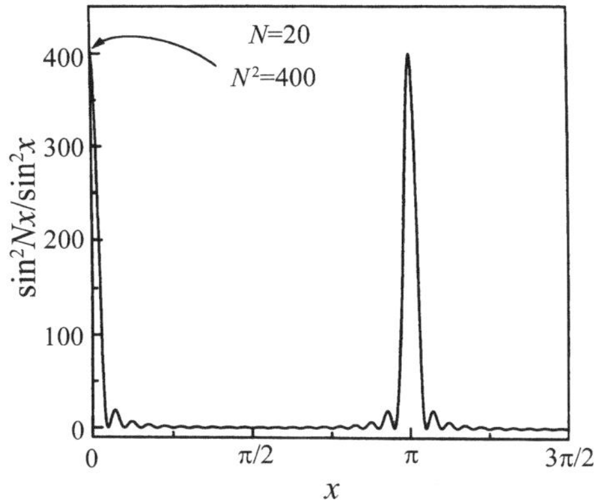

Fig. 2 Variations of the Laue function ($N = 20$)

Question 5.17 The function $y = \frac{\sin^2 Nx}{\sin^2 x}$ can be approximated by $y = N^2 \frac{\sin^2 \phi}{\phi^2}$ in the close vicinity of $x = 0$, where $\phi = Nx$.

(1) Derive an equation for providing the full peak width at half maximum ordinate given by this function.
(2) Derive an equation for providing the area under the peak given by this function.
(3) Discuss "what is the dependence of N," with respect to the maximum ordinate, of the full peak width at half maximum ordinate and of the area under the peak.

Answer 5.17

(1) Referring to the schematic diagram of Fig. 1, we compute the full width at half maximum ordinate of the following function

$$y = N^2 \frac{\sin^2 \phi}{\phi^2}$$

The purpose is to find ϕ, which gives $y = N^2$ and $y = N^2/2$ when $x = 0$.

$$N^2 \frac{\sin^2 \phi}{\phi^2} = \frac{N^2}{2}$$

$$\sin^2 \phi - \frac{\phi^2}{2} = 0$$

$$\left(\sin \phi + \frac{\phi}{\sqrt{2}}\right)\left(\sin \phi - \frac{\phi}{\sqrt{2}}\right) = 0$$

Since ϕ is also close to zero in the vicinity of $x = 0$, $\sin \phi$ and ϕ are the same signs and we find $\sin \phi + \frac{\phi}{2} \neq 0$.

Therefore, $\sin\phi - \frac{\phi}{\sqrt{2}} = 0$ so that we may use the approximation of $\sin\phi = x - \frac{x^3}{3!}$

$$\phi - \frac{\phi^3}{6} - \frac{\phi}{2} = 0$$
$$\frac{\phi^2}{6} - \left(\frac{\sqrt{2}-1}{\sqrt{2}}\right) = 0$$
$$\phi^2 = 3\sqrt{2}(\sqrt{2}-1) = 6 - 3\sqrt{2}$$
$$\phi = \pm\sqrt{6-3\sqrt{2}}$$

Therefore, it will be $x = \pm\frac{\sqrt{6-3\sqrt{2}}}{N}$, and the corresponding peak width Δx can be obtained as follows. $\Delta x = \frac{2\sqrt{6-3\sqrt{2}}}{N}$.

(2) Referring to the schematic diagram of Fig. 2, the area under the peak is estimated as follows. In the present case, the purpose is to find x, which gives $y = N^2 \frac{\sin^2 Nx}{\sin^2 x} = 0$.

From the condition $(\sin^2 Nx)/x^2 = 0$, we find $\sin Nx = 0$.

$$x = \frac{m\pi}{N} \quad (m \text{ is integer})$$

It will be $x = \pm\frac{\pi}{N}$ as $m = \pm 1$ in the close vicinity of $x = 0$. It is also found in this case that $\phi = \pm\pi$.

On the contrary, since $dx = d\phi/N$ from the definition $\phi = Nx$, we obtain the area under the peak in the following equation.

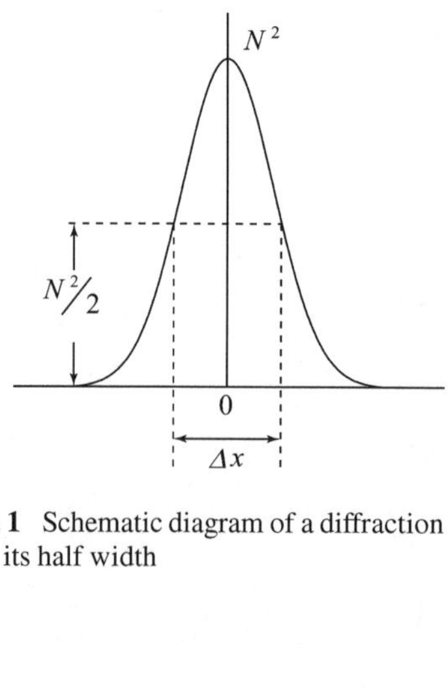

Fig. 1 Schematic diagram of a diffraction peak and its half width

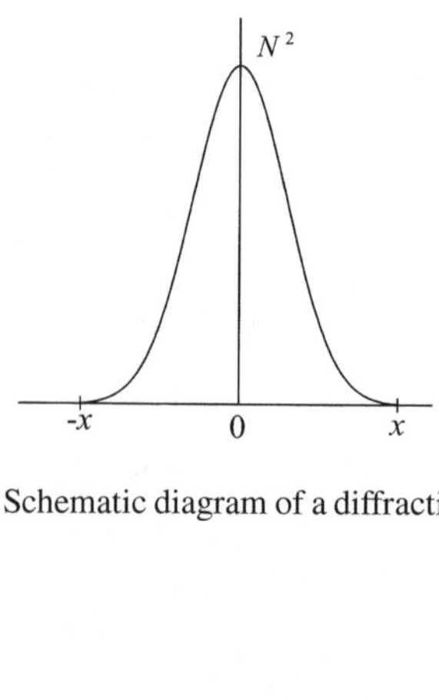

Fig. 2 Schematic diagram of a diffraction peak

$$\int_{-\pi/N}^{\pi/N} N^2 \frac{\sin^2 Nx}{N^2 x^2} \mathrm{d}x = \int_{-\pi}^{\pi} N^2 \frac{\sin^2 \phi}{\phi^2} \frac{\mathrm{d}\phi}{N} = 2N \int_0^{\pi} \frac{\sin^2 \phi}{\phi^2} \mathrm{d}\phi$$

Note: $\int_0^{\pi} \frac{\sin^2 \phi}{\phi^2} \mathrm{d}\phi$ is a constant that is independent on N.

(3) According to the above-mentioned results, the area under the peak is proportional to N, whereas the full peak width at half maximum ordinate is found to be proportional to N^{-1}. In addition, the maximum of the peak can be computed at the condition $x = 0$ ($\phi = 0$), and it is given in the following equation.

$$\lim_{\phi \to 0} N^2 \frac{\sin^2 \phi}{\phi^2} = N^2 \lim_{\phi \to 0} \left(\frac{\sin \phi}{\phi} \right)^2 = N^2$$

Therefore, the maximum of the peak is dependent on N^2.

Question 5.18 Answer the following questions concerning the measurement of the diffraction intensity in the symmetrical transmission mode for a thin slab of crystalline powder sample of mass per unit area M'. Suppose the incident beam and diffracted beams form equal angles with the surface of the slab sample, and the incident beam intensity is expressed by $I_0 A_0 = P_0$.

(1) Derive an equation providing the integrated intensity P' per unit length of the diffraction circle at a distance R from the thin slab sample.
(2) Discuss what value of M' will maximize P'.

Note: The integrated intensity P from the effective irradiated volume V is given by the following equation.

$$P = I_0 \left(\frac{e^4}{m_e^2 c^4} \right) \frac{V \lambda^3 p F^2}{4 v_a^2} \left(\frac{1 + \cos^2 2\theta}{2 \sin 2\theta} \right)$$

where λ is the wavelength, p is the multiplicity factor, F is the structure factor, and v_a is the volume of unit cell. The term given in parenthesis is the Lorentz polarization factor of a crystalline powder sample for an unpolarized incident X-ray beam I_0.

Answer 5.18

(1) The circumference of the Debye ring so as to form a cone of half apex angle θ located at distance R from a sample is $2\pi R \sin \theta$. Therefore, we find the following equation for the integrated intensity (see Fig. 5.4).

$$P' = \frac{P}{2\pi R \sin \theta} = \frac{I_0}{16\pi R} \left(\frac{e^4}{m_e^2 c^4} \right) \frac{V \lambda^3 p F^2}{v_a^2} \left(\frac{1 + \cos^2 2^{\theta}}{\sin \theta \sin 2\theta} \right) \qquad (1)$$

Here, V is the effective irradiated volume of a thin slab. By considering the geometrical relationship shown in Fig. 1 for the symmetrical transmission mode and setting the cross-sectional area to A_0, it can be replaced by the following equation.

$$V = \int_{t=1}^{t} \mathrm{e}^{(-\mu t \sec\theta)} A_0 \sec\theta \mathrm{d}x = A_0 t \sec\theta \mathrm{e}^{(-\mu t \sec\theta)} \tag{2}$$

From (1) and (2), we obtain.

$$P' = \frac{I_0 A_0}{16\pi R}\left(\frac{e^4}{m_e^2 c^4}\right)\frac{\lambda^3 p F^2}{v_a^2}\left(\frac{1+\cos^2 2\theta}{\cos\theta\sin^2\theta}\right) t\mathrm{e}^{(-\mu t \sec\theta)} \tag{3}$$

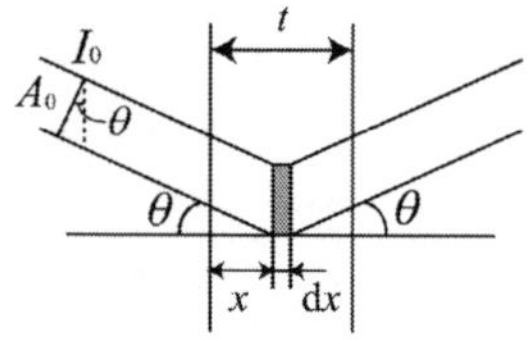

Fig. 1 Geometry for the symmetrical transmission mode

Since t corresponds to the thickness of the slab the sample is replaced with $t = M'/\rho_0$, where M' is mass and ρ_0 is density per unit area of the sample,

$$P' = \frac{I_0 A_0}{16\pi R}\left(\frac{e^4}{m_e^2 c^4}\right)\frac{\lambda^3 p F^2}{v_a^2}\left(\frac{1+\cos^2 2\theta}{\cos\theta\sin^2\theta}\right) \frac{m}{\rho_0}\mathrm{e}^{-\frac{\mu m}{\rho_0}\sec\theta} \tag{4}$$

(2) Setting $g = \frac{M'}{\rho_0}\mathrm{e}^{-\frac{\mu m}{\rho_0}\sec\theta}$, P' has its maximum if g is maximized, so we need to find the condition where the derivative of g is zero.

$$\begin{aligned} g' &= \frac{1}{\rho_0}\mathrm{e}^{-\frac{\mu M'}{\rho_0}\sec\theta} - \frac{\mu M'}{\rho_0}\sec\theta\mathrm{e}^{-\frac{\mu M'}{\rho_0}\sec\theta} \\ &= \frac{1}{\rho_0}\left(1 - \frac{\mu M'}{\rho_0}\sec\theta\right)\mathrm{e}^{-\frac{\mu M'}{\rho_0}\sec\theta} \end{aligned} \tag{5}$$

Based on this relationship, the maximum value of P' will be obtained at $M' = \rho_0/(\mu\sec\theta)$. However, keep in mind that it is actually not so easy to control M' using this relationship.

Chapter 6
Symmetry Analysis for Crystals and the Use of the International Tables

6.1 Symmetry Analysis

A crystal may be defined as a solid composed of atoms arranged on a regular three-dimensional lattice and such periodicity in the atomic distribution features their structure. The geometry of atomic distributions in crystals is known to be characterized by the repetition, such as lattice translation (see Chap. 2). In addition to lattice translations, we find reflection and rotation. In these cases, an object is brought into coincidence with itself by reflection in a certain plane; rotation upon around a certain axis; or reflection in a certain plane. The repetition of a pattern by specific rules characterizes all symmetry operations and their fundamental points are given below.

If a symmetry operation leaves a locus, such as a point, a line, or a plane unchanged (i.e., same atomic position), this locus is referred to as the symmetry element. For any operation excluding lattice translation for space group, the symmetry operation belongs to one of four cases: inversion (i) expressed by a change from (x, y, z) to $(-x, -y, -z)$; rotatory-inversion ($\bar{n}$); reflection (m: a mirror plane) expressed by a change from (x, y, z) to $(-x, y, z)$; and rotation (n, a rotation axis) expressed by a change $(360°/n)$ about an axis. In crystals, we may conclude that only one-, two-, three-, four-, and sixfold rotation axes can be accepted. Of course, the symmetry operations may be linked with one another. The symmetry operation called inversion relates a pairs of objects which are equidistant from and on opposite sides of a central point (called inversion center). That is, only a single point remains unchanged. Whereas, rotatory-inversion is one of the compound symmetry operations and it is frequently called roto-inversion. This operation is produced by the combination of a rotation of $(360°/n)$ around a certain axis followed by inversion through a point on the axis as a symmetry center.

As shown in Fig. 6.1, the rotatory-inversion operation provides five cases denoted by symbols $\bar{1}$, $\bar{2}$, $\bar{3}$, $\bar{4}$, and $\bar{6}$. However, the following three points are noteworthy. (1) Since the rotatory-inversion operation of $\bar{1}$ is a rotation of 360° followed by inversion through a point on the onefold roto-inversion axis, it is identical to inversion (i), simply called center of symmetry or inversion center. (2) The rotatory-inversion operation of $\bar{2}$ is represented by rotation through an angle of 180° followed by inversion to take one point into an equivalent one. However, these two points are

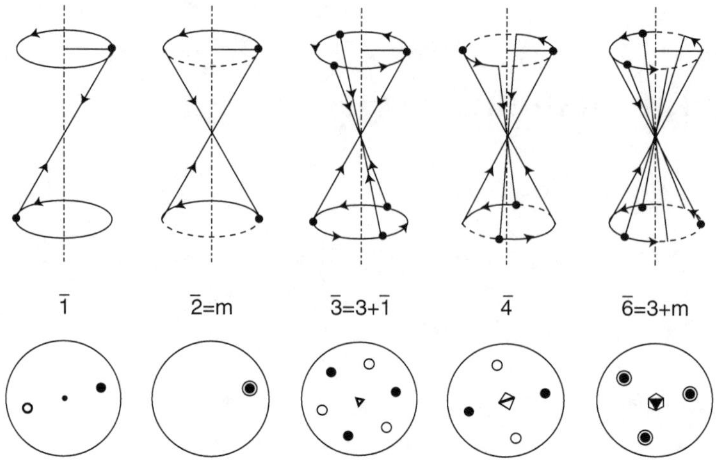

Fig. 6.1 Examples of rotatory-inversion operation

also related to one another by reflection in a plane normal to the axis, so that the rotatory-inversion of $\bar{2}$ is identical to a mirror reflection (m). (3) As easily seen in Fig. 6.1, successive applications of the rotation-inversion operation of $\bar{3}$ alter a point into altogether six equivalent positions. This variation can be reproduced by combining operations with a threefold rotation axis and inversion (i). Similarly, the rotation-inversion operation of $\bar{6}$ is also represented by combining operations with a threefold rotation axis and a twofold roto-inversion axis, or mirror plane perpendicular to the axis. These three points suggest that the rotatory-inversion operations except for $\bar{4}$ result in no new operation, so that $\bar{1}$, $\bar{2}$, $\bar{3}$, and $\bar{6}$ are not included in the independent symmetry operations.

For convenience, some additional details are given for understanding the rotatory-inversion operation of $\bar{4}$, using two different cases as an example. In Fig. 6.2a, we easily find the results obtained by symmetry operations of fourfold rotation and inversion, because successive operations about the fourfold axis move a point from 1 to 2, 3, 4, and back to 1. On the other hand, the inversion center alters it from each of those positions to 7, 8, 5, and 6, respectively. This combination of symmetry operations results in a mirror plane normal to the axis. In this case, two individual symmetry operations are found to be linked which are themselves symmetry operations. Whereas, in Fig. 6.2b illustrating the results obtained by symmetry operation of fourfold rotation about an axis followed by inversion through a point on its axis. Successive applications of these operations move a point at 1 to 2, 3, 4, and back to 1. In this case, we can find neither the inversion center nor the fourfold rotation axis (see auxiliary points denoted by open circles in Fig. 6.2b. Namely, two symmetry operations in this case are made in sequence as a single matter referred to as a new symmetry operation, which is called compound symmetry operation.

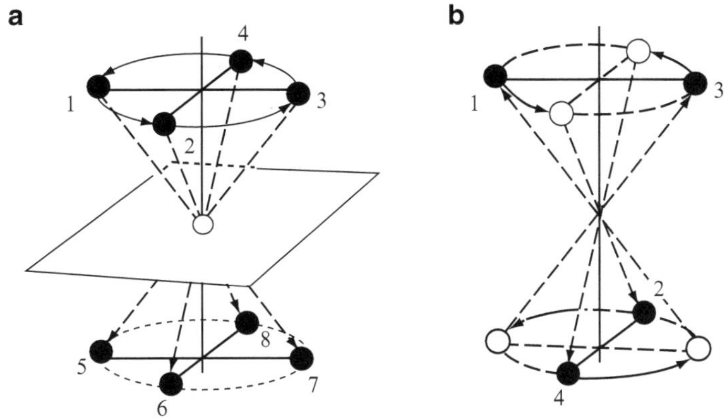

Fig. 6.2 Applications of the fourfold rotation and inversion. (**a**) Combination of independent symmetry operation, (**b**) Compound operation (roto-inversion)

In conclusion, the independent symmetry operations for discussing the symmetry of the three-dimensional atomic arrangement are eight: inversion (i), reflection (m), rotation (1, 2, 3, 4, 6), and rotatory-inversion of $\bar{4}$. This means that the whole periodic array observed in crystals can be covered by repeating the parallel translation (translational operation) of the structure derived from these eight kinds of symmetry element. In other words, as already mentioned in Chap. 2, there are seven crystal systems for classification, which consist of 14 kinds of Bravais lattices and a crystal is known to be classified into 32 point groups on the basis of eight kinds of symmetry element. In addition, when it is extended to include space groups, by adding point groups, Bravais lattices, screw axis, and glide reflection plane, it will be classified into 230 in total. For this analysis, we have to include two compound symmetry operations; rotation and translation (screw axis) and reflection and translation (glide plane). A brief description for screw axes and glide reflection planes is given below.

The symmetry operation of the so-called "screw rotation" consists of a rotation of $360°/n$ where n is 2, 3, 4, and 6 and a translation by a vector parallel to the axis. The screw axis is expressed by n_m and its operation is to translate by (m/n) times the length of a unit lattice vector along the direction of a rotation axis every one operation with respect to n-axis of rotation. All of the screw rotation axes are shown in Fig. 6.3 (see also Question 6.1). Although the direction of rotation itself is not so important in the screw rotation, the definition is illustrated in Fig. 6.4 using a right-handed axial system as an example. This case shows an operation which is a rotation around a c-axis from the a-axis toward the b-axis by an angle ϕ followed by a positive translation along a c-axis, called the motion of a right-handed screw.

The compound symmetry operation of a glide reflection consists of a reflection and a translation by the vector $\boldsymbol{q}_g$ parallel to the plane of reflection. For convenience, Fig. 6.5 shows a comparison of the operation of a glide plane with that of a mirror plane on a point lying off the planes.

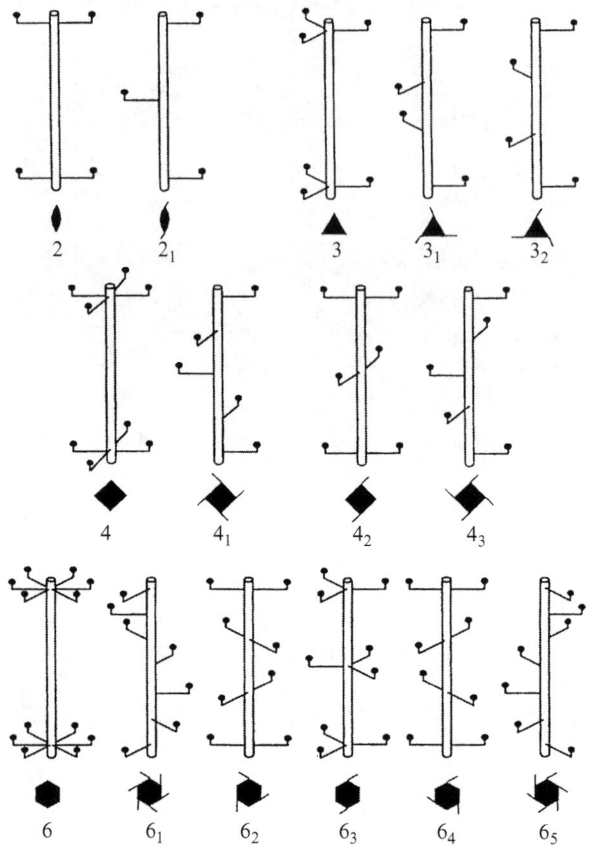

Fig. 6.3 Examples of possible screw axes

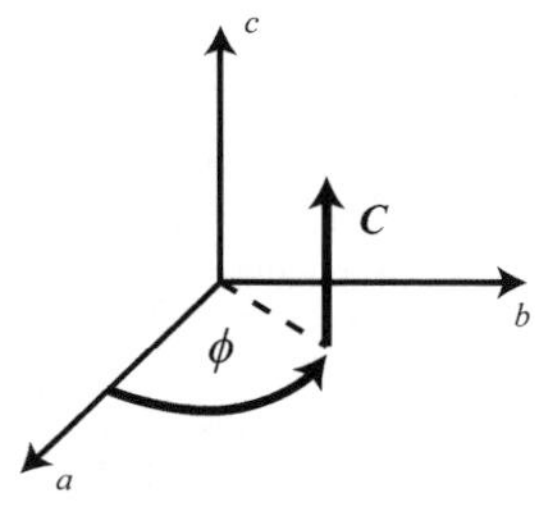

Fig. 6.4 A right-handed screw system

The description of this symmetry element may also be simplified by reference to the unit lattice vectors $\boldsymbol{a}$, $\boldsymbol{b}$, and $\boldsymbol{c}$. For example, with respect to a-glide plane perpendicular to the b-direction, the reflection operation is made through the glide plane and then displaced by the vector, corresponding to one-half of a lattice translation $\left(\frac{1}{2}\boldsymbol{a}\right)$, parallel to the glide plane. Similarly, we may obtain b- and c-glide planes perpendicular to one of the other directions. There is also a diagonal glide plane n by

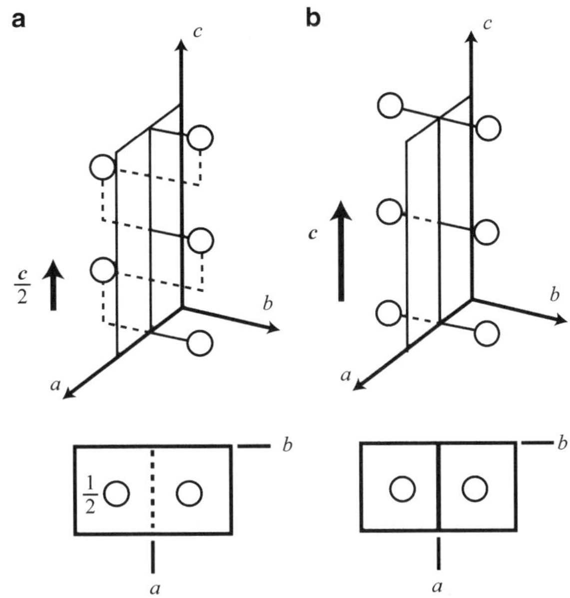

Fig. 6.5 Comparison of the operation of (**a**) a glide plane and (**b**) a mirror plane

translation of the diagonal direction. That is, the n-glide plane, if it is perpendicular to c, gives a glide component of $\left(\frac{1}{2}\boldsymbol{a} + \frac{1}{2}\boldsymbol{b}\right)$. Furthermore, we have one additional case; the diamond glide plane denoted by d which can be featured by one quarter $(1/4)$ of a lattice translation along the line parallel to the body-centered direction. Then, there are five kinds of glide reflection planes in all.

As mentioned previously, all crystal structures can be classified into seven crystal systems using parameters of three vectors $\boldsymbol{a}$, $\boldsymbol{b}$, and $\boldsymbol{c}$ (or those lengths a, b, and c) and the interaxial angles between them, α, β, and γ. The relationships between crystal systems and symmetry elements are summarized in Table 6.1.

The atomic distribution in crystals is characterized by its periodicity in a regular three-dimensional lattice and it is known to be classified into 32 point groups using eight symmetry elements (see Fig. 2.1). In addition, the periodicity in regular three-dimensional lattice can be analyzed by the concept of symmetry. Particularly, if we use eight kinds of symmetry element; "reflection (m)," "rotation (1, 2, 3, 4 and 6)," "inversion (i)," and "rotatory-inversion ($\bar{4}$)" together with eleven "screw axes" and five "glide planes," all the possible geometric arrangement of atoms in three-dimensional lattice space can be classified into 230 in all, called "space groups." This implies that the number of geometric arrangements with periodicity is limited in three-dimensional lattice space. In other words, any crystal can be described only by one of the 230 space groups. In addition, the real crystal structures are not evenly distributed over these 230 space groups. It is rather unevenly distributed, so that there are many space groups that do not represent any real crystal structure. The relationships of crystal systems with point group and space groups are summarized in Table 6.2.

Table 6.1 Seven crystal systems, symmetry elements, and relevant data

Crystal systems	Subunit	Minimum symmetry elements	Point groups	Number of rotation axes 2346	Number of mirror plane
Triclinic	$a \neq b \neq c$ $\alpha \neq \beta \neq \gamma \neq 90°$	None	$1, \bar{1}$	0000	0
Monoclinic	$a \neq b \neq c$ $\alpha = \gamma = 90° \beta \neq 90°$	One diad axis or mirror plane	2, m, 2/m	1000	1
Orthorhombic	$a \neq b \neq c$	Three orthogonal diad or inverse diad axis	$222, mm2, mmm$	3000	3
Tetragonal	$a = b \neq c$ $\alpha = \beta = \gamma = 90°$	One tetrad or inverse tetrad axis	$4, \bar{4}, 4/m, 422, 4mm, \bar{4}2m, 4/mmm$	4010	5
Cubic	$a = b = c$ $\alpha = \beta = \gamma = 90°$	Form triad axes	$23, m3, 432, \bar{4}3m\ m3m$	6430	9
Trigonal	$a = b \neq c$ $\alpha = \beta = 90°\ \gamma = 120°$ or $a = b = c$ $\alpha = \beta = \gamma \neq 90°$	One triad or inverse triad axis	$3, \bar{3}, 32, 3m, \bar{3}m$	3100	3
Hexagonal	$a = b \neq c$ $\alpha = \beta = 90°$ $\gamma = 120°$	One hexad or inverse hexad axis	$6, \bar{6}, 6/m622, 6mm, \bar{6}m2, 6/mmm$	6001	7

6.2 International Tables

It is not necessary to know all (230) the space groups individually, because we can easily get many of the most important information of the space groups from "International Tables for Crystallography Vol. A, Fifth Edition (2002)" published by the International Union of Crystallography (referred to as IUCr). Although two methods, Schönflies symbols and Hermann–Mauguin symbols have been used for describing the 32 point groups (see Question 6.2), the IUCr suggests the use of Hermann–Mauguin symbols. Note that Schönflies symbols are widely used in respect to spectroscopy, especially of organic molecules. This handbook is quite useful for determining or interpreting the structure of crystals of interest. Although the contents cover only 24 space groups frequently found in real crystals, a textbook is also available for beginners titled on Brief Teaching Edition of Volume A Space-group Symmetry, International Tables for Crystallography, Fifth Revised Edition, edited by T. Hahn, Kluwer Academic Publishers, Dordrecht, and Holland (2002). International Tables for Crystallography (often referred to as International Tables) provides the following information:

(1) Short space group symbol, Schönflies symbol, point group, crystal system, number of the space group, full space group symbol, and Patterson symmetry symbol.

Table 6.2 Crystal systems and relations to point groups and space groups

Crystal system	Point	Space groups
Triclinic	1	$P1$
	$\bar{1}$	$P\bar{1}$
Monoclinic	2	$P2, P2_1, C2$
	m	Pm, Pc, Cm, Cc
	$2/m$	$P2/m, P2_1/m, C2/m, P2/c, P2_1/c, C2/c$
Orthorhombic	222	$P222, P222_1, P2_12_12, P2_12_12_1, C222_1, C222, F222, I222, I2_12_12_1$
	$mm2$	$Pmm2, Pmc2_1, Pcc2, Pma2_1, Pca2_1, Pnc2_1, Pmn2_1, Pba2, Pna2_1,$
		$Pnn2, Cmm2, Cmc2_1, Ccc2, Amm2, Abm2, Ama2, Aba2, Fmm2,$
		$Fdd2, Imm2, Iba2, Ima2, Aba2, Fmm2, Fdd2, Imm2, Iba2, Ima2$
	mmm	$Pmmm, Pnnn, Pccm, Pban, Pmma, Pnna,$
		$Pmna, Pcca, Pbam, Pccn, Pbcm, Pnnm, Pmmn, Pbcn, Pbca, Pnma,$
		$Cmcm, Cmca, Cmmm, Cccm, Cmma, Ccca, Fmmm, Fddd,$
		$Immm, Ibam, Ibca, Imma$
Tetragonal	4	$P4, P4_1, P4_2, P4_3, I4, I4_1$
	$\bar{4}$	$P\bar{4}, I\bar{4}$
	$4/m$	$P4/m, P4_2/m, P4/n, P4_2/n, I4/m, I4_1/a$
	422	$P422, P42_12, P4_122$, $P4_12_12$, $P4_222$, $P4_22_12$
		$P4_322, P4_32_12, I422, I4_122$
	$4mm$	$P4mm, P4bm, P4_2cm, P4_2nm, P4cc, P4nc,$
		$P4_2mc, P4_2bc, I4mn, I4cm, I4_1md, I4cd$
	$\bar{4}2m$	$P\bar{4}2m, P\bar{4}2c, P\bar{4}2_1m, P\bar{4}2_1c, P\bar{4}m2, P\bar{4}c2,$
		$P\bar{4}b2, P\bar{4}n2, I\bar{4}m2, I\bar{4}c2, I\bar{4}2m, I\bar{4}2d$
	$4/mmm$	$P4/mmm, P4/mcc, P4/nbm, P4/nnc, P4/mbm,$
		$P4/mnc, P4/nmm, P4/ncc, P4_2/mmc, P4_2/mcm,$
		$P4_2/nbc, P4_2/nnm, P4_2/mbc, P4_2/mnm, P4_2/nmc,$
		$P4_2/ncm, I4/mmm, I4/mcm, I4_1/amd, I4_1/acd$
Trigonal–hexagonal	3	$P3, P3_1, P3_2, R3$
	$\bar{3}$	$P\bar{3}, R\bar{3}$
	32	$P312, P321, P3_112, P3_121, P3_212, P3_221, R32$
	$3m$	$P3m1, P31m, P3c1, P31c, R3m, R3c$
	$\bar{3}m$	$P\bar{3}1m, P\bar{3}1c, P\bar{3}m1, P\bar{3}c1, R\bar{3}m, R\bar{3}c$
	6	$P6, P6_1, P6_5, P6_3, P6_2, P6_4$
	$\bar{6}$	$P\bar{6}$
	$6/m$	$P6/m, P6_3/m$
	622	$P622, P6_122, P6_522, P6_222, P6_422, P6_322$
	$6mm$	$P6mm, P6cc, P6_3cm, P6_3mc$
	$\bar{6}\,m$	$P\bar{6}2m, P\bar{6}c2, P\bar{6}2m, P\bar{6}2c,$
	$6/mmm$	$P6/mmm, P6/mcc, P6_3/mcm, P6_3/mmc$
Cubic	23	$P23, F23, I23, P2_13, I2_13$
	$m\,\bar{3}$	$Pm\bar{3}, Pn\bar{3}, Fm\bar{3}, Fd\bar{3}, Im\bar{3}, Pa\bar{3}, Ia\bar{3}$
	432	$P432, P4_232, F432, F4_132, I432, P4_332, P4_132, I4_132$
	$\bar{4}3m$	$P\bar{4}3m, F\bar{4}3m, I\bar{4}3m, P\bar{4}3n, F\bar{4}3c, I\bar{4}3d$
	$m\bar{3}m$	$Pm\bar{3}m, Pn\bar{3}n, Pm\bar{3}n, Pn\bar{3}m, Fm\bar{3}m, Fm\bar{3}c,$
		$Fd\bar{3}m, Fd\bar{3}c, Im\bar{3}m, Ia\bar{3}d$

(2) Projection of the symmetry elements of the space group along special axes (high symmetry). The origin is in the upper left corner.
(3) Projection of a general position.
(4) Information for the selection of the origin.
(5) Asymmetric unit.
(6) Symmetry operations of the space group.
(7) General and special positions, multiplicity Wyckoff letter, site symmetry, and coordinates of equivalent positions.

Patterson symmetry is the symmetry of the Patterson function for Fourier transformation and Wyckoff letter provides the equivalent positions in a unit cell. More details are obtained from the International Tables for Crystallography.

The first alphabetical capital letter is to show the lattice symbol of the Bravais lattices (***P, F, I, A, B, C,*** and ***R***) as summarized in Table 6.3 and next three characters indicate symmetry elements related to the particular orientation in crystal systems as summarized in Table 6.4. For example, we find ***Cmm*2** (orthorhombic) in the International Tables, Vol.A, p.238. This ***Cmm*2** shows that space lattice is base-centered (***C***) and next three characters of ***mm*2** inform us the symmetry elements with respect to directions of [100], [010], and [001], respectively. That is, this orthorhombic space lattice shows mirror plane *m*, perpendicular to both ***a***- and ***b***-axes and a twofold rotation axis along the ***c***-axis. For another example, $\boldsymbol{P12_1/c1}$ (monoclinic) found in the International Tables, Vol.A, p.184 shows that space lattice is primitive (simple) and it has the $\mathbf{2_1}$ screw axis parallel to ***b***-axis and the ***c***-glide plane which is perpendicular to the $\mathbf{2_1}$ screw axis. Since the description of space groups is generally used in a simplified form as much as possible, so-called short space group symbol, for example, $\boldsymbol{P12_1/c1 \rightarrow P2_1/c}$ and $\boldsymbol{F4/m\bar{3}\,2/m \rightarrow Fm\bar{3}\,m}$ (see Vol. A, p. 688). Then, the users are requested to get familiar with the relationships between symbols and crystal systems including image of atomic positions. Practice makes perfect (see Questions 6.1–6.6).

Table 6.3 Number and coordinates of the lattice points in the unit cells of Bravais lattices

Lattice symbols	No. of lattice points in a unit cell	Coordinates of lattice points in a unit cell
P	1	0,0,0
A	2	$0,0,0\ ;\ 0,\frac{1}{2},\frac{1}{2}$
B	2	$0,0,0\ ;\ \frac{1}{2},0,\frac{1}{2}$
C	2	$0,0,0\ ;\ \frac{1}{2},\frac{1}{2},0$
I	2	$0,0,0\ ;\ \frac{1}{2},\frac{1}{2},\frac{1}{2}$
R	3	$0,0,0\ ;\ \frac{2}{3},\frac{1}{3},\frac{1}{3}\ ;\ \frac{1}{3},\frac{2}{3},\frac{2}{3}$
F	4	$0,0,0\ ;\ \frac{1}{2},\frac{1}{2},0\ ;\ \frac{1}{2},0,\frac{1}{2}\ ;\ 0,\frac{1}{2},\frac{1}{2}$

Table 6.4 The order of Hermann–Mauguin symbols and their relation to directions in a crystal

Crystal systems	1st index	2nd index	3rd index
Triclinic	None		
Monoclinic	[010](*b-axis*)* [001](*c-axis*)*		
Orthorhombic	[100]	[010]	[001]
Tetrogonal	[001]	$\left\{\begin{matrix}[100]\\ [010]\end{matrix}\right\}$	$\left\{\begin{matrix}[1\bar{1}0]\\ [110]\end{matrix}\right\}$
Trigonal Referred to hexagonal axes	[001]	$\left\{\begin{matrix}[100]\\ [010]\\ [\bar{1}\bar{1}0]\end{matrix}\right\}$	
Trigonal Referred to rhombohedral axes	[111]	$\left\{\begin{matrix}[1\bar{1}0]\\ [01\bar{1}]\\ [\bar{1}01]\end{matrix}\right\}$	
Hexagonal	[001]	$\left\{\begin{matrix}[100]\\ [010]\\ [\bar{1}\bar{1}0]\end{matrix}\right\}$	$\left\{\begin{matrix}[1\bar{1}0]\\ [120]\\ [\bar{2}\bar{1}0]\end{matrix}\right\}$
Cubic	$\left\{\begin{matrix}[100]\\ [010]\\ [001]\end{matrix}\right\}$	$\left\{\begin{matrix}[111]\\ [1\bar{1}\bar{1}]\\ [\bar{1}1\bar{1}]\\ [\bar{1}\bar{1}1]\end{matrix}\right\}$	$\left\{\begin{matrix}[1\bar{1}0] & [110]\\ [01\bar{1}] & [011]\\ [\bar{1}01] & [101]\end{matrix}\right\}$

* Orthogonal axis = Unique axis.

The International Tables do not cover only Vol.A focusing on symmetry of space groups, but also Vol.B published in 2001 covering information about the reciprocal lattice, structure factor, Fourier transform, and others including structural analysis by diffuse scattering, dynamical theory, and its applications. One can also find Vol.C, Third Edition in 2004 providing mathematical, physical, and chemical tables including absorption coefficients and X-ray atomic scattering factors. This volume includes sample preparation techniques, methods for the determination of lattice parameters, refining techniques for the structure determination. Further, Vol. D published in 2003 Physical Properties of Crystals, Vol.E in 2002, Sub-periodic Groups, Vol. F in 2001 Crystallography of Biological Macromolecules, and Vol.G in 2005, Definition and Exchange of Crystallographic Data. There is also A1, "Symmetry relations between space groups" published in 2004. It should be suitably selected, depending on the purpose.

6.3 Solved Problems (8 Examples)

Question 6.1 Explain screw axes and glide planes which are important for analyzing the three-dimensional regular array in crystal lattice.

Answer 6.1 A three-dimensional periodic array in crystals is known to be reproduced by the infinite repetition of identical structural units, but we find the following problem.

When supposing an operation around a point brings it to self-coincidence, it is difficult to distinguish between the result obtained by the one cycle operation and that where it returns to the original position from another lattice point separated with several cycles. For this purpose, the operations of rotation and translation may be linked with one another. This is particularly true for the space groups of centered lattices. That is, it is necessary to introduce screw rotation and glide reflection.

The screw axis may be denoted by n_m and its operation is to translate by (m/n) times the length of a unit lattice vector along the direction of a rotation axis every one operation about the n-axis of rotation. As a specific example, a comparison is made in Fig. 1 using the relationships between twofold rotation axis and twofold screw rotation axis. In the operation of twofold screw rotation axis (2_1), the point alters from A to B due to a translation operation applied by half of a unit lattice translation along the direction parallel to the long-axis after the 180° rotation as similar to the twofold rotation axis case and if the same operation is repeated, the point A does not return to its starting point. This operation does not come into coincidence with itself. The point A moves the point corresponding to the position applied by a unit lattice translation along the direction parallel to the long axis.

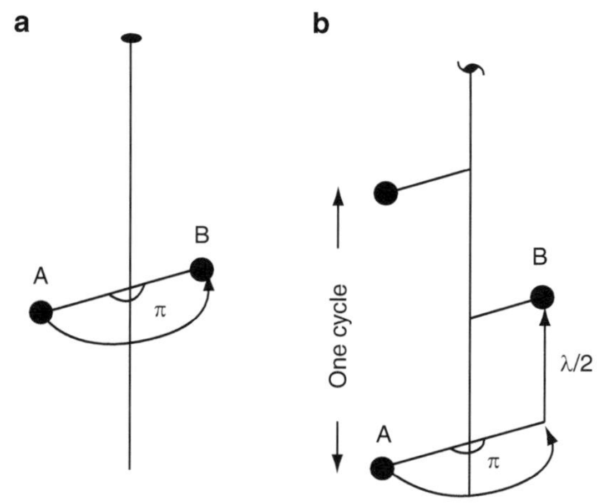

Fig. 1 Twofold rotation axis (**a**) and twofold screw rotation axis (**b**)

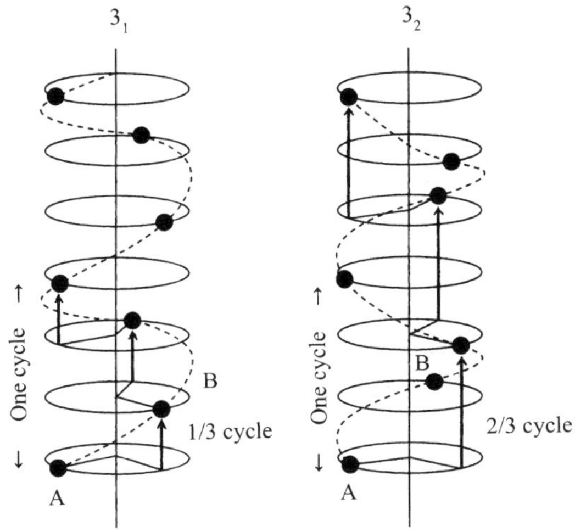

Fig. 2 Threefold screw axes

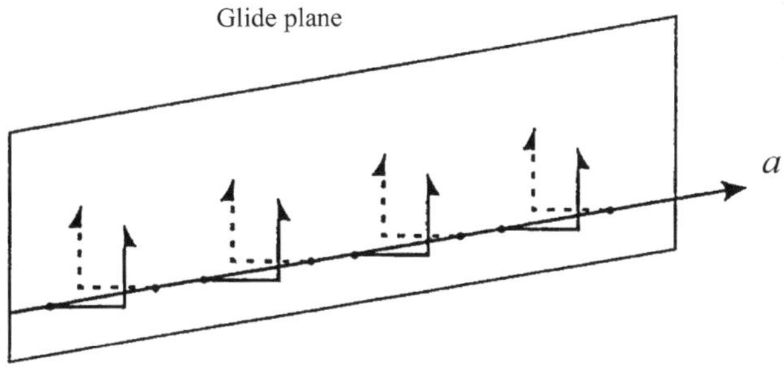

Fig. 3 Example of a glide reflection operation denoted by "a-glide plane"

In the operation of threefold screw rotation axis, a translation operation is made by 1/3 of a unit lattice translation along the direction of a rotation axis after every 120° rotation. However, as shown in Fig. 2, keep in mind that there are two ways like the treads of a spiral staircase, clockwise and counterclockwise. In order to distinguish these two cases, the threefold screw rotation axis describes clockwise (3_1) and counterclockwise (3_2). It may be added that there are three fourfold screw rotation axes and five sixfold screw rotation axes (see Fig. 6.3) and screw rotation axes are only allowed in crystals parallel to those directions possibly accepted for rotation axes in the corresponding point group.

Glide plane is divided into three categories; axial glide plane and diagonal and diamond glide plane. Reflection across the so-called mirror plane is followed by translation parallel to the plane by one-half of the length of a unit lattice translation vector. For example, "a-glide plane" is the case where it projects on a mirror surface in pairs and a translation operation by ($\frac{1}{2}\boldsymbol{a}$) is made along the direction parallel to the mirror surface (see Fig. 3). Similarly, there are axial b-glide plane and c-glide plane.

In other words, glide planes are designated by symbols suggesting the relationships of their glide components to the unit lattice vectors.

The n-glide plane is characterized by a translation operation along the diagonal direction. In addition, the diamond glide plane denoted by d-glide plane features one-quarter (1/4) of a lattice translation along the line parallel to the body-centered direction. Symbols and relevant information of these glide planes including a mirror plane are summarized in Table 1.

Table 1 Symbols and their relevance for glide planes including a mirror plane

Symbols	Graphic symbols		Glide plane direction and translation	Limiting conditions
m		⊥ paper	—	—
	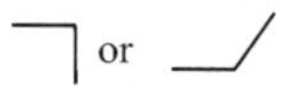	// paper	—	—
a		⊥ paper	$(h0l)$ $a/2$	$h0l : h = 2n$
	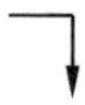	// paper	$(hk0)$ $a/2$	$hk0 : h = 2n$
b		⊥ paper	$(0kl)$ $b/2$	$0kl : k = 2n$
		// paper	$(hk0)$ $b/2$	$hk0 : k = 2n$
c		{ ⊥ paper ⊥ paper	$(0kl)$ $c/2$ $(h0l)$ $c/2$	$0kl : l = 2n$ $h0l : l = 2n$
n		{ ⊥ paper ⊥ paper	$(0kl)$ $(b + c)/2$ $(h0l)$ $(c + a)/2$	$0kl : k + l = 2n$ $h0l : l + h = 2n$
		// paper	$(hk0)$ $(a + b)/2$	$hk0 : h + k = 2n$
d		{ ⊥ paper ⊥ paper	$(0kl)$ $(b \pm c)/4$ $(h0l)$ $(c \pm a)/4$	$0kl : k + l = 4n$ $h0l : l + h = 4n$
		// paper	$(hk0)$ $(a \pm b)/4$	$hk0 : h + k = 4n$

Question 6.2 A crystal structure is known to be characterized by points in an infinite three-dimensional regular array. Explain the geometry of crystals using symmetry elements. Keep in mind the viewpoint of symmetry given in a crystal structure, not the mathematical issue.

Answer 6.2 In crystals, a three-dimensional periodic arrangement of atoms is always present and such nature is represented by points. The concept of a lattice corresponding to a three-dimensional periodic arrangement of points is a purely mathematical subject, but it is well accepted as the space lattice. In this case, we use lattice translation for the repetition operation. However, we find other repetition methods called symmetry operations. For example, it is relatively easy to understand the rotation operation as illustrated in Figs. 1 and 2. An operation is required to bring a point into coincidence with itself, such as rotation about an axis and reflection in a plane. When a symmetry operation gives a locus, such as a point, a line, or a plane which is left unchanged by this operation, the locus is referred to as the symmetry element. Note that symmetry may be defined as that spatial property of a pattern or body by which the pattern or body can be brought from a starting state to another indistinguishable state by a certain operation. A symmetry element is also considered to a geometrical entity (point, line, or plane) in a pattern or body, A point is a dimensionless entity in space with coordinates that specify its position. An axis is a line joining two points and a plane is formed by two intersecting lines.

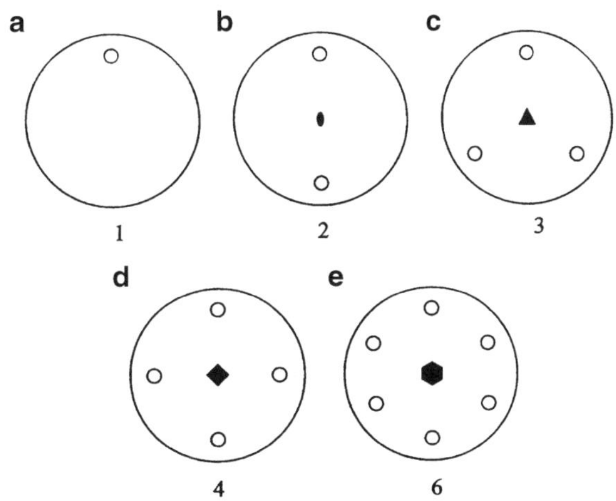

Fig. 1 Point groups having only rotation as symmetry element

Seven crystal systems, each related to the type of unit cell, are combined with 32 point groups which are associated with elements of symmetry in the unit cell itself. Note that the symmetry elements enable us to represent all the possible point arrangements, although a real crystal is a single, unrepeated object. Anyway, with respect to the components of symmetry element, we may suggest a center of symmetry, $\bar{1}$, (or a roto-inversion center), a mirror plane, glide planes, rotation axes, screw axes, and inversion axes. For example, the point which does not move in the inversion operation is called "inversion center." The operation-related mirror plane is given as follows. Any point on one side of a mirror plane is matched with an equivalent point on the other side at the same distance from the plane along a line

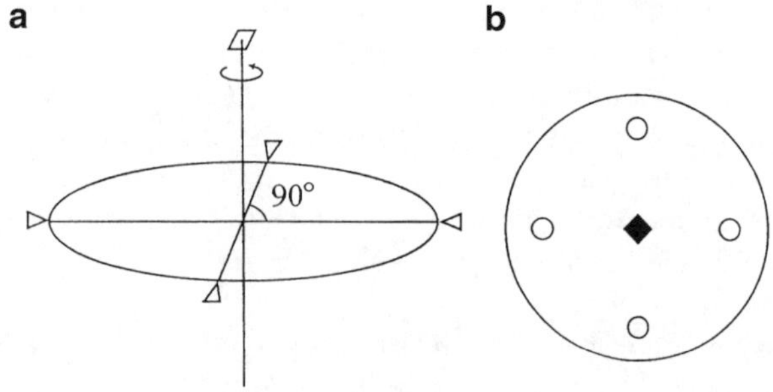

Fig. 2 Fourfold rotation axis (**a**) and its notation (**b**)

normal to it. In other words, the mirror reflection is to match any point or any object with the case where it is reversed after rotating 180° about a twofold axis perpendicular to the target plane. Similarly, the inversion operation can be combined with other axes of rotation called "rotatory-inversion" and this case is also called "onefold rotatory-inversion."

Considering all these factors, we may conclude for symmetry in the following. For the symmetry elements allowed in a three-dimensional periodic arrangement of points, we find ten different ways of "rotation and rotatory inversion" without translation; namely, the rotation axes of one, two, three, four, and sixfold rotation axes, as expressed by n and the rotatory inversion (roto-inversion) axes of $\bar{2}, \bar{3}, \bar{4}, \bar{6}$ denoted by $\bar{n}$. However, it is rather stressed here that there are only EIGHT independent symmetry elements by excluding threefold rotatory-inversion axis ($\bar{3}$) and sixfold rotatory-inversion axis ($\bar{6}$).

In other words, it is found out that there are 32 point groups for covering a three-dimensional periodic arrangement of points in space lattice using these eight symmetry elements and their combinations. Some key points for point groups are summarized as follows.

(1) Point groups where only rotation is recognized as a symmetry element (five cases: 1, 2, 3, 4, and 6).
(2) Point groups where only axis of rotatory-inversion is recognized as symmetry element (five cases: $\bar{1}$, $\bar{2} = m$, $\bar{3}$, $\bar{4}$, and $\bar{6} = 3/m$).
(3) Point groups where n-fold rotation axis is perpendicular to twofold rotation axis (four cases: 222, 32, 422, and 622).
(4) Point groups which have a mirror plane perpendicular to the n-fold rotation axis (three cases: $2/m$, $4/m$, and $6/m$).
(5) Point groups which have a mirror plane parallel to the n-fold rotation axis (four cases: $2mm$, $3m$, $4mm$, and $6mm$).
(6) Point groups of 222, 32, 422 and 622 (corresponding to the No.3 cases), when further considering a mirror plane perpendicular to the n-fold rotation axis (four cases: mmm, $\bar{6}m2$, $4/mmm$ and $6/mmm$).
(7) Point groups of 222 and 32 (corresponding to the No.3 case), when further considering a mirror planes so as to bisect an angle formed by twofold rotation axes parallel to the plane of drawing (two cases: $\bar{4}2m$ and $\bar{3}m$).

(8) Point group where four threefold rotation axes mutually intersect to make the tetrahedral angle of 109.471° (one case only: 23). Similarly, point group where four threefold rotation axes mutually intersect to make the octahedral angle of 70.529° (one case only: 432).

(9) Point groups of 23 and 432 (corresponding to the No.8 case), when further considering several mirror planes (three cases: $m3$, $\bar{4}3m$, and $m3m$).

In order to represent the 32 point groups, we find two methods, Schöenflies and Hermann–Mauguin symbols. Although Schönflies symbols are widely used in the field of spectroscopy, especially of organic molecules, the use of Hermann–Mauguin symbols become popular in crystallography, because of the IUCr suggestion. If need, we can use the results summarized in Table 1. It may also be noteworthy that the abbreviation form is generally used in cases possibly described by combination of higher symmetry. For example, in the point group of $4\bar{3}2$ for a cube described by a combination of symmetry operations of fourfold, threefold, and twofold axes, the best description for symmetry is given by two mirror planes; one is $4/m$ perpendicular to fourfold axis and another is $2/m$ perpendicular both to threefold rotatory-inversion axis $\bar{3}$ and twofold axis. In this case, the original full-notation is $\frac{4}{m}\bar{3}\frac{2}{m}$, but the abbreviation form of m$\bar{3}$m is widely employed. Some helpful information can be obtained from the results of Table 2.

The following information may be convenient in the symmetry operations for crystallography. When including translation, it is necessary to consider "screw axes" with translation to the direction of rotation and its axis and a mirror plane and a "glide plane" with translation parallel to it. Keeping these factors in mind, it is necessary to stipulate the direction of an axis as well as the direction of the translational operation, so that the combination of eleven screw axes and five glide planes as listed in Table 3 may be linked with one another.

Table 1 Hermann–Mauguin and Schöenflies symbols for the 32 crystallographic point groups

Harmann–Mauguin symbols	Schöenflies symbols	Harmann–Mauguin symbols	Schöenflies symbols	Harmann–Mauguin symbols	Schöenflies symbols
1	C_1	4 2 2	D_4	$6/m$	C_{6h}
1	C_i	$4mm$	C_{4v}	6 2 2	D_6
2	C_2	$\bar{4}\ 2\ m$	$D_{2\alpha}$	$6\ m\ m$	C_{6v}
m	C_s	$4/mmm$	D_{4h}	$\bar{6}\ m\ 2$	D_{3h}
$2/m$	C_{2h}	3	C_3	$6/mmm$	D_{6h}
2 2 2	D_2	3	C_{3i}	2 3	T
$mm\ 2$	C_{2v}	3 2	D_3	$\bar{m}\ 3$	T_h
mmm	D_{2k}	$3\ m$	C_{3v}	432	O
4	C_4	$3\ m$	$D_{3\alpha}$	$4\ 3\ m$	T_α
4	S_4	6	C_6	$m\ 3\ m$	O_h
$4/m$	C_{4h}	6	D_{3h}		

Table 2 The 32 point groups and their relation to the crystal systems

Crystal system	Point groups	
Triclinic	$\bar{1}$	1
Monoclinic	$2/m$	m,2
Orthorhombic	$2/m\ 2/m\ 2/m$ (mmm)	$mm2$, 222
Tetragonal	$4/m\ 2/m\ 2/m$ $(4/mmm)$	$\bar{4}2m$, $4mm$, 422 $4/m$, $\bar{4}$, 4
Trigonal	$\bar{3}2/m$ $(\bar{3}m)$	$3m$, 32, $\bar{3}$, 3
Hexagonal	$6/m\ 2/m\ 2/m$ $(6/mmm)$	$\bar{6}m2$, $6mm$, 622 $6/m$, $\bar{6}$, 6
Cubic	$4/m, \bar{3}\ 2/m$ $(m\bar{3}m)$	$\bar{4}3m$, 432, $2/m\ \bar{3}$, 23 $(m\bar{3})$

( ): abbreviated symbols.

Table 3 Symbols and their relevance of symmetry elements including translation

Symbol	Symmetry elements	Graphical symbol	Translation
2_1	2-fold screw		$c/2, a/2$ or $b/2$
3_1	3-fold screw		$c/3$
3_2			$2c/3$
4_1	4-fold screw		$c/4$
4_2			$2c/4$
4_3			$3c/4$
6_1	6-fold screw		$c/6$
6_2			$2c/6$
6_3			$3c/6$
6_4			$4c/6$
6_5			$5c/6$
a, b	Glide plane	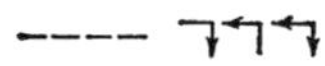	Translation parallel to the plane of paper ($a/2, b/2$ etc.)
c			Translation perpendicular to the plane of paper ($c/2$ etc.)
n	Diagonal glide plane		$(a+b)/2$ etc.
d	Diamond glide plane		$(a+b)/4$ etc.

Question 6.3 The mathematical concept of space groups is used for structural analysis of crystals. Explain the space groups using a monoclinic system as an example.

Answer 6.3 Usually more than one symmetry element is present in crystals. For crystals, only the rotational values $n = 1, 2, 3, 4,$ and 6 are permitted and 32 crystallographic point groups are generated. Point groups are the symmetry groups of a finite body, but space groups should be described as an infinite set of symmetry elements. That is, space groups provide the symmetry not only of crystal lattices but also of crystal structures.

When 32 point groups are analyzed and arranged on the basis of degrees of rotation axis, all crystals are known to be classified into one of the seven crystal systems. Furthermore, the periodic sequence found in crystals has been systematically analyzed using all symmetry operations possible; four symmetry operations of mirror reflection, rotation, inversion and rotatory-inversion and eleven screw axes and five translational operations of glide planes and their combinations and then we obtain the 230 space groups. Combinations of these 230 space groups are proved from purely mathematical point of view. In other words, any periodic arrangement found in crystals can be expressed by one of the 230 space groups.

In the monoclinic system, the lattices are characterized by two unit cell descriptions; simple lattice (P) which has only one lattice point in a unit cell and one centered lattice (C) which has a lattice point at the center of the ab-plane. Point groups are 2, m, and $2/m$. For example, let us consider the case of P and C with a point group of 2, we obtain four combinations such as $P2$, $C2$, $P2_1$, and $C2_1$. However, $C2$ is equivalent to $C2_1$, as readily seen from the results of Fig. 1, so that this combination provides three space groups denoted by $P2$, $C2$, and $P2_1$.

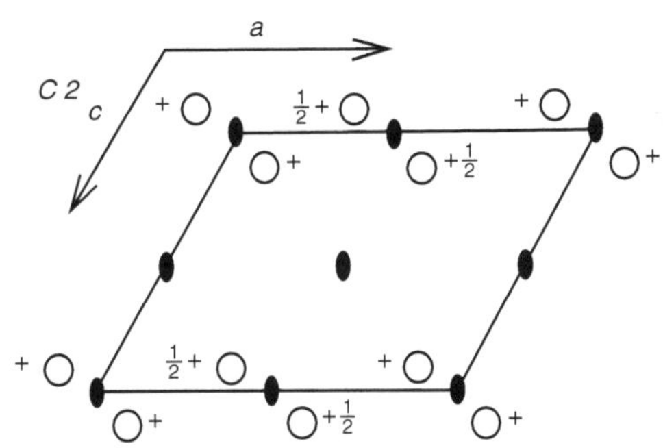

Fig. 1 The operation of the space group $C2$

Since one screw axis and one glide-plane (c-glide plane in this case) will be taken into consideration, we have to check the following five cases of 2_1, c, $2_1/m$, $2/c$, and $2_1/c$. Taking all these factors, the 13 combinations are found to be possible as the space lattices P and C of the monoclinic system. The results are summarized in Table 1 together with subgroups of $P2/m$ and $C2/m$. Note that we also find the relationships of $C2_1 = C2/m$, $C2_1/c = C2/c$, and $C2_1 = C2$.

All possible space groups of the monoclinic system can be derived in a slightly different way. Starting from the two monoclinic space groups of highest symmetry of $P2/m$ and $C2/m$. In $C2/m$, there are a-glide planes at (x, 1/4, z) and (x, 3/4, z) and 2_1-axes at (1/4, y, 0), (1/4, y, 1/2), (3/4, y ,0), and (3/4, y, 1/2).

The monoclinic subgroups for the point group $2/m$ are m and 2. Then, the replacement of point symmetry elements of 2 and m can be done by using 2_1 and a-glide plane, respectively. Since m is parallel to the plane denoted by (010), only a-, c-, and n-glide planes are possible. However, a different selection of the a- and c-axes will convert either a-glide or n-glide into c-glide plane. For this reason we need to take only the c-glide plane into account. Accordingly, the replacement of 2 and m by 2_1 and c results in the 13 monoclinic space groups as summarized in Table 1.

Table 1 Space and point groups for the monoclinic crystal system

Point groups	Space groups	
$2/m$	$P2/m$	$C2/m$
	$P2_1/m$	$C2_1/m^* \equiv C2/m$
	$P2/c$	$C2/c$
	$P2_1/c$	$C2_1/c^* \equiv C2/c$
m	Pm	Cm
	Pc	Cc
2	$P2$	$C2$
	$P2_1$	$C21^* \equiv C2$

When the point symmetry elements 2 and m is replaced by 2_1, a screw axis always appears between the twofold rotation axes, so that the symmetry element of 2_1 has been excluded in the centered-lattice (C) case. In other words, a- and n-glide planes occur in the C-centered space group case, so that the pairs of symbols $C2_1/m = C2/m$, $C2_1/c = C2/c$, and $C2_1 = C2$ give only a single space group each. Similarly, the same combination may be considered about other crystal systems and it is quite complicated as actual work. Since this point is purely handled as mathematical issue and the answer (restricted to 230 cases) has already come out, we should use the result. Detailed information of these 230 space groups are available in the International Tables, Volume A with the chart showing an equivalent positions in a unit cell.

For convenience, Table 2 shows the space group symbols for the 14 Bravais lattices.

Table 2 The space group symbols for the 14 Bravais lattices

<table>
<tr><th></th><th>P</th><th>C</th><th>I</th><th>F</th></tr>
<tr><td>Triclinic</td><td>$P\bar{1}$</td><td></td><td></td><td></td></tr>
<tr><td>Monoclinic</td><td>$P2/m$</td><td>$C2/m$</td><td></td><td></td></tr>
<tr><td>Orthorhombic</td><td>$P2/m\ 2/m\ 2/m$</td><td>$C2/m\ 2/m\ 2/m$</td><td>$I2/m\ 2/m\ 2/m$</td><td>$F2/m\ 2/m\ 2/m$</td></tr>
<tr><td>Tetragonal</td><td>$P4/m\ 2/m\ 2/m$</td><td></td><td>$I4/m\ 2/m\ 2/m$</td><td></td></tr>
<tr><td>Trigonal</td><td rowspan="2">$P6/m\ 2/m\ 2/m$</td><td colspan="3">$R\bar{3}2/m$</td></tr>
<tr><td>Hexagonal</td><td></td><td></td><td></td></tr>
<tr><td>Cubic</td><td>$P4/m\bar{3}2/m$</td><td></td><td>$I4/m\bar{3}2/m$</td><td>$F4/m\bar{3}2/m$</td></tr>
</table>

Question 6.4 Explain the Laue groups which are important as an indicator, when determining the space group of the structure of a desired substance.

Answer 6.4 The periodicity in regular three-dimensional lattice can be analyzed by the concept of symmetry, but one of our main interests in X-ray crystallography is to reveal the structure of the desired substances. This suggests "how to obtain information that the atomic distribution in a crystal of interest is described by one of the 230 space groups and by one of the 32 point groups with sufficient reliability."

Since the measured X-ray diffraction data enable us to provide information of the reciprocal lattices, the symmetry of reciprocal lattices is examined first and then the symmetry of crystals are extracted. The use of centrosymmetric point groups called "Laue groups" is known to be quite useful for this purpose. Figure 1 shows the experimental set-up for taking a Laue photograph on a flat-plate film using a white X-ray source. The crystal is stationary with respect to the X-ray beam, so that the crystal acts as a kind of filter for selecting the correct wavelength for each reflection under the Bragg law. The resultant spots lie on ellipses and all of which have one end of their major axis at the center of the photographic film. All spots on one ellipse arise through reflections from planes that lie in one and the same zone. A diffraction pattern formed by spots is centrosymmetric.

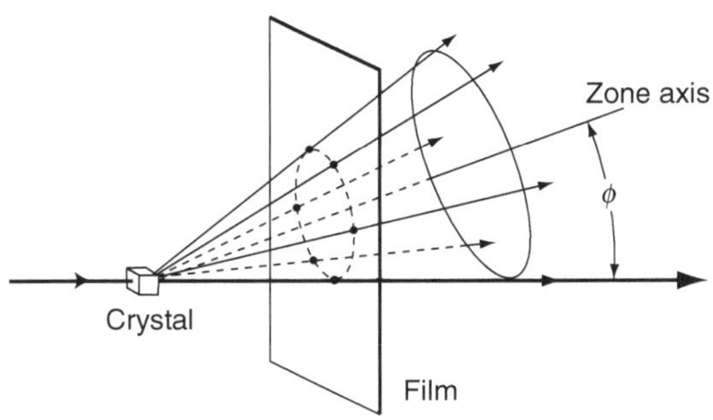

Fig. 1 Schematic diagram for the experimental condition for taking a Laue photograph on a flat-plane film (Transmission type)

Laue photographs using a white X-ray source providing information about symmetry of the weighted reciprocal lattice will reveal the presence of all the symmetry elements associated with the various point groups, but will add a center of symmetry (for noncentrosymmetric point groups). Namely, the arrangement of spots obtained on the Laue photograph exhibits only the symmetry that would be found from a crystal having the corresponding centrosymmetric point groups. There are only eleven possible symmetries and they are called "Laue groups." It may be added that the Laue group assigned to a crystal of interest gives the symmetry of the complete diffraction pattern from that crystal. Thus, the classification of the 32 point groups is possible by means of the Laue diffraction symmetry as shown in Table 1, where the symmetry of the Laue photographs on a flat-plate film can be described for directions of the incident X-ray beam normal to the crystallographic forms as listed. Note that point group projection symmetry corresponds to the symmetry of the projection of the general form of a point group on to a plane.

Table 1 Crystal systems, Laue groups, and Laue projection symmetry

Crystal systems	Point groups	Laue groups	Laue-projection symmetry normal to the given form		
			{100}	{010}	{010}
Triclinic	1, $\bar{1}$	$\bar{1}$	$\bar{1}$	$\bar{1}$	$\bar{1}$
Monoclinic	2, *m*, 2/*m*	2/*m*	*m*	2	*m*
Orthorhombic	222, *mm*2, *mmm*	*mmm*	2*mm*	2*mm*	2*mm*
			{001}	{100}	{110}
Tetragonal	4, $\bar{4}$, 4/*m*	4/*m*	4	*m*	*m*
	422, 4*mm*, $\bar{4}$ 2*m*, 4/*mmm*	4/*mmm*	4*mm*	2*mm*	2*mm*
			{0001}	$\{10\bar{1}0\}$	$\{11\bar{2}0\}$
Trigonal*	3, $\bar{3}$	$\bar{3}$	3	1	1
	32, 3*m*, $\bar{3}$*m*	$\bar{3}$*m*	3*m*	*m*	2
Hexagonal	6, $\bar{6}$, 6/*m*	6/*m*	6	*m*	*m*
	622, 6*mm*, $\bar{6}$*m*2, 6/*mmm*	6/*mmm*	6*mm*	2*mm*	2*mm*
			{100}	{111}	{110}
Cubic	23, *m*3	*m*3	2*mm*	3	*m*
	432, $\bar{4}$3*m*, *m*3*m*	*m*3*m*	4*mm*	3*m*	2*mm*

[a]Referred to hexagonal axis.

A rotation axis of crystal lattice becomes that of the corresponding reciprocal lattice and the distribution of the weighted reciprocal lattice points usually has a symmetry center. Therefore, a point group appeared in the reciprocal lattice points can be attributed to one of the Laue groups. In this respect, symmetry elements and equivalent positions of the Laue groups are very important fundamental information for structural analysis of crystals by X-ray diffraction. Such information is summarized in Fig. 2a using the method of projection.

These results are obtained in the following procedure. Let us consider the spherical surface passing through a symmetry center and mark the position where a symmetry element and this spherical surface intersect. Then, as shown in Fig. 2b, project the marked positions from right above on the equatorial plane of the sphere.

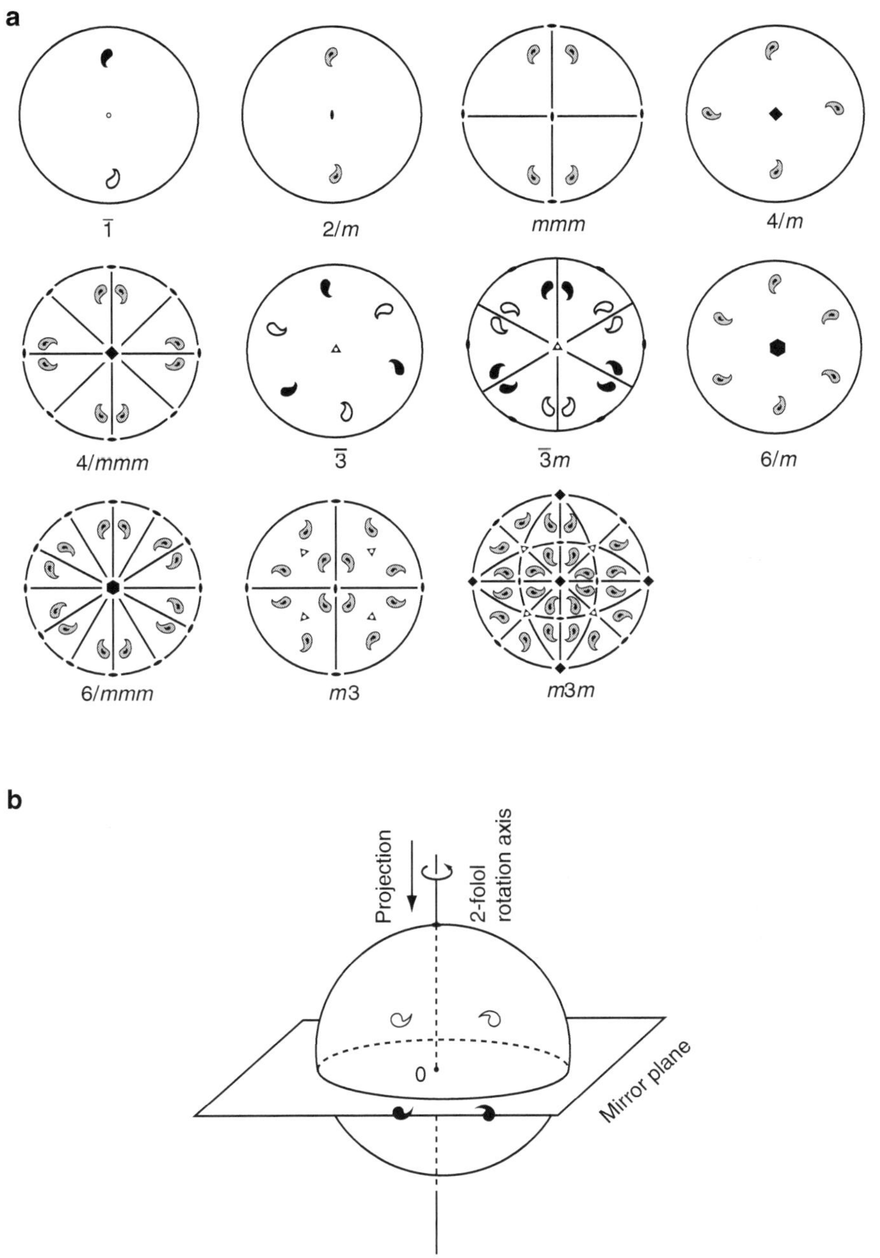

Fig. 2 Symmetry elements and equivalent positions of the Laue group (**a**). Example of projection using the 2/m case (**b**)

The equivalent positions are displayed by a huge comma, and its top surface is colored in white and its bottom in black. If the mirror plane is situated on the equatorial plane, two marks overlap each other upside down on the plane as they are projected. Such overlap is illustrated by black and white overlaid in the figure. Note that the circumference of a circle drawn by thick line indicates that a mirror plane is on the paper surface, the lines other than the circumference show the crossing section between the mirror plane and the sphere by projection.

Question 6.5 Show the variation of Hermann–Mauguin symbol, when the axes of the space group of ***Pnma*** (orthorhombic) are altered within a unit cell. In the same way, show the results of the space group given by ***Pna*** $\mathbf{2_1}$ (orthorhombic).

Answer 6.5 Let us consider the variation expressed by $\boldsymbol{abc} \rightarrow \boldsymbol{b\bar{a}c} \rightarrow \boldsymbol{\overline{ca}b} \rightarrow \boldsymbol{\bar{c}ba} \rightarrow \boldsymbol{bca} \rightarrow \boldsymbol{a\bar{c}b}$, when the starting axis is set to ***abc***. In this case, keep in mind that the origin, O, should be positioned at the top-left corner and further set ***b***-axis to abscissa (horizontal direction), ***a***-axis to ordinate (vertical direction), and ***c***-axis to perpendicular direction to the plane of the drawing in the right-handed system. As shown in Fig. 1, this corresponds to the case of watching a crystal from the right-hand side. While keeping this condition and if every rotat-

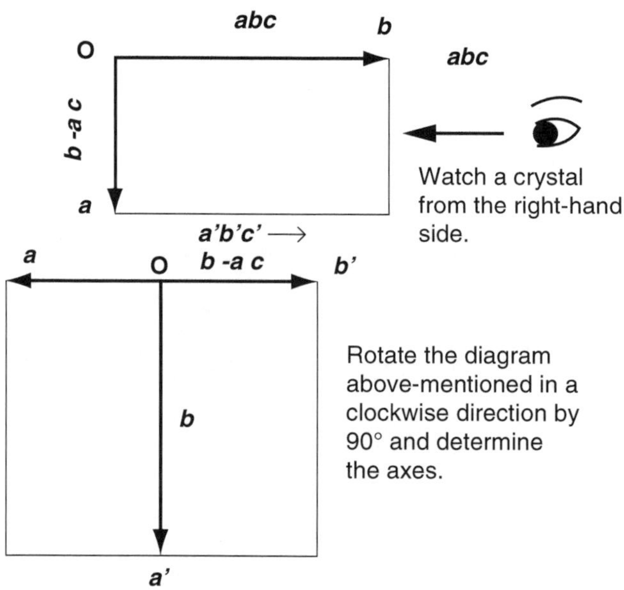

Fig. 1 Fundamentals for changing axes with respect to the orthorhombic system

ing the drawing clockwise by a quarter turn (90°) as shown in Fig. 1, decide axes of ***a***′, ***b***′, ***c***′. Such processes are summarized in Fig. 2. For convenience, Fig. 3 also shows the results using the conversion relationships between ***Pnma*** and ***Pbnm*** when changing the axes from ***abc*** in ***Pnma*** to $\overline{\boldsymbol{ca}}\boldsymbol{b}$ in ***Pbnm***, as an example.

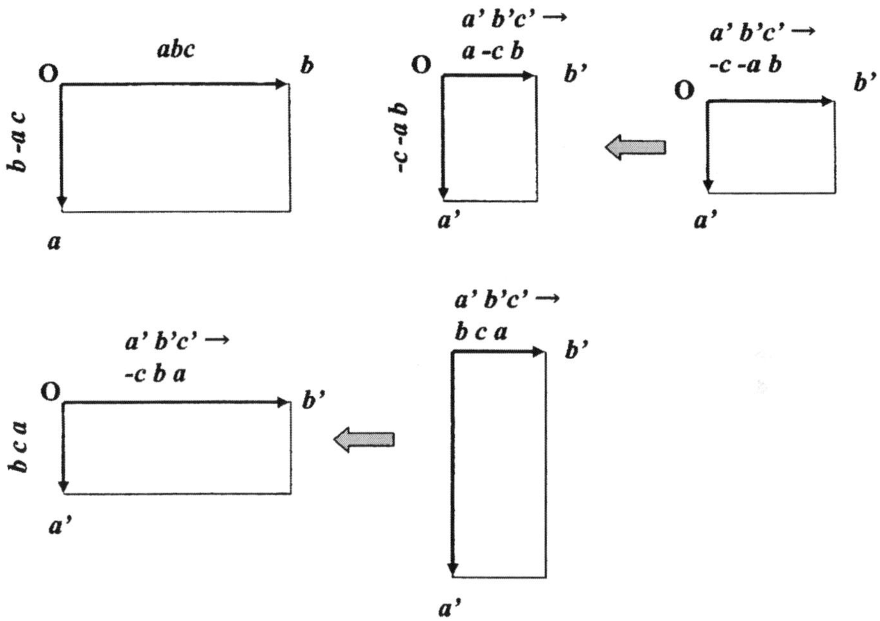

Fig. 2 Variations due to $abc \to b\bar{a}c \to \overline{ca}b \to \bar{c}ba \to bca \to a\bar{c}b$

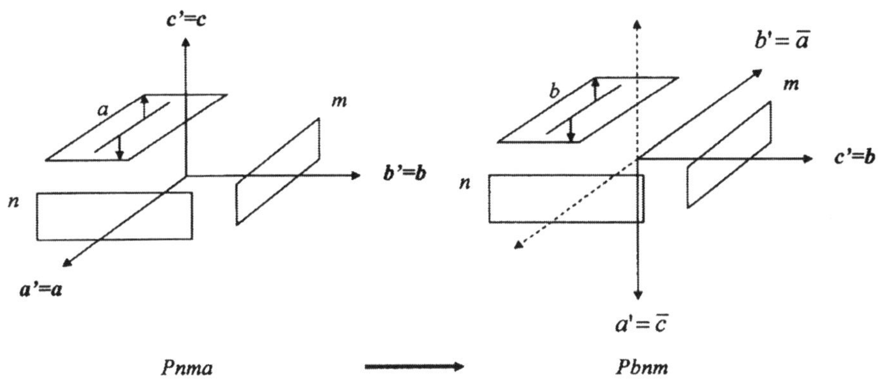

Fig. 3 The relationships between *Pnma* and *Pbnm* due to the change in axes of $abc \to \overline{ca}b$

The results are summarized as follows.

$$\boldsymbol{abc} \rightarrow \boldsymbol{b\bar{a}c} \rightarrow \boldsymbol{\overline{ca}b} \rightarrow \boldsymbol{\bar{c}ba} \rightarrow \boldsymbol{bca} \rightarrow \boldsymbol{a\bar{c}b}$$

$$\boldsymbol{Pnma, Pbnm, Pmcn, Pnam, Pmnb, Pcmn}$$

$$\boldsymbol{Pna2_1, P2_1nb, Pc2_1n, Pn2_1a, Pbn2_1, P2_1cn}$$

When the crystal axes are altered, we find changes for not only atomic coordinates (x, y, z) but also the basis vectors of reciprocal space $(\boldsymbol{a}^*, \boldsymbol{b}^*, \boldsymbol{c}^*)$ as well as the reflection indices (hkl) and lattice points (u, v, w). Since information about the axial transformation in each crystal system is provided using matrices in the International Tables, Volume A (see pp.77–89), some additional details are given below.

The general transformation of the coordinate system consists of two parts; a linear part and a shift of origin. The linear part suggests a variation of orientation or length or both of the basis vectors $\boldsymbol{a}, \boldsymbol{b}, \boldsymbol{c}$. Note that the shift vector is zero for a purely linear transformation. The 3 rows × 3 columns matrices $\boldsymbol{P}$ of the transformation from $\boldsymbol{a}, \boldsymbol{b}, \boldsymbol{c}$ to $\boldsymbol{a}', \boldsymbol{b}', \boldsymbol{c}'$ is given using row matrices $(\boldsymbol{a}, \boldsymbol{b}, \boldsymbol{c})$ as follows.

$$(\boldsymbol{a}', \boldsymbol{b}', \boldsymbol{c}') = (\boldsymbol{a}, \boldsymbol{b}, \boldsymbol{c})\boldsymbol{P} \tag{1}$$

$$= (\boldsymbol{a}, \boldsymbol{b}, \boldsymbol{c}) \begin{pmatrix} P_{11} & P_{12} & P_{13} \\ P_{21} & P_{22} & P_{23} \\ P_{31} & P_{32} & P_{33} \end{pmatrix} \tag{2}$$

Miller indices of a plane $(h\ k\ l)$ are also given in the same way.

$$(h', k', l') = (h, k, l)\,\boldsymbol{P} \tag{3}$$

Note that the Miller indices are usually made relative prime before and after the transformation.

If using the inverse matrices of $\boldsymbol{P}$ to $\boldsymbol{Q}(= \boldsymbol{P}^{-1})$, the transformation of reciprocal lattice axes, atomic coordinates, and lattice points will be given by the following equation using row matrices, $(\boldsymbol{a}^*/\boldsymbol{b}^*/\boldsymbol{c}^*)$, $(x/y/z)$, and $(u/v/w)$.

$$\begin{pmatrix} \boldsymbol{a}^{*\prime} \\ \boldsymbol{b}^{*\prime} \\ \boldsymbol{c}^{*\prime} \end{pmatrix} = \boldsymbol{Q} \begin{pmatrix} \boldsymbol{a}^* \\ \boldsymbol{b}^* \\ \boldsymbol{c}^* \end{pmatrix} \tag{4}$$

$$= \begin{pmatrix} Q_{11} & Q_{12} & Q_{13} \\ Q_{21} & Q_{22} & Q_{23} \\ Q_{31} & Q_{32} & Q_{33} \end{pmatrix} \begin{pmatrix} \boldsymbol{a}^* \\ \boldsymbol{b}^* \\ \boldsymbol{c}^* \end{pmatrix} \tag{5}$$

$$\left.\begin{aligned} (x'/y'/z') &= \boldsymbol{Q}(x/y/z) \\ (u'/v'/w') &= \boldsymbol{Q}(u/v/w) \end{aligned}\right\} \tag{6}$$

For convenience, a general comment is supplemented with respect to the case where the axes of a space group are changed within a unit cell. A space group symbol is usually described by four symbols such as ***Pnma*** and ***Cmca***, and they provide the relationship of each symmetry element with respect to the direction of a crystal axis. In other words, a space group symbol such as monoclinic and orthorhombic crystal systems is found to depend on the choice of a crystal axis. However, for crystal systems of relatively high symmetry the selection of a crystal axis does only a small variation in the space group symbol, because the selection of an axis is rather limited. In addition, since the triclinic crystal system has no axial symmetry element, the space group symbol is limited to only $\boldsymbol{P1}$ and $\boldsymbol{P\bar{1}}$.

If $\boldsymbol{b}$-axis is fixed in a monoclinic crystal system, the selection of other two axes is limited to a two-dimensional plane. Thus, it is relatively easy to understand that three kinds of notations given by $\boldsymbol{P2_1/c}, \boldsymbol{P2_1/a}$, and $\boldsymbol{P2_1/n}$ are the same space groups. Whereas, it is not optional how to select an axis as for the orthorhombic crystal system characterized by the three axes which are at right angles to one another (mutually perpendicular). Nevertheless, there are 3 ways of selecting $\boldsymbol{a}$-axis and 2 ways for $\boldsymbol{b}$-axis after fixing $\boldsymbol{a}$-axis, which results in six combinations even if the selection is limited to the right-handed system. Note that the space group symbols related to the axis are just three and the first being the notation to represent centered lattices, the second being the order of appearance of the axis symbol or plane one, and the third being the forward direction of a glide plane. This is applicable to the unit cell transformations. For example, when the directions of three axes in space are numbered as 1, 2, and 3 and each direction is named by $\boldsymbol{a}$, $\boldsymbol{b}$, and $\boldsymbol{c}$, respectively, one can obtain six combinations as summarized in Table 1.

Table 1 Transformations of the crystal axes in the orthorhombic system

	1	2	3	$3m31$
(1)	*a*	*b*	*c*	***Cmca***
(2)	*a*	*c*	$-b$	***Bmab***
(3)	*b*	*c*	*a*	***Abma***
(4)	*b*	*a*	$-c$	***Ccmb***
(5)	*c*	*a*	*b*	***Bbcm***
(6)	*c*	$-b$	*a*	***Acam***

Some additional details for this combinations are given. Let us temporarily assume that the case (1) of Table 1 is given by ***Cmca*** which represents the symbol of space lattice, the first symmetry element, the second symmetry element, and third symmetry element. The symbol of ***Cmca*** is, of course, a temporary one, so that it depends on the method "how to give name an axis." For this reason, $\boldsymbol{3m31}$ related to the directions of three fixed axes 1, 2, and 3 is provided in the top right end of Table 1. As setting is made in this way, six variations will be decided in turn and the resultant symbols will be obtained as shown in the right-end column of Table 1.

For example, since the axis of 3 is given as ***a*** for the case (3), its symbol will be ***Abma*** with ***A*** for the space lattice, ***b*** for the first symmetry element which is the direction of the glide plane for ***a***-axis, ***m*** for the second symmetry element as ***b***-axis matches the axis of 1 and then no change, and ***a*** for the third symmetry element which corresponds to the direction of the glide plane for ***c***-axis.

More information can be obtained from the International Tables for Crystallography.

Question 6.6 Obtain the structure factor of F_{hkl} and $|F_{hkl}|^2$ and show the condition where the diffraction intensity can be observed in the following three cases:

(1) The fourfold rotation axis along the a_3-axis in a unit cell.
(2) The 4_1 screw axis along the a_3-axis in a unit cell.
(3) The 4_2 screw axis along the a_3-axis in a unit cell.

Answer 6.6

(1) The atomic positions in a unit cell with the fourfold rotation axis are given by x, y, z; $\bar{y}, x, z$; $\bar{x}, \bar{y}, z$; and $y, \bar{x}, z$. Then, the structure factors F_{hkl} and $|F_{hkl}|^2$ in this case are computed in the following equations.

$$\begin{aligned}F_{hkl} &= f\mathrm{e}^{2\pi \mathrm{i}lz}\left\{\mathrm{e}^{2\pi \mathrm{i}(hx+ky)} + \mathrm{e}^{2\pi \mathrm{i}(kx-hy)} + \mathrm{e}^{-2\pi \mathrm{i}(hx+ky)} + \mathrm{e}^{-2\pi \mathrm{i}(kx-hy)}\right\}\\ &= 2f\mathrm{e}^{2\pi \mathrm{i}lz}\left\{\cos 2\pi(hx+ky) + \cos 2\pi(kx-hy)\right\}\\ &= 4f\mathrm{e}^{2\pi \mathrm{i}lz}\cos\pi\{(h+k)x-(h-k)y\}\cos\pi\{(h-k)x+(h+k)y\}\end{aligned}$$

$$|F_{hkl}|^2 = 16f^2\cos^2\pi\{(h+k)x-(h-k)y\}\cos^2\pi\{(h-k)x+(h+k)y\}$$

Therefore, there is no condition where the diffraction intensity cannot be detected with respect to the atoms in a unit cell with the fourfold rotation axis.

(2) The atomic positions in a unit cell with the 4_1-screw axis are given by x, y, z; $\bar{y}, x, z+\frac{1}{4}$; $\bar{x}, \bar{y}, z+\frac{1}{2}$; and $y, \bar{x}, z+\frac{3}{4}$. The structure factors for this case are computed as follows.

$$\begin{aligned}F_{hkl} = f\mathrm{e}^{2\pi \mathrm{i}lz}\{&\mathrm{e}^{2\pi \mathrm{i}(hx+ky)} + \mathrm{e}^{\mathrm{i}\pi l/2}\mathrm{e}^{2\pi \mathrm{i}(kx-hy)}\\ &+\mathrm{e}^{\mathrm{i}\pi l}\mathrm{e}^{-2\pi \mathrm{i}(hx+ky)} + \mathrm{e}^{\mathrm{i}3\pi l/2}\mathrm{e}^{-2\pi \mathrm{i}(kx-hy)}\}\end{aligned}$$

Here, let us consider the case of $l = 4n$,

$$F_{hkl} = f\mathrm{e}^{2\pi \mathrm{i}lz}\left\{\mathrm{e}^{2\pi \mathrm{i}(hx+ky)} + \mathrm{e}^{2\pi \mathrm{i}(kx-hy)} + \mathrm{e}^{-2\pi \mathrm{i}(hx+ky)} + \mathrm{e}^{-2\pi \mathrm{i}(kx-hy)}\right\}$$

$$|F_{hkl}|^2 = 16f^2\cos^2\pi\{(h+k)x-(h-k)y\}\cos^2\pi\{(h-k)x+(h+k)y\}$$

There is no condition where the diffraction intensity cannot be detected, along the way similar to the fourfold rotation axis case.

On the other hand, let us consider the case of $l = 4n \pm 1$, one obtains the following results.

$$F_{hkl} = 2fe^{2\pi ilz}\{\mathrm{i}\sin 2\pi(hx+ky) \mp \sin 2\pi(kx-hy)\}$$
$$|F_{hkl}|^2 = 4f^2\left\{\sin^2 2\pi(hx+ky) + \sin^2 2\pi(kx-hy)\right\}$$

In addition, the following results are obtained for the case of $l = 4n + 2$.

$$F_{hkl} = 4fe^{2\pi ilz}\sin\pi\{(h+k)x-(h-k)y\}\sin\pi\{(h-k)x+(h+k)y\}$$
$$|F_{hkl}|^2 = 16f^2\sin^2\pi\{(h+k)x-(h-k)y\}\sin^2\pi\{(h-k)x+(h+k)y\}$$

It is noteworthy from these two results that the diffraction intensity for a peak whose Miller indices are given by $(0\ 0\ l)$ can be observed only under the condition $l = 4n$.

(3) The structure factor F_{hkl} is computed from the atomic positions in a unit cell with the 4_2-screw axis; x, y, z; $\bar{y}, x, z+\frac{1}{2}$; $\bar{x}, \bar{y}, z$; $y, \bar{x}, z+\frac{1}{2}$ in the following form.

$$F_{hkl} = 2fe^{2\pi ilz}\left\{\cos 2\pi(hx+ky) + e^{\mathrm{i}\pi l}\cos 2\pi(kx-hy)\right\}$$

When considering $l = 2n$,

$$F_{hkl} = 2fe^{2\pi ilz}\{\cos 2\pi(hx+ky) + \cos 2\pi(kx-hy)\}$$

$$|F_{hkl}|^2 = 4f^2\{\cos 2\pi(hx+ky) + \cos 2\pi(kx-hy)\}^2$$

When considering $l = 2n + 1$,

$$F_{hkl} = 2fe^{2\pi ilz}\{\cos 2\pi(hx+ky) - \cos 2\pi(kx-hy)\}$$

$$|F_{hkl}|^2 = 4f^2\{\cos 2\pi(hx+ky) - \cos 2\pi(kx-hy)\}^2$$

In conclusion, the diffraction intensity for a peak whose Miller indices are given by $(0\,0\,l)$ can be observed only under the condition $l = 2n$.

Question 6.7 Let us consider the monoclinic space group denoted by $\boldsymbol{P2_1/c}$, showing the primitive lattice with the 2_1-fold screw axis parallel to $\boldsymbol{b}$-axis, the c-glide plane being perpendicular to $\boldsymbol{b}$-axis with glide of $\frac{c}{2}$. Answer the following questions.

(1) Illustrate the symmetry of this space lattice such as a symmetry center and show both general position and special positions.
(2) Find the extinction condition.

Answer 6.7

(1) Assuming the 2_1-fold screw axis and the glide plane being perpendicular to ***b***-axis as shown in Fig. 1. In addition, the center of point symmetry (a symmetry center) is indicated by a solid circle at $\frac{1}{4}$ or $\frac{3}{4}$ along the c-axis. If the position of $(0, \frac{1}{4}, \frac{1}{4})$ is set to the symmetry center, we obtain the results of Fig. 2. Therefore, the general positions can be given as follows.

$$x, y, z; \bar{x}, \bar{y}, \bar{z}; \bar{x}, \frac{1}{2} + y, \frac{1}{2} - z; x, \frac{1}{2} - y, \frac{1}{2} + z$$

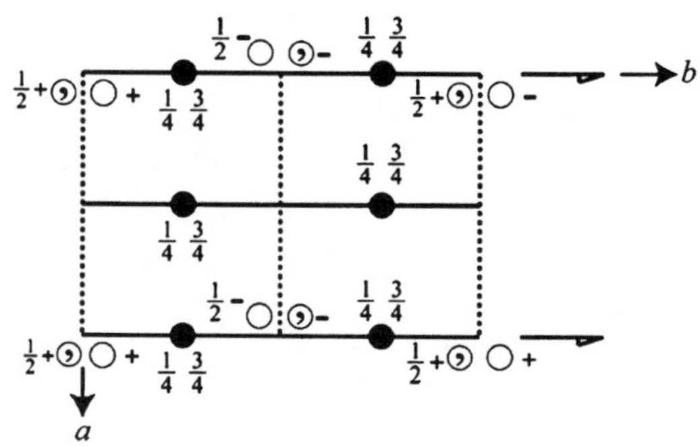

Fig. 1 Monoclinic space lattice. Solid circles indicate symmetry centers

The special positions corresponding to the symmetry center are also given as follows (see solid circles in Fig. 2)

$$(0\ \ 0\ \ 0) : \left(0\ \ \frac{1}{2}\ \ \frac{1}{2}\right), \left(\frac{1}{2}\ \ 0\ \ 0\right) : \left(\frac{1}{2}\ \ \frac{1}{2}\ \ \frac{1}{2}\right)$$

$$\left(0\ \ 0\ \ \frac{1}{2}\right) : \left(0\ \ \frac{1}{2}\ \ 0\right), \left(\frac{1}{2}\ \ 0\ \ \frac{1}{2}\right) : \left(\frac{1}{2}\ \ \frac{1}{2}\ \ 0\right)$$

(2) In order to find the extinction condition, the following formulas for trigonometric functions were used.

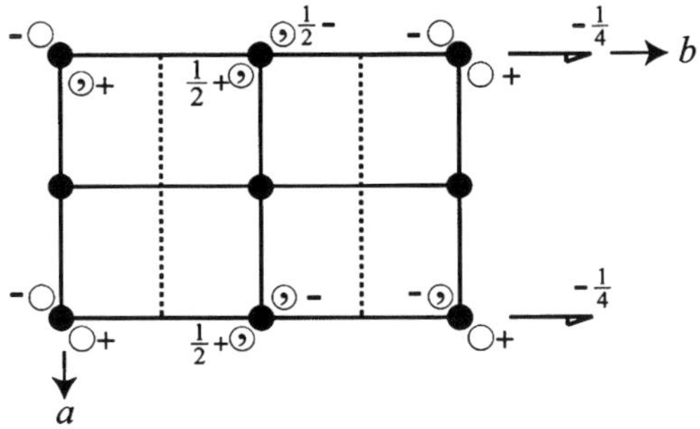

Fig. 2 Monoclinic space lattice when setting a symmetry center to $\left(0, \frac{1}{4}, \frac{1}{4}\right)$. Solid circles indicating a symmetry center at 0 or $\frac{1}{2}$ along the c-axis

$$\begin{cases} \sin A + \sin B = 2\sin\dfrac{A+B}{2}\cos\dfrac{A-B}{2} \\ \cos A + \cos B = 2\cos\dfrac{A+B}{2}\cos\dfrac{A-B}{2} \\ \cos A - \cos B = 2\sin\dfrac{A+B}{2}\sin\dfrac{A-B}{2} \end{cases}$$

The structure factor is computed as follows.

$$\begin{aligned} \frac{F}{f} &= e^{2\pi i(hx+ky+lz)} + e^{2\pi i(-hx-ky-lz)} \\ &\quad + e^{2\pi i\{-hx+k(\frac{1}{2}+y)+l(\frac{1}{2}-z)\}} + e^{2\pi i\{hx+(\frac{1}{2}-y)k+(\frac{1}{2}+z)l\}} \\ &= \cos 2\pi(hx+ky+lz) - i\sin 2\pi(hx+ky+lz) \\ &\quad + \cos 2\pi(-hx-ky-lz) - i\sin 2\pi(-hx-ky-lz) \\ &\quad + \cos 2\pi\left\{-hx + \left(\frac{1}{2}+y\right)k + \left(\frac{1}{2}-z\right)l\right\} \\ &\quad - i\sin 2\pi\left\{-hx + \left(\frac{1}{2}+y\right)k + \left(\frac{1}{2}-z\right)l\right\} \\ &\quad + \cos 2\pi\left\{hx + \left(\frac{1}{2}-y\right)k + \left(\frac{1}{2}+z\right)l\right\} \\ &\quad - i\sin 2\pi\left\{hx + \left(\frac{1}{2}-y\right)k + \left(\frac{1}{2}+z\right)l\right\} \\ &= 2\cos(2\pi, 0)\cos 2\pi\frac{2(hx+ky+lz)}{2} \\ &\quad + 2\cos 2\pi\frac{k+l}{2}\cos 2\pi(hx-ky+lz) \end{aligned}$$

$$-\mathrm{i}\sin 2\pi\frac{k+l}{2}\cos(hx-ky+l)$$

$$= 2\cos 2\pi(hx+ky+lz) + 2\cos 2\pi\frac{k+l}{2}\cos 2\pi(hx-ky+lz)$$

(i) In the case of $k + l = 2n$,

$$\frac{F}{f} = 4\cos 2\pi(hx+lz)\cos 2\pi ky$$

The extinction does not take place.

(ii) In the case of $k + l = 2n + 1$,

$$\frac{F}{f} = -4\sin 2\pi(hx+lz)\sin 2\pi ky$$

The extinction takes place under the condition of $h = l = 0$ or $k = 0$.

Question 6.8 Information about the space group ***Pnma***, No.62 (orthorhombic) can be obtained from the International Tables for Crystallography, Volume A, page 298-299. Explain the key points.

Answer 6.8 The symbol ***Pnma*** described by a shortened form (notation) of Hermann–Mauguin method describes that Bravais lattice is a simple lattice(P) with three symmetry elements ***mnm*** with respect to the direction of [100], [010], and [001] (see Table 6.4). Namely, this orthorhombic has mirror planes perpendicular to these directions. The symbol of D_{2h}^{16} provides the Schöenflies' description and No.62 corresponds to the number allocated in the 230 space groups. Next, $\boldsymbol{P}$ $2_1/n/2_1/m/2_1/a$ is full expression of three symmetry elements and they are as follows. $2_1/n$ shows the 2_1 screw axis and a diagonal plane of glide reflection perpendicular to it for the direction of $\boldsymbol{a}$-axis, the $2_1/m$ indicates the 2_1 screw axis and a mirror plane vertical to it for the direction of $\boldsymbol{b}$-axis, and the $2_1/a$ shows the 2_1 screw axis and a mirror plane perpendicular to it for the direction of $\boldsymbol{c}$-axis, respectively. The space group of the Patterson function is given by ***Pmmm***. The Patterson function corresponds to the Fourier transform of the square of the structure factor and it is widely used to obtain a map of interatomic distances in the unit cell. In other words, the Patterson function is a superposition of peaks derived from all atomic pairs in the unit cell directly related to the measured diffraction data. (Refer to other textbooks for details of the Patterson function, for example, M.M. Woolfson, An Introduction to X-ray Crystallography, 2nd Edition, Cambridge University Press (1997).)

Four figures are the so-called standard setting of the space group or the space-group tables and first three figures are characterized by the Hermann–Mauguin symbols in the headline. The space-group tables for each orthorhombic space group is known to consist of three projections of the symmetry elements along the $\boldsymbol{c}$-axis (upper left), the ***a***-axis (lower left), and the ***b***-axis (upper right), in addition to the general position diagram. For example, with respect to the upper left corner as its origin, the projection is made along the ***c***-axis, setting ***a***-axis and ***b***-axis to the horizontal (abscissa) and the vertical (ordinate) axes, respectively. For convenience, diagrams for the standard setting are shown in Fig. 1 using the orthorhombic and tetragonal space groups as an example, where G = general position diagram. Note that all these figures are described in the right-handed coordinate system.

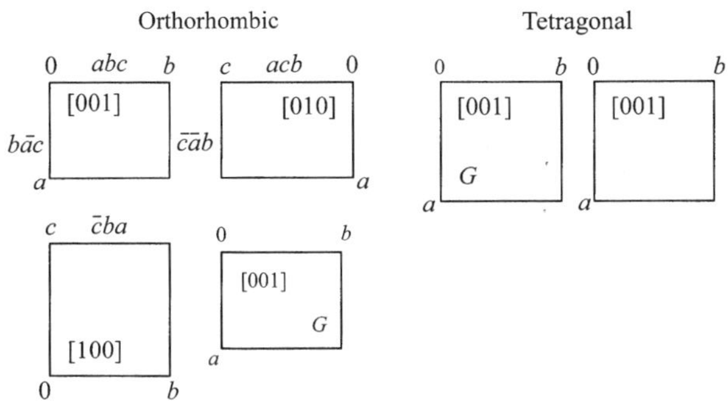

Fig. 1 Diagrams for the standard setting as described in the space-group tables; orthorhombic and tetragonal space groups

For each orthorhombic space group, there are six different ways of assigning the labels ***a, b, c*** to the three orthorhombic symmetry directions. These settings correspond to the six conversions of the labels of the axes including identity conversion. Three space group symbols written on the horizontal axes in three figures of projection are ***Pmnb***, ***Pbnm***, and ***Pmcn*** which are corresponding to the axial conversions of ***abc*** (fundamental axis), $\boldsymbol{a\bar{c}b}$, and $\boldsymbol{\bar{c}ba}$. Similarly, ***Pmnb***, ***Pbnm***, and ***Pmcn*** are related to the axial conversions of $\boldsymbol{b\bar{a}c}$, $\boldsymbol{\bar{c}\bar{a}b}$, and ***bca*** in the vertical axes case. In addition, the numerical value 1/4 placed beside the symmetry element symbols shows the inner height of a unit cell in the projection direction.

The figure at the lower right is corresponding to the general position diagram, which is given only in the projection along ***c***-axis. It may also be suggested that this figure shows the equivalent positions in the same projection as the upper left and provides information about what arrangement of atoms in the general positions will be possible in a unit cell. Both marks ◯ and ⨀ show the equivalent positions and ◯ and ⨀ are related by a mirror plane. They are called enantiomorphs, so that ◯ represents the left-handed system, if ⨀ is given in a right-handed system.

The signs and numerical values, $+, -$ and $\frac{1}{2}+$, $\frac{1}{2}-$ placed beside ◯ are the coordinates to the projection direction of the equivalent positions, and the present case suggests $+z$, $-z$, $\frac{1}{2}+z$, $\frac{1}{2}-z$.

Some other information are summarized as follows.

Origin: The determination of crystal structures is facilitated by the selection of a suitable origin. The line of "Origin" provides the origin selected in the space group table. According to the International Tables, the position of symmetry center is set to the origin for space groups with a symmetry center and the position of the highest site symmetry becomes the origin for space groups without a symmetry center. In the space group ***Pnma***, the origin is set to a symmetry center denoted by $\bar{\mathbf{1}}$ on the twofold screw axis (as given by "on **1** $\mathbf{2_1}$ **1**") which is equal to the ***b***-axis.

Asymmetric unit: An asymmetric unit of a space group is considered to be the smallest closed part of space from which by using all symmetry operations, the whole space can be filled. In other words, the line of "Asymmetric unit" shows that the independent area in a unit cell is given in the region; $0 \leq x \leq \frac{1}{2}$, $0 \leq y \leq \frac{1}{4}$, and $0 \leq z \leq 1$ in ***Pnma***.

Symmetry operations: The geometric description of the symmetry operations is given in the space-group tables under the heading "Symmetry operations." These information give a link between the space group diagrams and the general positions. The line of "Symmetry operations" shows the symmetry elements as well as positions related to symmetry operations. In the space group ***Pnma***, symmetry operations of (1)–(8) correspond to eight equivalent positions.

Generators: The line of "Generators" provides all symmetry operations required to generate all equivalent positions of the general positions from coordinates x, y, z. For example, $t(1, 0, 0)$, $t(0, 1, 0)$, and $t(0, 0, 1)$ indicate the translational operation that moves the coordinates x, y, z described by (1) of general positions to the directions of ***a***,- ***b***-, and ***c***-axes by one unit cycle. In the space group ***Pnma***, although symmetry operations of (1)–(8) are cited, all equivalent positions can be generated by operations of (1), (2), (3), and (5). Accordingly, the operations of (4), (6), (7), and (8) are excluded.

Positions: The column of "Position" more explicitly called Wyckoff positions provide information of the equivalent positions when considering site symmetry and are defined as a group of crystalline positions. The following information (a) to (e) classified into general positions and special positions are provided in the space group table.

(a) Multiplicity: This is the number of equivalent points per unit cell and keep in mind it differs from the number of the equivalent lattice planes in one plane of a form called multiplicity factor. The multiplicity of the general position is equal to the order of the point group to which the space group under consideration belongs. On the other hand, the multiplicity of the special position is given by the divisor of multiplicity of the general position. For example, the number of equivalent point in the special position 4*c* is one half of the general equivalent position. This is attributed to the condition that one special position is formed by overlapping two general equivalent positions.
(b) Wyckoff letter: This is simply a coding scheme for the Wyckoff positions. Usually, the notation is made due to the higher degree of site symmetry, starting with *a* at the bottom column for the special position and upward in the alphabetical order traced back to the general position. For example, the general position of

atomic position in the space group ***Pnma*** is $8d$ and its special positions are $4c$, $4b$, $4a$ in order.

(c) Site symmetry: The site symmetry groups of the different points of the same special position are symmetrically equivalent subgroups of the space group and then all points of one special position can be described by the same site symmetry symbol. The column of "Site symmetry" provides the symmetry which the atomic position has. There are two ways in the space group ***Pnma*** where 4 atoms occupy 8 symmetry center sites and the special positions $4a$ and $4b$ correspond to them. There is the mirror symmetry (m) in the special position $4c$. In addition, in order to clarify symmetry directions, the irrelevant axial directions are shown by dots such as $.m.$ Here this $.m.$ indicates a mirror plane perpendicular to the ***b***-axis.

(d) Coordinates: The sequence of the coordinate triplets is based on the Generators. For centered space groups, the centering translations such as $(0, 0, 0)+$ $\left(\frac{1}{2}\frac{1}{2}\frac{1}{2}\right)+$ are given above the coordinate triplets. The symbol $+$ indicates that the components of the centering translations have to be added to the listed coordinate triplets for obtaining a complete Wyckoff position. Coordinates corresponding to ***a***-, ***b***-, and ***c***-axes with a parallelepiped as a unit cell are referred to as x, y, z. The length of this unit cell is normalized as 1. When one atom exists in the space group ***Pnma***, we find eight atoms without exception at the following positions in a unit cell.
(x, y, z), $\left(-x+\frac{1}{2}, -y, z+\frac{1}{2}\right)$, $\left(-x, y+\frac{1}{2}, -z\right)$, $\left(x+\frac{1}{2}, -y+\frac{1}{2}, -z+\frac{1}{2}\right)$, $(-x, -y, -z)$, $\left(x+\frac{1}{2}, y, -z+\frac{1}{2}\right)$, $\left(x, -y+\frac{1}{2}, z\right)$, $\left(-x+\frac{1}{2}, y+\frac{1}{2}, z+\frac{1}{2}\right)$, where $-x$ is $\bar{x}$.

(e) Reflection conditions: Information of the extinction rule is given for the case where atoms are located at the general positions. For example, $0kl : k+l = 2n$ (n is integer) shows, with respect to the $0kl$ reflection, that when $k+l$ is an odd number, its crystal structure factor is zero and if $k+l$ is an even number, it is not zero. On the other hand, when atoms are located only at the special positions, new information of the extinction rule appears in the column listed as Special for every Wyckoff sign, in addition to the conditions given for the general positions. For example, if an atom is only at the $4a$ position, the structure factor becomes zero, when $h+l$ or k is an odd number.

Symmetry of special projections provides information corresponding to two-dimensional space groups and for example, this is used to project the crystal structure with respect to the direction perpendicular to the reciprocal lattices using the two-dimensional intensity data of a zero layer. In each space group table, three different projections are given with respect to the direction indicated by "Along" which is the projection to the plane perpendicular to this direction. Projections depend on crystal system and such information is as follows. Projections are made to the directions of ***c***-, ***a***-, and ***b***-axes for triclinic, monoclinic, and orthorhombic systems. Similarly, the directions of ***c***- and ***a***-axes and [110] for tetragonal system, ***c***- and ***a***-axes and [210] for hexagonal system, [111], $[1\bar{1}0]$, and $[2\bar{1}\bar{1}]$ for trigonal system and [001], [111], and [110] for cubic system, respectively, are used for the projections. Following the projection direction, information related to the plane groups generated by the projection of the space groups are provided as the Hermann–Mauguin symbol, $p2gm$, at the present case. In the following line, the relationship with the basic axes $\boldsymbol{a}'$, $\boldsymbol{b}'$ and in the line after that, the origin of plane groups, for example, $(0, 0, z)$ is given using the unit cell coordinates of the space group.

Chapter 7
Supplementary Problems (100 Exercises)

Exercise 1.1 When accelerating an electron by 1 kV, compute the values of energy, momentum and wavelength using the de Broglie relation.

Exercise 1.2 Calculate the values of mass absorption coefficient of gallium arsenide (GaAs) and barium titanate ($BaTiO_3$) for Cu-Kα radiation.

Exercise 1.3 Air composition can be considered to be mostly 80mass%N_2 and 20mass%O_2 and its density is $1.29 \times 10^{-3} \mathrm{Mg/m^3} = \mathrm{g/cm^3}$ at normal temperature (273.15 K) and pressure (101,325 Pa). If Cu-Kα radiation passes through 360 mm of normal air, by what percentage is the intensity reduced. Perform a similar calculation for the Fe-K radiation.

Exercise 1.4 3.138×10^{21} atoms are found to be included in 0.5 g of a metallic substance consisting of a single element. Calculate the atomic weight of this substance.

Exercise 1.5 What is the relationship between the wavelengths λ_K of absorption edge found in each atom at the characteristic energies and the critical excitation voltage V_K for the case where the K-shell electron is removed. Estimate the excitation voltage of Mo-Kα radiation assuming that the wavelength of K-absorption edge of molybdenum is 0.06198 nm.

Exercise 1.6 In the photoelectric absorption process where a photoelectron is released from an atom, priority is given to inner-shell electrons; K-shell electrons are released before L-shell electrons. The recoil of atom is necessarily produced in the photoelectric absorption process, but its energy variation is known to be negligibly small (see Question 1.6). In other words, photoelectric absorption by free electrons does not occur. Explain why this is so using the law of conservation of momentum. When a lead plate was irradiated with X-rays with an energy of 150 keV, K-shell photoelectric absorption occurred at the surface of the plate. In this case, the speed of the photoelectrons was found to be 1.357×10^8 m/s. Calculate the K-absorption edge of lead, assuming that no energy loss occurs.

Exercise 1.7 When a tungsten plate was irradiated by X-rays with the energy of 150 keV, the photoelectric absorption of K-shell was made at the surface of plate and the photoelectron was ejected. Assuming that there is no energy loss in this photoelectric absorption process and the K absorption edge of tungsten is $E_K^W = 69.52$ keV, find the velocity of photoelectron, at ejection.

Exercise 1.8 Compute the energy of Kα radiation emitted from tungsten (atomic number 74), under the condition that Moseley's law is obeyed and the shielding constant for generating the Kα radiation is set to zero and the Rydberg constant is 1.097×10^7 (m^{-1}).

Exercise 1.9 In the operation of X-ray tube, if setting applied voltage to V, excitation voltage to V_K and current to i, the number of photons of the characteristic radiation emitted, I_K, is approximated by the following equation, called "Storm formula".

$$I_K = B_S i(V - V_K)^n$$

Where B_S is a proportionality constant and $n = 1.5$ is widely used. The value of a proportionality constant is set to 4.25×10^8 (As$^{-1} \cdot$ sr^{-1}) and the excitation voltage of tungsten is given by 69.5 keV. When a tungsten X-ray tube is operated at 100 kV and 1 mA, answer the following questions.

(1) Compute the number of photons of the characteristic X-rays released per unit solid angle using the Storm formula.
(2) In this case, the average energy of the characteristic X-rays was also found to be 60.7 keV. Compute the intensity.

Exercise 1.10 The transmission rate was found to be 0.65, when an iron plate was irradiated by X-rays with the energy of 100 keV. The mass absorption coefficient and density of iron for X-rays with the energy of 100 keV are 0.215 cm^2/g and 7.87 g/cm^3, respectively.

(1) Estimate the thickness of this iron plate.
(2) Check the amount of intensity loss in this iron plate as a function of the incident X-rays.

Exercise 1.11 The thickness of a substance required to make intensity of the incident X-rays down to 50% is called "the half value layer". Calculate the thickness of a layer required to decrease the incident X-ray intensity to 10% when the half value layer of iron to X-rays with a certain energy is known to be 0.041 mm.

Exercise 1.12 The minimum energy required to release an electron from an atom in a solid substance is called the work function. During experiments involving light exposure of the tungsten used for the filament of an X-ray tube, the threshold wavelength for photoelectron emission is found to be 274.3 nm. Calculate the work function of the metal. On the other hand, applying a reverse voltage to the filament can prevent electron emission. Calculate the reverse voltage required to prevent electron emission when tungsten is irradiated by light with a wavelength of 100 nm.

Exercise 1.13 The wavelengths of Kα and Kβ radiation for magnesium are 0.9890 and 0.9521 nm, respectively. With reference to the K-Shell, describe the atomic core level of the L- and M-shells in magnesium.

Exercise 2.1 Illustrate the planes of $(\bar{1}10), (11\bar{1}), (102), (020), (2\bar{2}0), (11\bar{2})$ and the directions of $[001], [100], [110], [210], [122], [1\bar{2}0]$ for a cubic lattice.

Exercise 2.2 Identify the closest packing plane for bcc and fcc structures and estimate the atomic density on these planes.

Exercise 2.3 Graphite, sometimes called "black lead", is known to have a layered structure in which carbon atoms occupy the vertex positions of a regular (equilateral) hexagon. Draw the unit cell in the plane of this layered structure.

Exercise 2.4 Potassium has a bcc structure with a lattice parameter of a = 0.520 nm.

(1) Compute the interatomic distances of the 1st and 2nd nearest neighbors.
(2) Estimate the coordination numbers of the 1st and 2nd nearest neighbors.
(3) Estimate the density based on the crystal structure.

Exercise 2.5 It is known that the hexagonal close-packed lattice will be materialized in the case of $c/a = \sqrt{8/3} = 1.633$, and magnesium is close to this ideal case, because of $c/a = 1.624$. The values of density and molar mass of magnesium are $(1.74 \times 10^6\ \text{g/m}^3)$ and 24.305 g, respectively.

(1) Compute the volume of a unit cell.
(2) Estimate the lattice parameter a as well as the 1st nearest neighbor interatomic distance.

Exercise 2.6 Typical crystal structures of metallic elements are known to be fcc, hcp and bcc. Compute the packing fraction of these three structures. For comparison, compute the packing fraction of simple cubic structure.

Exercise 2.7 CaS and MgS have a NaCl-type structure. The radii of Ca^{2+} and Mg^{2+} ions are 0.099 and 0.065 nm, respectively. On the other hand, the radius of S^{2-} ion is known to be 0.182 nm. Describe the stability of CaS and MgS crystals using the difference in size of positive and negative ions.

Exercise 2.8 Aluminum has a fcc structure with a lattice parameter of a = 0.40497nm. Compute the interatomic distances in the planes (100) and (111).

Exercise 2.9 The coordination polyhedra can be defined as a group of the nearest neighbor atoms surrounding a central atom. The number of such nearest neighbors is called the coordination number. If the coordination number decreases, the volume of the coordination polyhedra will decrease. Calculate the reduction percentage in atomic radius when the coordination number changes from 12 to 8 or from 12 to 4.

Exercise 2.10 Supposing that a body-centered cubic (bcc) structure is filled with hard spheres, obtain the maximum radius of the sphere which can fit into octahedral and tetrahedral voids.

Exercise 2.11 Iron exhibits an α-phase at temperatures below 1,183 K and a δ-phase at temperatures above 1673 K. Both of these phases are characterized by a bcc structure, whereas the γ-phase found at temperatures between 1,183 and 1,673 K has a fcc structure. It is also known that only the γ-phase exhibits relatively high solubility for carbon.

(1) Explain the difference in carbon solubility of these phases from a structural point of view.
(2) Titanium is known to have a hcp structure at room temperature. Explain the difference in formation of titanium hydride and titanium carbide.

Exercise 2.12 Sodium chloride is known to contain four molecules in a unit cell. The molecular weight of sodium chloride per mole and the value of density at 298 K are 58.44 g and 2.164×10^6 g/m^3, respectively.

(1) Compute the specific volume and molecular volume of sodium chloride crystal.
(2) Estimate the distance between sodium ion and chlorine ion in sodium chloride crystal.

Exercise 2.13 With respect to caesium chloride crystal, (1) estimate the specific volume and molecular volume. (2) Also find the distance between cesium ion and chlorine ion. The molecular weight of causium chloride per mole and the density value at 298 K are 168.5 g and 3.970×10^6 g/m^3, respectively.

Exercise 2.14 Potassium bromide (KBr) has a NaCl-type structure. The ionic radii of K^+ and Br^- are 0.133 and 0.195 nm, respectively. The molecular weight of KBr is 119.00 g.

(1) Determine the lattice parameter a and the density of a KBr crystal by assuming the additivity of ionic radii.
(2) Find the ratio of r^+/r^- required to prevent the direct contact of anions.
(3) Potassium halides also have a NaCl-type structure. For the same halogen, the lattice parameter of rubidium halide is 0.028 nm larger than that of potassium halide. Estimate the ionic radius of Rb^+ from this difference.

Exercise 2.15 In the zinc blende (ZnS) structure, one element occupies the corner and the face-centered positions of a unit cell, and the other element occupies the tetrahedral positions of the diamond structure. In addition, the ionic radii of Zn^{2+} and S^{2-} are 0.074 and 0.184 nm, respectively.

(1) Find the nearest-neighbor coordination number of Zn^{2+} and S^{2-}.
(2) Estimate the angle formed by a S^{2-} and consecutive nearest-neighbor Zn^{2+} ions.
(3) Estimate the radius ratio (r^+/r^-) required to avoid direct contact between anions (S^{2-}) by assuming that positive and negative ions touch each other.
(4) Give a possible reason why ZnS does not have a NaCl-type structure.

Exercise 2.16 A unit cell of a caesium chloride (CsCl) crystal with a lattice parameter of $a = 0.4123$ nm has a Cl^- ion at each vertex and Cs^+ ion at the center of cube. The molecular weight of CsCl is 168.36 g. Compute the distance between Cs^+ and Cl^- ion and the density of this ionic crystal.

Exercise 2.17 The simplest crystal structure for zinc sulfide is the diamond-like cubic structure. The diamond structure may be visualized as follows; four atoms are added every other one to the tetrahedral site which exists at the center of eight "octants" contained in a unit cell of fcc structure. Explain the relationships among the face-centered cubic, diamond and zinc sulfide structures by illustrating their unit cells and the atomic arrangements along the vertical-axis for zinc sulfide structure. Note: An octant is the name of a small cubic unit cell.

Exercise 2.18 In annealing twins of a metallic element with fcc structure, it is likely to find the deformation in the (111) plane so as to form twin structure in the $[11\bar{2}]$ direction. Illustrate the movement of each atom during such deformation with twin structure.

Exercise 2.19 Draw the atomic arrangement in unit cells of fcc, hcp and bcc lattices which are the typical crystal structures of metallic elements. Show atomic arrangement at different distances along z-axis (vertical axis).

Exercise 2.20 Draw the layer-by-layer atomic arrangement in a unit cell along the vertical (z-) axis for of diamond and NaCl structures.

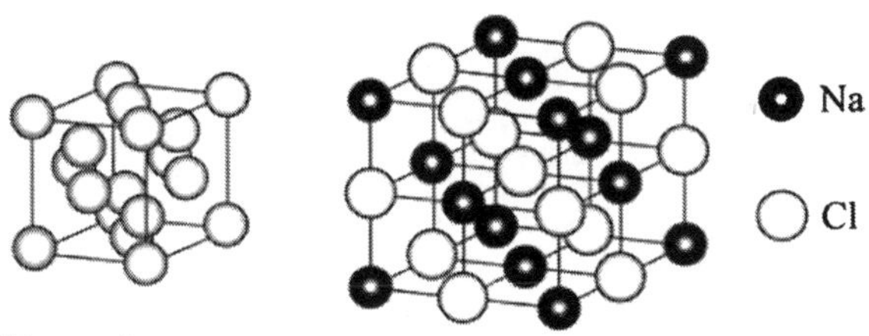

Diamond structure NaCl structure

Exercise 2.21 Draw the layer-by-layer atomic arrangement in a unit cell along the vertical (z-) axis for zinc blende (ZnS), fluorite (CaF_2) and rutile (TiO_2) structures.

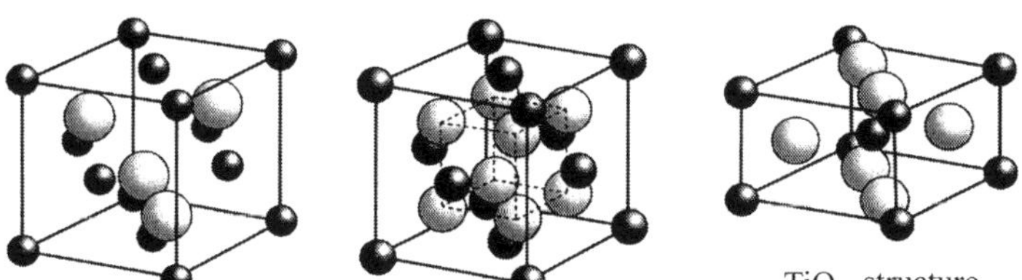

ZnS structure CaF_2 structure TiO_2 structure

Exercise 2.22 A four-index system ($H\ K\ i\ L$) is often used to identify planes in a hexagonal lattice, in addition to the usual three-index ($h\ k\ l$) system. A similar method is frequently used for describing directions.

(1) Using the three-index system, planes, (100), (010) and (001) may appear equivalent to one another. However, show that (001) is not equivalent to other two planes by the use of four indices. Further, show that (110) is equivalent to $(1\bar{2}0)$.
(2) Show that the directions [100], [010], [001] are not equivalent by the use of four indices. Also demonstrate that the $[10\bar{1}0]$ direction is perpendicular to the $(10\bar{1}0)$ plane.

Exercise 2.23 The lattice points having the symmetry elements characteristic of the rhombohedral system may also be referred to a hexagonal cell. Explain the transformation between rhombohedral axes and hexagonal axes.

Exercise 2.24 Find the geocentric angle between Sendai, Japan (38° of north latitude, 141° of east longitude) and Seattle, U.S.A. (47° of north latitude, 123° of west longitude).

Exercise 2.25 Draw the standard stereographic projection of a cubic crystal when projected on the plane of (001) or (011).

Exercise 2.26 Show the direction equivalent to perpendicular axis (1) and hexagonal axis (2) in the stereographic projection.

Exercise 2.27 Show the stereographic projection of two cases; (1) tetragonal pyramid and (2) tetragonal dipyramid. Also show the angular coordinates with respect to a certain plane in the tetragonal pyramid case.

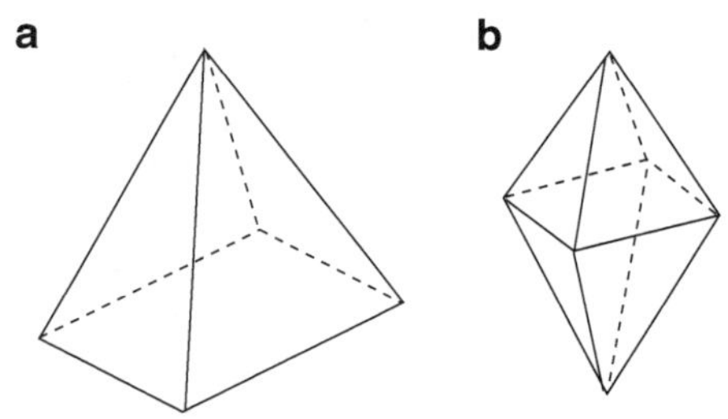

Exercise 2.28 Explain how to determine the position of a rotating axis of a cube using the Wulff net.

Exercise 3.1 Find the number of molecules and electrons contained in one cubic meter (m^3) water (H_2O) at 273 K.

Exercise 3.2 Differential cross section $d\sigma_e/d\Omega$ of one free electron for coherent X-ray scattering, which is generated from the area per unit solid angle in the direction of scattering angle θ, is given by the following equation.

$$\frac{d\sigma_e}{d\Omega} = \frac{r_e}{2}(1 + \cos^2\theta) \qquad (m^2/sr)$$

where r_e is the classical electron radius (2.8179×10^{-15}) m and sr is steradian.

(1) Estimate the value of $d\sigma_e/d\Omega$ in the direction of scattering angle 45° with respect to unit solid angle and unit scattering angle.
(2) Estimate the values of $d\sigma_e/d\Omega$ in the area from scattering angle zero to π radian (180°) and show the results in the graphical form.

Exercise 3.3 Compute the number of photons produced per unit area by the incoherent scattering, when the X-ray beam with 100 keV energy and the number of photons per unit area is given by $N_x = 2 \times 10^{12}$ (m^2) transmits through 20 mm thick water layer. Note that the value per electron of the incoherent scattering cross-section for the incident X-rays with energy of 100 keV is $\sigma_e = 70.01 \times 10^{-30}$ m^2.

Exercise 3.4 Compute how much the scattering intensity from one electron is at the distance of 0.01 m away from an electron, assuming that the polarization factor is negligible. Next estimate the number of electrons per unit mass for magnesium. The atomic number of magnesium is 12 and its molar mass is 24.305 g. In addition, confirm the scattering intensity from 1 g of Mg, at the distance of 0.01 m.

Exercise 3.5 Let us consider that X-rays with the wavelength of 0.01 nm collide with a free electron and produce incoherent scattering at scattering angle of 60°.

(1) Find the wavelength of the scattered photon.
(2) Estimate the energy of the scattered photon in keV.

Exercise 3.6 The incoherent scattering was obtained at 180° the scattering angle; the energy of the recoil electron was found to be 30 keV. Compute the scattered photon energy, by considering conservation law of both energy and momentum. Note that if the momentum of an electron is set to p, the rest mass to m_e and the speed of light to c, the relationship of $(E/c)^2 - p^2 = (m_e c)^2$ shall be approved with respect to the total energy of an electron E.

Exercise 3.7 Calculate the so-called Compton wavelength accompanying the Compton scattering of an electron.

Exercise 3.8 The electron density of a hydrogen atom is given as follows.

$$\rho = \frac{(e^{-2r/a})}{\pi a^2} \qquad a = 0.053 \text{ nm} = 0.53 \text{ Å} \qquad \int \rho dV = 1$$

(1) Derive the atomic scattering factor f and the incoherent scattering intensity per atom $i(M)$ as a function of $(\sin\theta/\lambda)$.
(2) Calculate the values of f and $i(M)$ at values of $(\sin\theta/\lambda) = 0.0$, 0.2 and 0.4.

Exercise 3.9 Let us consider that X-rays with an energy of 100 keV collide with a free electron and are incoherently scattered. Find the energy of the scattered photon under the condition that the recoil angle (ϕ) is zero.

Hint: A value can not be assigned to $1/\tan\phi$ because of the zero in the denominator, so use that $2\theta \to 2\pi$ as $\phi \to 0$, where 2θ is the scattering angle.

Exercise 3.10 Cuprous chloride (CuCl) has ZnS type structure and the molecular weight per mole is 99.00 g and the density is 4.135×10^6 g/m^3 at 298 K. When the Mo-Kα radiation with the wavelength of $\lambda = 0.07107$ nm is used, find the angle at which a strong peak corresponding to the reflection from the (111) plane will be observed.

Exercise 3.11 The molar mass of magnesium oxide having NaCl type structure is 40.30 g and the density is 3.58×10^6 g/m^3 at 298 K.

(1) Compute the values of $(\sin\theta/\lambda)$ at which peaks corresponding to the reflection from the planes of (100), (110) and (111) may appear.
(2) Explain the condition for detecting these three peaks by the crystallographic structure factor.

Exercise 3.12 AgCl is known to have NaCl type structure. Which of the following indices are allowed in the X-ray diffraction pattern? 100, 010, 001, 110, 101, 011, 111, 200, 020, 002, 120, 102, 012, 210, 201, 021, 220, 202, 022, 211, 121, 112, 221, 212, 122 and 222.

Exercise 3.13 Uranium is known to have an orthorhombic crystal lattice with four atoms in a unit cell. When the length of the unit cell is a standard unit, the positions of the four atoms $u\,v\,w$ are expressed as $0y\frac{1}{4}; 0-y\frac{3}{4}; \frac{1}{2}\frac{1}{2}+y\frac{1}{4}$ and $\frac{1}{2}\frac{1}{2}-y\frac{3}{4}$, where y expresses an arbitrary location.

(1) Find the Bravais lattice of uranium in addition to the structure factor F_{hkl}.
(2) Determine the plane indices for which the diffracted intensity cannot be observed.

Exercise 3.14 In the calcium fluorite (CaF_2) structure, atoms occupy the position of $\frac{3}{4}\frac{3}{4}\frac{3}{4}$, which is not filled in the so-called zinc blende (ZnS) structure.
That is, Ca occupies the positions of $000; \frac{1}{2}\frac{1}{2}0; 0\frac{1}{2}\frac{1}{2}; \frac{1}{2}0\frac{1}{2}$, and F occupies the positions of $\frac{1}{4}\frac{1}{4}\frac{1}{4}; \frac{3}{4}\frac{3}{4}\frac{1}{4}; \frac{1}{4}\frac{3}{4}\frac{3}{4}; \frac{3}{4}\frac{1}{4}\frac{3}{4}; \frac{3}{4}\frac{3}{4}\frac{3}{4}; \frac{1}{4}\frac{1}{4}\frac{3}{4}; \frac{3}{4}\frac{1}{4}\frac{1}{4}; \frac{1}{4}\frac{3}{4}\frac{1}{4}$.

(1) Obtain the structure factor F_{hkl} of CaF_2 and also express it using the structure factor $F(fcc)$ of a face-centered cubic lattice.

(2) Determine the plane indices for which the diffracted intensity can and cannot be observed.

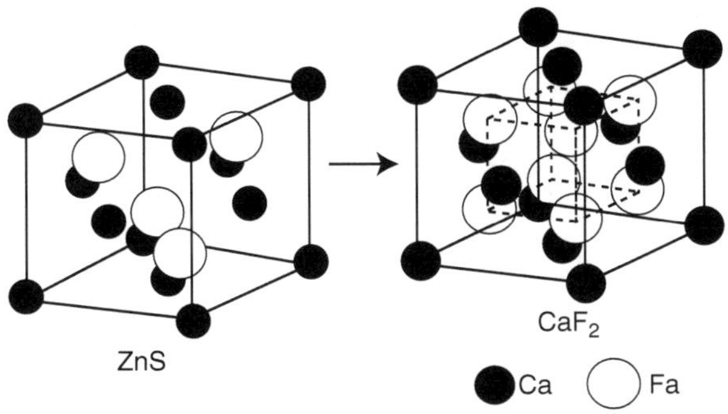

Exercise 3.15 Graphite has hexagonal crystal lattice which contains four atoms per unit cell at positions of 000; $\frac{1}{3}\frac{2}{3}0$; $00\frac{1}{2}$ and $\frac{2}{3}\frac{1}{3}\frac{1}{2}$. Show that the structure factor is given by the following equations. In addition, find the condition of $h\ k\ l$ in which the diffracted intensity cannot be observed.

$$l = \text{even}: \qquad F = 4f\cos^2\left\{\pi\left(\frac{h+2k}{3}\right)\right\}$$

$$l = \text{odd} \qquad F = \mathrm{i}2f\sin\left\{2\pi\left(\frac{h+2k}{3}\right)\right\}$$

Exercise 3.16 Explain characteristic features of the structure of caesium chloride (CsCl) crystal. In addition, estimate the diffracted intensity values for peaks from the (100), (110) and (111) planes by assuming that the atomic scattering factor can be approximated by the atomic number of each element, that is 55 for Cs and 17 for Cl, respectively.

Exercise 4.1 Derive an equation of absorption factor for measuring the diffraction intensity from a slab sample with thickness t by the symmetry-transmission method.

Exercise 4.2 Thermal expansion coefficient of copper is 16.6×10^{-6}/K. Compute how much the temperature control is required for a sample in order to obtain the lattice parameter of copper within the uncertainty of ± 0.00001 nm at 293 K.

Exercise 4.3 The lattice parameter of aluminum at room temperature is $a = 0.4049$ nm and its Debye–Waller factor is $B_{\mathrm{T}} = 8.825 \times 10^{-3}$ nm^2. Let us consider the intensity measurements of aluminum for peaks corresponding to the 111, 311 and 420 planes. What is the intensity at room temperature as a percentage of that at absolute zero Kelvin.

Exercise 4.4 A diffraction pattern of a powder crystalline sample with a very small volume was recorded using a Debye–Scherrer camera with a flat-plate film positioned at a distance D from the specimen and perpendicular to the incident beam. Find the intensity P' per unit length of the diffraction circle on the flat-plate film as a function of D.

Exercise 4.5 For two metallic samples, the following numerical data (Tables A and B) were obtained from X-ray diffraction patterns using Cu-Kα radiation ($\lambda = 0.1542$ nm). By considering the fact that fcc or hcp-type feature is observed in the diffraction pattern, determine the crystal structure and also estimate the lattice parameter.

Table A

	2θ	d(A)	I / I_0
1	43.16	2.0761	100
2	50.28	1.8148	48
3	73.97	1.2816	26
4	89.86	1.0917	24
5	95.05	1.0453	7

Table B

	2θ	d(A)	I / I_0
1	36.30	2.4751	53
2	38.97	2.3114	40
3	43.22	2.0935	100
4	54.32	1.6890	37
5	70.07	1.3430	48
6	70.61	1.3341	35
7	77.04	1.2380	6
8	82.11	1.1739	26
9	83.70	1.1556	4
10	86.53	1.1249	16
11	89.90	1.0913	11
12	94.92	1.0464	6

Exercise 4.6 For two unknown samples, the following numerical data (Tables A and B) were obtained from X-ray diffraction patterns using Cu-Kα radiation ($\lambda = 0.1542$ nm). By assuming that the unknown sample is a single phase, identify each sample using the Hanawalt method.

Table A

	2θ	d(A)	I / I_0
1	36.90	2.436	20
2	42.77	2.114	100
3	62.14	1.494	55
4	74.50	1.274	5
5	78.51	1.218	15
6	93.95	1.055	5
7	105.74	0.967	2
8	109.92	0.942	15

Table B

	2θ	d(A)	I / I_0
1	35.83	2.506	85
2	41.74	2.164	100
3	60.45	1.532	60
4	72.33	1.307	35
5	76.05	1.252	20
6	90.77	1.083	10
7	101.68	0.994	15
8	105.47	0.969	25

Exercise 4.7 X ray diffraction pattern (Fig. A) and the relevant numerical data (Table A) for the unknown sample were obtained by Cu-Kα radiation. Identify the sample by applying the Hanawalt method.

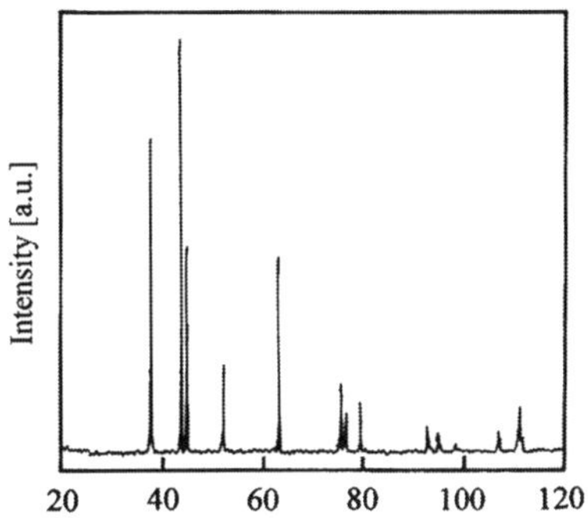

Fig.1

Table A

	2θ	d (A)	I/I_0
1	37.32	2.410	70
2	43.34	2.088	100
3	44.53	2.035	60
4	51.92	1.761	30
5	62.89	1.478	55
6	75.38	1.261	15
7	76.31	1.248	10
8	79.40	1.207	10
9	92.76	1.065	10
10	94.97	1.046	5
11	98.33	1.019	5
12	107.34	0.957	5
13	111.10	0.935	15

Exercise 4.8 X ray diffraction pattern (Fig. A) and the relevant numerical data (Table A) for the unknown sample were obtained by Cu-Kα radiation. Identify the sample by applying the Hanawalt method.

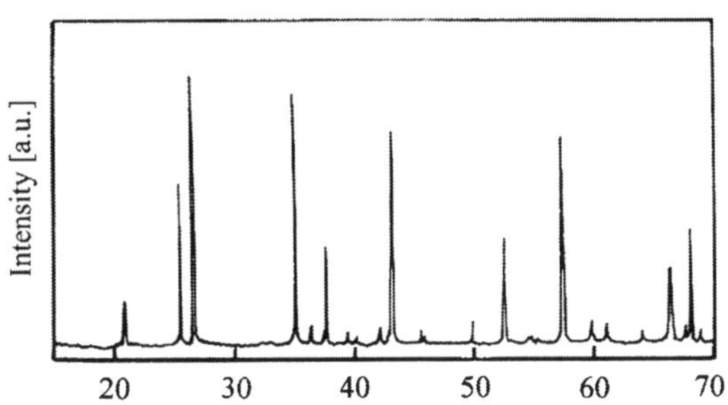

Fig. A

Table A

	2θ	d (A)	I/I_1		2θ	d (A)	I/I_1
1	20.80	4.271	17	13	50.25	1.816	10
2	25.59	3.482	62	14	52.54	1.742	41
3	26.66	3.344	100	15	54.88	1.673	3
4	35.16	2.553	95	16	55.31	1.661	2
5	36.56	2.458	7	17	57.42	1.605	80
6	37.77	2.382	36	18	59.96	1.543	8
7	39.46	2.284	5	19	61.32	1.512	7
8	40.27	2.240	1	20	64.10	1.453	2
9	42.42	2.131	6	21	66.51	1.406	28
10	43.36	2.087	90	22	67.71	1.384	4
11	45.78	1.982	2	23	68.21	1.375	40
12	46.15	1.967	2	24	68.44	1.371	3

Exercise 4.9 A diffraction experiment was carried out on a powder sample of silicon using Cu-Kα radiation. The structure of silicon is the same as that of diamond crystal lattice. In the high angle region, the peak splitting attributed to the difference between Kα_1 and Kα_2 is clearly observed in Fig. A and the relevant angular data are summarized in Table A for the (440), (531), (620), and (533) planes. Compute the lattice parameter using the extrapolation method.

Table A

	Radiation	2θ (degree)
1	Kα_1	106.71
2	Kα_2	107.11
3	Kα_1	114.10
4	Kα_2	114.56
5	Kα_1	127.55
6	Kα_2	128.13
7	Kα_1	136.92
8	Kα_2	137.67

Cu- Kα_1=0.154056 nm
Cu- Kα_2=0.154439 nm

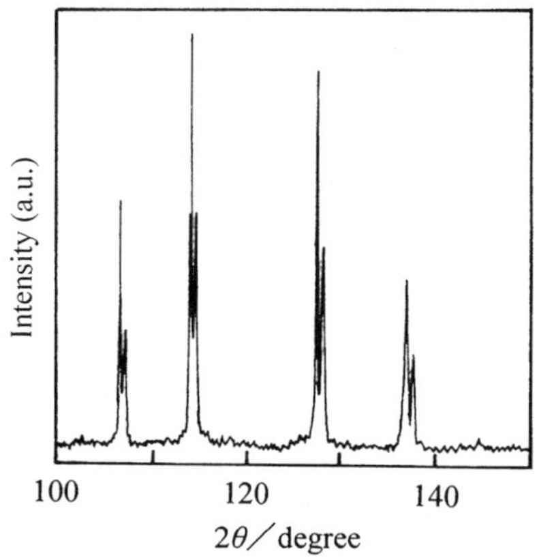

Fig. A

Exercise 4.10 Ultra fine particles of iron were produced by the evaporation method and their average sizes are estimated 25, 50, 90 and 120 nm using the laser particle analyzer. Assuming that the resultant particles are strain free and the peak broadening detected in the measurement with Cu-Kα radiation (λ = 0.15406 nm) arises only from variation of size of the crystallites, estimate the peak width (the value of FWHM) of the (110) plane possibly detected by using the Scherrer equation.

Exercise 4.11 The peak broadening as well as the decrease in peak height was observed, during a diffraction experiment by Cu-Kα ($\lambda = 0.1542$ nm) radiation on the cold worked aluminum sample. The results are summarized in Table A where the FWHM values for peaks corresponding to the (111), (200), (220), and (311) planes. The aluminum sample was fully annealed in order to remove the stress and strain induced by cold-work and similar measurements were done for comparison. The results are summarized in Table B with the corresponding FWHM values. Compute the average size of crystallites in the cold-worked aluminum sample.

Table A

	hkl	2θ (degree)	FWHM (degree)
1	111	38.47	0.188
2	200	44.72	0.206
3	220	65.13	0.269
4	311	78.27	0.303

Table B

	hkl	2θ (degree)	FWHM (degree)
1	111	38.47	0.102
2	200	44.70	0.065
3	220	65.10	0.089
4	311	78.26	0.091

Exercise 4.12 The diffraction experiment using Cu-Kα ($\lambda = 0.1542$ nm) radiation was carried out for the mixed powder sample of MgO and CaO and two or more clearly separated diffraction peaks were obtained. According to the preliminary analysis, the diffraction peak of the (111) plane of MgO is found to overlap with that of the (200) plane of CaO near $2\theta = 37°$ (see Fig. A) and it is difficult to separate. Other peaks can be identified with each component. The measured integrated intensities of the corresponding peaks are summarized in Table A. Calculate the contents of MgO and CaO. Both of MgO and CaO have NaCl type structure

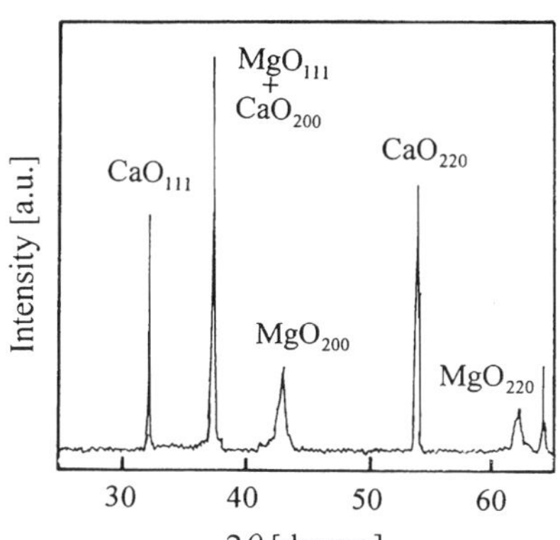

Fig. A

and the lattice parameters are 0.42112 nm for MgO and 0.48105 nm for CaO (see Appendix A.9).

Table A

	2θ (degree)	Intensity I*	hkl
1	32.18	73.9	CaO(111)
2	37.17	-	MgO(111)+CaO(200)
3	42.90	123.1	MgO(200)
4	53.85	106.0	CaO(220)
5	62.27	74.5	MgO(220)
6	64.16	-	CaO(311)

* counts per unit area and unit time

Exercise 4.13 The diffraction experiment using Cu-Kα line ($\lambda = 0.1542$ nm) radiation was made for a mixed powder sample of Si containing Cu and four clearly separated diffraction peaks were obtained. These four peaks can be attributed to peaks of the (111) and (220) planes of Si and the (111) and (200) planes of Cu. The integrated intensities of these four peaks are summarized in Table A. Calculate the contents of Cu and Si using the direct method.

Table A

	2θ (degree)	Integrated intensity I	hkl
1	28.40	162.3	Si(111)
2	43.28	359.7	Cu(111)
3	47.31	87.2	Si(220)
4	50.43	120.4	Cu(200)

Exercise 4.14 Slags used in steel-making process dissolve various elements in the glassy phase and the melilite component is known to be frequently involved. After griding this slag sample in a ball mill, the diffraction peak corresponding to the (211) plane of the melilite component was observed at about 32° scattering angle, as shown in Fig. A. With increasing milling time, the peak broadening as well as the decrease in peak height was observed as summarized in Table A. Compute the variation in average size of crystallites by assuming that measured FWHM values arise only from change of the crystallite size.

Table A

Milling time (hour)	Melilite (211)	
	2θ degree	FWHM degree
0	31.24	0.059
1	31.25	0.105
2	31.27	0.272
4	31.30	0.319
6	31.31	0.483

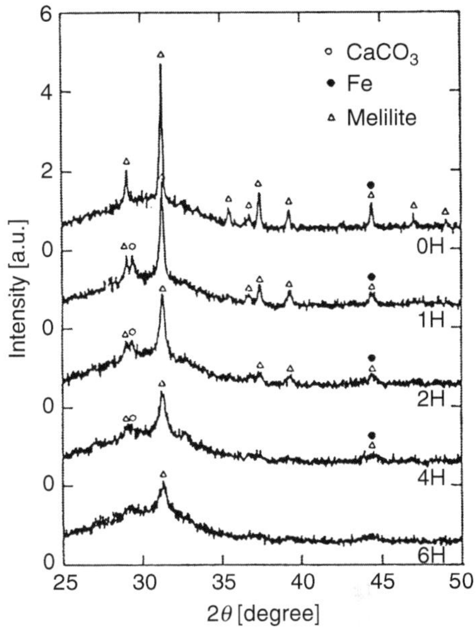

Fig. A

Exercise 5.1 For a rectangular (or orthogonal) coordinate system *XY*, let us consider that the unit vectors, taken from a point O along the positive directions of the individual axes, are e_x and e_y. If the coordinates of two vectors from O to the points P_1 and P_2 are given by $P_1(x_1, y_1)$ and $P_2(x_2, y_2)$, obtain the relationship for the coordinates $P(x, y)$ which divides the distance between P_1 and P_2 with the ratio of $m : n$.

Exercise 5.2 In a rectangular coordinate system *XY*, when the angle between a straight line and the positive X-axis is θ, the angle to the positive Y-axis is given by $\{(\pi/2) - \theta\}$. Accordingly, the two directional cosines are $\lambda = \cos\theta$ and $\mu = \cos\{(\pi/2) - \theta\} = \sin\theta$, respectively. In this case, $m = (\mu/\lambda) = (\sin\theta/\cos\theta) = \tan\theta$ is called the directional coefficient of the straight line. Obtain the relationships of the directional coefficients and directional cosines when a straight line $\boldsymbol{q}_1$ is perpendicular to a straight line $\boldsymbol{q}_2$, where their directional coefficients are given by m_1 and m_2, respectively.

Exercise 5.3 Let us consider that a two-dimensional lattice is expressed by the basic vectors $\boldsymbol{a} = 2\boldsymbol{e}_x$ and $\boldsymbol{b} = \boldsymbol{e}_x + 2\boldsymbol{e}_y$. Obtain the primitive vectors of the reciprocal-lattices, $\boldsymbol{A}^*$ and $\boldsymbol{B}^*$.

Exercise 5.4 Show that the magnitude of the reciprocal-lattice vector $\boldsymbol{b}_1$ is equal to the reciprocal of a spacing of the (100) plane.

Exercise 5.5 Real crystal lattice vectors $\boldsymbol{a}_1$ and $\boldsymbol{a}_2$ are parallel to the drawing plane, whereas $\boldsymbol{a}_3$ is perpendicular to the drawing plane. When the lengths of the real crystal-lattice vectors are $|\boldsymbol{a}_1| = 3.0, |\boldsymbol{a}_2| = 2.0, |\boldsymbol{a}_3| = 1.0$ and the angle formed by $\boldsymbol{a}_1$ and $\boldsymbol{a}_2$ is $\alpha_{12} = 60°$, solve the following problems.

(1) Draw the real crystal-lattice vectors $\boldsymbol{a}_1$ and $\boldsymbol{a}_2$, and also the reciprocal-lattice vectors $\boldsymbol{b}_1$ and $\boldsymbol{b}_2$ for the above conditions.
(2) Draw the (110), (210) and (310) planes in the real crystal-lattice based on the definition of Miller indices *hkl*, and also show $\boldsymbol{H}_{110}, \boldsymbol{H}_{210}$ and $\boldsymbol{H}_{310}$ in the reciprocal-lattice when $\boldsymbol{H}_{hkl} = h\boldsymbol{b}_1 + k\boldsymbol{b}_2 + l\boldsymbol{b}_3$.

Exercise 5.6 Let us consider that a crystallographic direction may be expressed by $\boldsymbol{A}_{uvw} = u\boldsymbol{a}_1 + v\boldsymbol{a}_2 + w\boldsymbol{a}_3$ and the direction perpendicular to the (hkl) plane by $\boldsymbol{H}_{hkl} = h\boldsymbol{b}_1 + k\boldsymbol{b}_2 + l\boldsymbol{b}_3$. If the angle between $\boldsymbol{A}_{uvw}$ and $\boldsymbol{H}_{hkl}$ is ϕ, express $\cos\phi$ for the orthorhombic system in terms of $u, v, w, h, k, l, \boldsymbol{a}_1, \boldsymbol{a}_2$ and $\boldsymbol{a}_3$.

Exercise 5.7 It is known that the trigonal (rhombohedral) lattice unit cell is the primitive unit cell for the fcc structure. Similarly, the orthorhombic lattice unit cell is the primitive lattice unit cell for the hexagonal structure. Find the method of conversion for the equivalent plane between these two primitive unit cells using the fact that only one reciprocal-lattice vector can have a real physical meaning.

Exercise 5.8 For a face-centered cubic lattice, answer the following questions.

(1) Obtain the primitive (translation) vectors of the reciprocal-lattices.
(2) Show the first Brillouin zone.

Exercise 5.9 Obtain the so-called Bragg equation or Laue equation by using the condition that the strongest scattering amplitude from crystals is obtained when the scattering angle is in agreement with the reciprocal-lattice vector.

Exercise 5.10 Show that a given formula for the summation of the real crystal lattice vectors is not zero only if the vector $\boldsymbol{q}$ is equal to the reciprocal-lattice vector.

$$G(\boldsymbol{q}) = \sum_n \mathrm{e}^{-2\pi i \boldsymbol{q}\cdot\boldsymbol{r}_n}$$

Exercise 5.11 If the $\sin\theta/\lambda$ value is very small, show that both the atomic scattering factor f and the incoherent scattering intensity $\boldsymbol{i}(M)$ are given by a parabolic function of $\sin\theta/\lambda$.

Exercise 5.12 In the classical theory, the amplitude of vibration of atomic electrons is considered to be continuously and infinitely variable. On the other hand, from a quantum-mechanical point of view, the angular frequency is that of an X-ray photon with just enough energy to eject the electron from the atom. This is the case when the wavelength of the incident X-rays is close to the absorption edge and it is known as anomalous scattering (or resonance scattering) phenomenon. Considering this phenomenon as the interaction between X-rays and a harmonic oscillator with an attenuation term, explain the variation of the atomic scattering factor f.

Exercise 5.13 With respect to a diffraction phenomenon from one-dimensional slit called the Fraunhofer diffraction which is attributed only to the aperture width L, obtain the diffraction intensity at the point P which is at distance R sufficiently far from the slit.

Exercise 5.14 Obtain the diffraction intensity produced from a two-dimensional lattice described by an interval a repeated m times in the x-direction and an interval b repeated n times in the y-direction, respectively. Here, m and n take sufficiently large values.

Exercise 5.15 Let us consider a one-dimensional lattice in which identical scattering centers are at positions $\boldsymbol{r}_n = n\boldsymbol{a}$ (n: integer). The scattering amplitude G in this one-dimensional lattice is proportional to $G = \sum e^{-n\boldsymbol{a}\cdot\boldsymbol{q}}$, and the sum over m lattice points is given by.

$$G = \sum_{n=0}^{m-1} e^{-in\boldsymbol{a}\cdot\boldsymbol{q}} = \frac{1 - e^{-im(\boldsymbol{a}\cdot\boldsymbol{q})}}{1 - e^{-i(\boldsymbol{a}\cdot\boldsymbol{q})}}$$

(1) Obtain the diffraction intensity I.
(2) In this case, the maximum value of the diffraction intensity can be obtained when $\boldsymbol{a}\cdot\boldsymbol{q}$ is given by integer multiples of 2π. Let us consider the case of a small change in $\boldsymbol{q}$ such that $\boldsymbol{a}\cdot\boldsymbol{q} = 2\pi\boldsymbol{h} + \delta(\delta > 0)$. Calculate the minimum value of δ for which the intensity becomes zero.

Exercise 5.16 In the model system of carbon tetrachloride, four chlorine atoms are located at the positions of regular tetrahedron and the carbon atom occupies the center of the tetrahedron. By applying Debye's equation, obtain the approximate expression for the diffraction intensity from carbon tetrachloride molecules constructed with a total of five atoms with two distinct elements as the scatterer.

Exercise 5.17 Consider diffraction from a single crystal which N_1a, N_2a, and N_3a primitive cubic lattices arranged along the three coordinate axis, $< 100 >$ direction, with the lattice parameter of $a = 0.25$ nm. Explain features of the reciprocal-lattices for the following three cases with different numbers of unit cells along the three directions.

(1) $N_1 = N_2 = N_3 = 10^4$
(2) $N_1 = N_2 = 10^4, N_3 = 10$ (Reciprocal-lattice of a thin film layer)
(3) $N_1 = 10^4, N_2 = N_3 = 10$ (Reciprocal-lattice of a peace of string sample)

Exercise 5.18 Calcium fluorite (CaF_2) has a face-centered cubic lattice containing four (CaF_2) per unit cell and the lattice parameter is $a = 0.5463$ nm. The atomic positions, 000 for Ca, $\frac{1}{4}, \frac{1}{4}, \frac{1}{4}, \frac{3}{4}, \frac{3}{4}, \frac{3}{4}$ for F and other positions can be obtained by the face-centering translations.

(1) Obtain the structure factor F.
(2) Compute the values of the squares of the scattering factor F^2 for peaks corresponding to the planes 111 and 222 using the numerical values of f from Appendix A.3.
(3) Compute the F^2 values including the anomalous dispersion factors f and f'' for Ca.

Exercise 6.1 When the axes of the orthorhombic space group ***Pbcn*** are altered within the same unit cell, show the changes that occur to Hermann–Mauguin symbols. Repeat for the orthorhombic space group ***Cmca***.

Exercise 6.2 Show the coordinates of the equivalent positions when the symmetry operation is made for the general positions x, y, and z expressed by $\boldsymbol{C}2/c$ (International Tables: No.15, c-axis is taken on a unique axis) and $\boldsymbol{P}2_1 2_1 2_1$ (International Tables: No.19).

Exercise 6.3 If the (100) plane is set to a glide plane, explain the effect of the glide-reflection operation of $c/2$, with respect to the (100) plane, on the structure factor F_{hkl} and $|F_{hkl}|^2$. In addition show the conditions which enable us to detect the diffraction intensity.

Exercise 6.4 Explain the method for determining the screw axis including translational operation or a plane of gliding-reflection by applying the extinction law to diffraction peaks.

Exercise 6.5 Rutile-type titanium dioxide (TiO_2) belongs to the tetragonal system containing two TiO_2 formula units per unit cell and the atomic positions are given as;

$$\text{Ti}: \quad 000 \quad : \quad \frac{1}{2}\frac{1}{2}\frac{1}{2}$$

$$\text{O}: \quad uu0 \quad : \quad -u-u0 \quad : \quad \frac{1}{2}-u u+\frac{1}{2}\frac{1}{2} \quad : \quad u+\frac{1}{2}\frac{1}{2}-u\frac{1}{2}$$

(1) Obtain the structure factor F by considering the equivalent lattice points.
(2) Obtain the conditions for which F becomes zero without reference to the value of u.
(3) Explain which of a simple tetragonal or body-centered tetragonal lattice is a Bravais lattice.

Exercise 6.6 Supposing an incident beam is reflected from a plane $h_1k_1l_1$, it is possible to coincide this beam with the direction of the reflected beam from another plane $h_2k_2l_2$.

(1) Show that the direction of the second reflection is equivalent to that due to direct reflection from a plane $h_3k_3l_3$ such that:

$$h_3 = h_1 + h_2, k_3 = k_1 + k_2, l_3 = l_1 + l_2$$

(2) When double reflections from the $(\bar{1}\bar{1}1)$ and $(h_2k_2l_2)$ planes of a diamond crystal with a lattice parameter of $a = 0.3567$ nm are produced using Cu-Kα ($\lambda = 0.1542$ nm) radiation, it is possible to obtain a so-called pseudo-reflection from the (222) plane. Identify such $(h_2k_2l_2)$ planes. In addition, show the direction of the incident beam by the directional cosine S_0 along three orthogonal axes a_1, a_2 and a_3.

Exercise 6.7 Following data give the lattice parameters and the basis of the rutile structure of titanium dioxide (TiO_2). Draw both the perspective drawing and the projection on the $(x, y, 0)$ plane. In addition, show a stereographic projection.

Lattice	Basic atom position	Space groups	Atom positions		
Tetragonal P	Ti : $0, 0, 0$		a	Ti : $0, 0, 0$	
$a = 0.459$ nm	$\frac{1}{2}, \frac{1}{2}, \frac{1}{2}$			$\frac{1}{2}, \frac{1}{2}, \frac{1}{2}$	
	O : $0.3, 0.3, 0$	$P4_2/mnm$	f	O : $x, x, 0$	
$c = 0.296$ nm	$0.8, 0.2, \frac{1}{2}$			$\frac{1}{2}+x, \frac{1}{2}-x, \frac{1}{2}$	$x = 0.3$
	$0.2, 0.8, \frac{1}{2}$			$\frac{1}{2}-x, \frac{1}{2}+x, \frac{1}{2}$	
	$0.7, 0.7, 0$			$\bar{x}, \bar{x}, 0$	

Exercise 6.8 Pyrite is known to have NaCl type structure with four FeS_2 formula units per unit cell. The lattice parameter is $a = 0.5408$ nm. In addition, the space group symmetry suggests the following atomic positions;

$$\text{Fe} : 000\,;\ 0\frac{1}{2}\frac{1}{2}\,;\ \frac{1}{2}0\frac{1}{2}\,;\ \frac{1}{2}\frac{1}{2}0$$

$$\text{S} : u, u, u\,;\ \frac{1}{2}+u, \frac{1}{2}-u, \bar{u};$$

$$\bar{u}, \bar{u}, \bar{u}\,;\ -\frac{1}{2}-u, -\frac{1}{2}+u, u;$$

$$\bar{u}, \frac{1}{2}+u, \frac{1}{2}-u\,;\ \frac{1}{2}+u, \bar{u}, \frac{1}{2}+u$$

$$\bar{u}, -\frac{1}{2} - u\ ;\ -\frac{1}{2} + u\ ;\ -\frac{1}{2} + u, u, -\frac{1}{2} - u,$$

(1) Obtain the structure factor F for the unmixed $h\ k\ l$ by excluding the anomalous dispersion factors.
(2) For a powder sample of pyrite, the integrated intensities of the diffraction peaks for the (111) and (200) planes were found to be 69.2 and 277.5 in arbitrary units, respectively. Compute the value of u from the ratio of the integrated intensities of two peaks, when considering only the u-range between 0.30 and 0.40.

Exercise 6.9 Show the rotatory inversion operation of $\bar{5}$ and $\bar{10}$ on a general pole of the stereographic projection.

Exercise 6.10 With respect to the space group $\boldsymbol{P}2_1/c$, which is frequently found in real crystals, information can be obtained from the International Tables for Crystallography, Volume A, page 184–185. Explain the key points.

Exercise 6.11 Information on the space group $\boldsymbol{Cmm2}$, which is frequently found in real crystals, can be obtained from the International Tables for Crystallography, Volume A, page 238–239. Explain the key points.

Chapter 8
Solutions to Supplementary Problems

Exercise 1.1

$$E = 1.602 \times 10^{-16}\,\text{J},\ p = 1.708 \times 10^{-23}\,\text{kg}\cdot\text{m/s},\ \lambda = 3.879 \times 10^{-11}\,\text{m}$$

Exercise 1.2

$$\left(\frac{\mu}{\rho}\right)_{\text{GaAs}} = 68.62\,(\text{cm}^2/\text{g}),\ \left(\frac{\mu}{\rho}\right)_{\text{BaTiO}_3} = 234.8\,(\text{cm}^2/\text{g})$$

Exercise 1.3

$$32\%(\text{Cu} - \text{K}\alpha),\ 54\%(\text{Fe} - \text{K}\alpha)$$

Exercise 1.4

$$5.95\,\text{g}$$

Exercise 1.5

$$V_{\text{K}}(\text{kV}) = \frac{1.240}{\lambda_{\text{K}}(\text{nm})},\ \text{For Mo-K}\alpha,\ V_{\text{K}} = 20.01\,\text{kV}$$

Exercise 1.6

$$E_{\text{K}}^{\text{Pb}} = 88\,\text{keV}$$

Exercise 1.7

$$1.510 \times 10^8\,\text{m/s}$$

Exercise 1.8

$$55.9\,\text{keV}$$

Exercise 1.9

Photon number I_{K} : $1.28 \times 10^{11}[\text{mAs}^{-1}\cdot\text{sr}^{-1}]$, Intensity I : 1.24×10^{-3} (J/mAs·sr)

Exercise 1.10

$$2.55\,\text{mm}$$

Exercise 1.11

$$0.14\,\text{mm}$$

Exercise 1.12

Work function 4.52 eV, Reverse voltage V_i : 7.88 Volt

Exercise 1.13

1290 eV (L-shell) and 48 eV (M-shell) higher than the K-shell energy.

Exercise 2.2

bcc : 1.06, fcc : 1.15

Exercise 2.3

Rhombuses obtained by drawing lines between the centers of regular hexagons.

Exercise 2.4

1. $r_1 = 0.450\,\text{nm}$, $r_2 = 0.520\,\text{nm}$
2. $n_1 = 8$, $n_2 = 6$
3. Density : $0.923 \times 10^6\,\text{g/m}^3$

Exercise 2.5

1. $46.39 \times 10^{-30}\,\text{m}^3$
2. $a = 0.3207\,\text{nm} \qquad r_1 = a$

Exercise 2.6

fcc : 0.7405, hcp : 0.7405, bcc : 0.6802, simple cubic : 0.5236

Exercise 2.7

See the coordination number of 6 in Question 2.15. The sphere radius of octahedron formed by six S^{2-} ions is about 0.075 nm which is smaller than the ionic radius of Mg^{2+} ion.

Exercise 2.8

0.2864 and 0.4050 nm for (100) plane, 0.2864 nm for (111) plane.

Exercise 2.9

Reduction percentage from 12 to 8: 26.8% and from 12 to 4: 77.5%

Exercise 2.10

Octahedral void: $0.155r$, Tetrahedral void: $0.291r$

Exercise 2.11

1. Tetrahedral void in fcc: 0.107 nm ($0.414r$), Octahedral void in fcc: 0.058 nm ($0.225r$). Tetrahedral void in bcc: 0.072 nm ($0.291r$). Carbon atom (size: 0.14∼0.15 nm) is found to be relatively easy to occupy the tetrahedral void in fcc structure.
2. Packing condition of hcp in the ideal state is equal to the fcc case.

Exercise 2.12

1. 0.4621×10^{-6} (m^3/g), 27.01×10^{-6} (m^3/mol)
2. 0.2820 nm

Exercise 2.13

1. 0.2519×10^{-6} (m^3/g), 42.41×10^{-6} (m^3/mol)
2. 0.3577 nm

Exercise 2.14

1. $a = 0.656$ nm, Density $= 2.80 \times 10^6$ g/m^3
2. $r^+/r^- = 0.414$
3. Ionic radius of Rb^+: 0.147 nm

Exercise 2.15

1. 4
2. $\theta = 109.48°$
3. $r^+/r^- = 0.225$
4. $r^+/r^- = 0.402$ for ZnS. Since the radius of Zn^{2+} cation is smaller than the maximum radius fitting into the vacant space formed by NaCl structure, the direct contact between anions of S^{2-} is possible, so that the structure is unstable.

Exercise 2.16

Distance : 0.3571 nm, Density : 3.989×10^6 g/m^3

Exercise 2.18

When applying the annealing process to metals with fcc structure after cold-working, we can frequently observe the twin crystal growth, called "annealing twins." Note that "deformation twins" are also known in metals with bcc or hcp structure. In order to explain the formation of the annealing twin structure, the positions of about 67% (=2/3) of total atoms are unchanged and we take a uniform shear motion of the (111) layer to the $[11\bar{2}]$ direction into account. This enables us to move each layer only the quantity proportional to the distance of twins.

Exercise 2.22

Three indices (hkl)	↔ Four indices $(HKiL)$	Three indices $[uvw]$	↔ Four indices $[UVtW]$
(100)	$(10\bar{1}0)$	[100]	$[2\bar{1}\bar{1}0]$
(010)	$(01\bar{1}0)$	[010]	$[\bar{1}2\bar{1}0]$
(001)	(0001)	[001]	[0001]
$(0\bar{2}0)$	$(0\bar{2}10)$		
(110)	$(11\bar{2}0)$		

Exercise 2.23

The relationship between (hkl) for rhombohedral structure and $(HKiL)$ for the hexagonal case is $-H + K + L = 3k$ (k : integer). If $(-H + K + L)$ is not given by integer multiple of 3, the system is considered to be hexagonal.

Exercise 2.24

68°

Exercise 2.28

In cubic, the zone pole and the corresponding zone circle are at right angles to one another, so that the angle between the planes of two zone circles is equal to that formed by two poles of the corresponding zones.

Exercise 3.1

$N_m = 3.34 \times 10^{28}$ molecules/m^3, $N_e = 3.34 \times 10^{29}$ electrons/m^3

Exercise 3.2

1. $\frac{d\sigma_e}{d\Omega} = 5.96 \times 10^{-30}$ m^2/sr. Estimate the differential cross section using the relation of $d\Omega = 2\pi \sin\theta d\theta$. $\frac{d\sigma_e}{d\Omega} = 2.65 \times 10^{-30}$ m^2/rad at ($\sin 45° = 0.7071$).
2. The values of differential cross section per unit solid angle are obtained as a function of angle θ. Note that the value at 90° is found half of the 180° case.

θ (degree)	$\cos^2\theta$	$\frac{d\sigma_e}{d\Omega}$	θ (degree)	$\cos^2\theta$	$\frac{d\sigma_e}{d\Omega}$
0	1.0	7.94×10^{-30}	100	0.030	4.09×10^{-30}
30	0.750	6.95×10^{-30}	130	0.413	5.61×10^{-30}
90	0	3.97×10^{-30}	180	1.0	7.94×10^{-30}

Exercise 3.3

If the number of electrons in water layer is set to N_e, the number of photons per unit area N_{ip} due to the incoherent scattering is given as $N_{ip} = \sigma_e \cdot N_e \cdot N_x = 6.95 \times 10^9 \text{m}^{-2}$.

Exercise 3.4

Intensity scattered from one electron : $I_e = 7.95 \times 10^{-26} I_0$, Intensity scattered from electrons of 1 g Mg : $I'_e = 0.024 I_0$, so that it is measurable.

Exercise 3.5

(1) $\lambda = 0.0112\,\text{nm}$, (2) $E = 110.7\,\text{keV}$

Exercise 3.6

$E = 73.815\,\text{keV}$

Exercise 3.7

$\Delta\lambda = 0.002426(1 - \cos 2\theta)\,\text{nm}$

Exercise 3.8

1.

$$a = 0.53\,\text{Å} \text{ and } k = 4\pi \sin\theta/\lambda$$

$$f = f_e = \frac{16}{(a^2k^2 + 4)^2} = \frac{1}{\{1 + (1.06\pi \sin\theta/\lambda)^2\}^2}$$

$$i(M) = 1 - f_e^2 = 1 - \frac{1}{\{1 + (1.06\pi \sin\theta/\lambda)^2\}^4}$$

2.

$\sin\theta/\lambda$	f	$i(M)$
0.0	1.0	0
0.2	0.48	0.77
0.4	0.13	0.98

Exercise 3.9

$E = h\nu = 71.7\,\text{keV}$

Exercise 3.10

$\theta = 6.52°$

Exercise 3.11

For (100) : $0.119 \times 10^{10}\,\mathrm{m}^{-1}$, For (110) : $0.168 \times 10^{10}\,\mathrm{m}^{-1}$ and For (111) : $0.206 \times 10^{10}\,\mathrm{m}^{-1}$. Intensity is not detected if (hkl) is mixed.

Exercise 3.12

Four indices : 111, 200, 220 and 222 are allowed.

Exercise 3.13

$$F^2 = 4f_{\mathrm{U}}2\left\{1 + \cos\left(\frac{\pi(h+k)}{2}\right)\right\} \times \{1 + \cos\pi(-4yk + l)\}$$

$$F = 0, \text{if } (h+k) \text{ is odd number.}$$

Exercise 3.14

$$F = \left[f_{\mathrm{Ca}} + 2f_{\mathrm{F}}\cos\frac{\pi}{2}(h+k+l)\left\{2\cos\frac{\pi}{2}(h+k+l) - 1\right\}\right] F(fcc)$$

h, k, l	$h+k+l$	F
Mixed	–	0
Unmixed	$4n$	$4(f_{\mathrm{Ca}} + 2f_{\mathrm{F}})$
Unmixed	$4n \pm 1$	$4f_{\mathrm{Ca}}$
Unmixed	$4n \pm 2$	$4(f_{\mathrm{Ca}} - 2f_{\mathrm{F}})$

Exercise 3.15

$F = 0$, when $l = 2n + 1, h = 2k = 3n$

Exercise 3.16

$$|F_{100}|^2 = (55 - 17)^2 = 38^2, \qquad |F_{111}|^2 = (55 - 17)^2 = 38^2$$

$$|F_{110}|^2 = (55 + 17)^2 = 72^2$$

Exercise 4.1

$$A = \frac{\sec\theta}{\mathrm{e}^{-\mu_s t_s(1-\sec\theta)}}$$

Exercise 4.2

$$\Delta T = \pm 1.67\mathrm{K}$$

Exercise 4.3

111 : 92.3%, 311 : 74.4%, 420 : 58.4%

Exercise 4.4

$$P' = \frac{I}{2\pi D \tan 2\theta} = \frac{I_0}{2\pi D}|F|^2 p \left(\frac{1+\cos^2 2\theta}{\sin^2\theta\cos\theta\tan 2\theta}\right)$$

Exercise 4.8

Note: Not only three d values, but also eight d values are used. When using the Hanawalt method, preference of d values is suggested in comparison to those of the relative intensity ratio.

Exercise 4.9

$a = 0.54305$ nm (in average), $a = 0.54302$ nm (least-squares method).

Exercise 4.10

0.34° for $t = 25$ nm, 0.17° for $t = 50$ nm, 0.10° for $t = 90$ nm and 0.07° for $t = 120$ nm.

Exercise 4.11

93 nm (Hall method), 110 nm (least-squares method).

Exercise 4.12

$$\frac{c_{\mathrm{CaO}}}{c_{\mathrm{MgO}}} = \frac{I_{\mathrm{CaO}}}{I_{\mathrm{MgO}}} \times \frac{R_{\mathrm{MgO}}}{R_{\mathrm{CaO}}} = \frac{106.0}{74.5} \times \frac{0.97\times 10^7}{1.34\times 10^7} = 1.03$$

$$c_{\mathrm{MgO}} = 0.49 \quad \text{and} \quad c_{\mathrm{CaO}} = 0.51$$

Exercise 4.13

$$\frac{c_{\mathrm{Cu}}}{c_{\mathrm{Si}}} = \frac{I_{\mathrm{Cu}}}{I_{\mathrm{Si}}} \times \frac{R_{\mathrm{Si}}}{R_{\mathrm{Cu}}} = \frac{359.7}{162.3} \times \frac{1.74\times 10^7}{16.87\times 10^7} = 0.228$$

$$c_{\mathrm{Si}} = 0.81 \quad \text{and} \quad c_{\mathrm{Cu}} = 0.19$$

Exercise 4.14

$$B_r = 2\Delta\theta(\mathrm{FWHM}) = \frac{0.9\lambda}{\varepsilon\cos\theta} \rightarrow \varepsilon = \frac{0.9\lambda}{B_r\cos\theta}$$

Milling time hour	2 θ degree	cosθ	ϵ (µm)
1	31.25	0.9630	0.095
2	31.27	0.9630	0.031
4	31.30	0.9629	0.026
6	31.31	0.9629	0.017

Exercise 5.1

$$x = \frac{nx_1 + mx_2}{m+n} \quad y = \frac{ny_1 + my_2}{m+n}$$

Exercise 5.2

The angle formed by the straight lines $\boldsymbol{q}_1$ and $\boldsymbol{q}_2$ is given by $\cos\alpha = \lambda_1\lambda_2 + \mu_1\mu_2$. If these two lines are mutually perpendicular, we find $\cos\alpha = 0$. From the relationships of $\lambda_1^2 + \mu_1^2 = 1$ and $\sin^2\alpha + \cos^2\alpha = 1$, one obtains $\sin\alpha = (\lambda_2\mu_1 - \lambda_1\mu_2)$ and $\tan\alpha$. The direction coefficients of two straight lines are expressed by $m_1 = \frac{\mu_1}{\lambda_1}$ and $m_2 = \frac{\mu_2}{\lambda_2}$ respectively, then one obtains;

$$\tan\alpha = \frac{\frac{\mu_1}{\lambda_1} - \frac{\mu_2}{\lambda_2}}{1 + \frac{\mu_1\mu_2}{\lambda_1\lambda_2}} = \frac{m_1 - m_2}{1 + m_1 m_2}$$

The required condition is given by $1 + m_1 m_2 = 0$.

Exercise 5.3

$$A^* = \frac{\boldsymbol{b}\times\boldsymbol{c}}{\boldsymbol{a}\cdot(\boldsymbol{b}\times\boldsymbol{c})} = \frac{2\boldsymbol{e}_x - \boldsymbol{e}_y}{4} = \frac{1}{2}\boldsymbol{e}_x - \frac{1}{4}\boldsymbol{e}_y$$
$$B^* = \frac{\boldsymbol{c}\times\boldsymbol{a}}{\boldsymbol{a}\cdot(\boldsymbol{b}\times\boldsymbol{c})} = \frac{2\boldsymbol{e}_y}{4} = \frac{1}{2}\boldsymbol{e}_y$$

Exercise 5.4

$$|b_1| = \left|\frac{\boldsymbol{a}_2\times\boldsymbol{a}_3}{\boldsymbol{a}_1\cdot(\boldsymbol{a}_2\times\boldsymbol{a}_3)}\right| = \frac{\boldsymbol{a}_2\times\boldsymbol{a}_3}{|\boldsymbol{a}_1||\boldsymbol{a}_2\times\boldsymbol{a}_3|\cos\theta} = \frac{1}{\boldsymbol{a}_1\cos\theta} \to \frac{1}{d_{100}}$$

Exercise 5.5

$$\boldsymbol{b}_1 = \frac{\boldsymbol{a}_2\times\boldsymbol{a}_3}{\boldsymbol{a}_1\cdot\boldsymbol{a}_2\times\boldsymbol{a}_3} = \frac{1}{3\sqrt{3}}\left(\sqrt{3},-1,0\right), \quad \boldsymbol{b}_2 = \frac{\boldsymbol{a}_3\times\boldsymbol{a}_1}{\boldsymbol{a}_1\cdot\boldsymbol{a}_2\times\boldsymbol{a}_3} = \frac{1}{3\sqrt{3}}(0,3,0)$$

Exercise 5.6

$$\cos\phi = \frac{hu + kv + lw}{\left[\frac{h^2}{a_1^2} + \frac{k^2}{a_2^2} + \frac{l^2}{a_3^2}\right]^{\frac{1}{2}} [u^2 a_1^2 + v^2 a_2^2 + w^2 a_3^2]^{\frac{1}{2}}}$$

Exercise 5.7

For vector $\boldsymbol{A}$, we obtain the following relations:

$$\boldsymbol{A} = m_{11}\boldsymbol{a} + m_{12}\boldsymbol{b} + m_{13}\boldsymbol{c}$$

$$(m_{11}\boldsymbol{a} + m_{12}\boldsymbol{b} + m_{13}\boldsymbol{c}) \cdot (h\boldsymbol{a}^* + k\boldsymbol{b}^* + l\boldsymbol{c}^*) = m_{11}h + m_{12}k + m_{13}l = H$$

Similarly, we obtain for vectors $\boldsymbol{B}$ and $\boldsymbol{C}$ and they are summarized as follows:

$$\begin{pmatrix} H \\ K \\ L \end{pmatrix} = \Delta \begin{pmatrix} h \\ k \\ l \end{pmatrix} = \begin{pmatrix} m_{11}m_{12}m_{13} \\ m_{21}m_{22}m_{23} \\ m_{31}m_{32}m_{33} \end{pmatrix} \begin{pmatrix} h \\ k \\ l \end{pmatrix}$$

Exercise 5.8

The unit vectors of reciprocal lattices $\boldsymbol{b}_1$, $\boldsymbol{b}_2$ and $\boldsymbol{b}_3$ are as follows:

$$\left.\begin{aligned} \boldsymbol{b}_1 &= \frac{\boldsymbol{a}_2 \times \boldsymbol{a}_3}{V} = \frac{1}{a}(\boldsymbol{e}_x + \boldsymbol{e}_y - \boldsymbol{e}_z) \\ \boldsymbol{b}_2 &= \frac{\boldsymbol{a}_3 \times \boldsymbol{a}_1}{V} = \frac{1}{a}(-\boldsymbol{e}_x + \boldsymbol{e}_y + \boldsymbol{e}_z) \\ \boldsymbol{b}_3 &= \frac{\boldsymbol{a}_1 \times \boldsymbol{a}_2}{V} = \frac{1}{a}(\boldsymbol{e}_x - \boldsymbol{e}_y + \boldsymbol{e}_z) \end{aligned}\right\}$$

The arbitrary reciprocal lattice vector $\boldsymbol{H}_{pqr}$ is given by the following:

$$\boldsymbol{H}_{pqr} = p\boldsymbol{b}_1 + q\boldsymbol{b}_2 + r\boldsymbol{b}_3 = [(p - q + r)\boldsymbol{e}_x + (p + q - r)\boldsymbol{e}_y + (-p + q + r)\boldsymbol{e}_z]$$

Then we obtain the shortest non-zero vectors which consist of eight $\{111\}$ planes of the equilateral hexagon and six $\{002\}$ planes of the square. They give the first Brillouin zone of a face-centered cubic lattice and this result is known to correspond to the Wigner Seitz unit cell of a body-centered cubic lattice.

Exercise 5.9

The scalar products of crystal lattice vector $\boldsymbol{R}_{pqr}$ and its reciprocal lattice vector $\boldsymbol{H}_{hkl}$ is known to be always an integer. If atoms in a crystal are located only at the lattice point $\boldsymbol{R}_{pqr}$, a scatterer may be set with $\boldsymbol{r}_n = \boldsymbol{R}_{pqr}$. Therefore, the condition which enables us to detect a diffraction peak with sufficient intensity is given by $\boldsymbol{q} = \boldsymbol{H}_{hkl}$. The absolute value of both sides of this equation is taken.

$$|\boldsymbol{q}| = |\boldsymbol{H}_{hkl}| \rightarrow \frac{2\sin\theta}{\lambda} = \frac{1}{d_{hkl}}$$

$$\therefore 2d_{hkl}\sin\theta = \lambda \qquad \text{(Bragg condition)}$$

When considering the reciprocal lattice vectors and the scalar products of crystal lattice vectors, we readily obtain a formula of Laue condition, such as $\boldsymbol{a}_1\cdot(\boldsymbol{s}-\boldsymbol{s}_0) = h\lambda$.

Exercise 5.10

In a sufficiently large real space lattice, the summation of the given formula is unchanged even if substituting $\boldsymbol{r}'_n = \boldsymbol{r}_n + \boldsymbol{n}$ for $\boldsymbol{r}_n$. Such particular relation is used.

$$G(\boldsymbol{q}) = \sum_n \mathrm{e}^{-2\pi\mathrm{i}\boldsymbol{q}\cdot\boldsymbol{r}_n} \Rightarrow (1 - \mathrm{e}^{-2\pi\mathrm{i}\boldsymbol{q}\cdot\boldsymbol{n}})G(\boldsymbol{q}) = 0$$

Exercise 5.11

Discussion may be possible if using Taylor's expansion of $f(x) = \frac{1}{(1+x^2)^2}$ and in the very small value of $\frac{\sin\theta}{\lambda}$ $\left(\frac{\sin\theta}{\lambda} \ll 1\right)$, the atomic scattering factor f_n can be approximated by $f_n \approx 1 - 2\left\{\left(\frac{2\pi a_n \sin\theta}{\lambda}\right)^2\right\}$.

Exercise 5.12

The anomalous dispersion terms are given by following equations (for details, refer to a monograph for the related subjects, such as R.W. James: Optical Principles of the Diffraction of X-rays, G. Bell & Sons, London (1954)).

$$f'(\omega) = -\frac{1}{2}\int\left(\frac{\mathrm{d}g_{oj}}{\mathrm{d}\omega_{jo}}\right)\omega_{jo}\left\{\frac{\omega_{jo}-\omega}{(\omega_{jo}-\omega)^2+\gamma_{oj}^2/4} + \frac{\omega_{jo}+\omega}{(\omega_{jo}+\omega)^2+\gamma_{oj}^2/4}\right\}\mathrm{d}\omega_{jo}$$

$$f''(\omega) = \frac{1}{2}\int\left(\frac{\mathrm{d}g_{oj}}{\mathrm{d}\omega_{jo}}\right)\omega_{jo}\frac{\gamma_{oj}/2}{(\omega_{jo}-\omega)^2+\gamma_{oj}^2/4}\mathrm{d}\omega_{jo}$$

Where ω is the photon energy and its subscript denotes the state of photon such as the initial (o) and the j-th scattering process. γ_{oj} shows the convoluted width of states of o and j and g_{oj} is the so-called characteristic oscillatory strength.

There are some methods of computing a function of $(\mathrm{d}g_{oj}/\mathrm{d}\omega_{jo})$, but the procedure of Cromer and Liberman's scheme using the relativistic wave function is probably the best at the present time. Information of the anomalous dispersion terms including mass absorption coefficient of various elements in the wide energy region is available in the SCM-Database,

URL: http://res.tagen.tohoku.ac.jp/˜waseda/scm/AXS/index.html

Exercise 5.13

Considering that γ is the angle between the vector showing the direction of propagation of the wave and z-axis and s_0 is the unit vector of the incident wave, respectively, the diffracted intensity I_P is given as follows.

$$I_P = L^2 \left\{ \frac{\sin(\pi L \cdot s_0 \sin \gamma)}{\pi L \cdot s_0 \sin \gamma} \right\}^2$$

Exercise 5.14

Considering the two-dimensional slit system arrayed repeatedly in $\boldsymbol{a}$-period to x-direction and $\boldsymbol{b}$-period to y-direction and the diffraction intensity is estimated if a small slit may be expressed by a wave vector $\boldsymbol{s}_x \cdot \boldsymbol{s}_y$ of each direction.

$$I = \left| \frac{\sin(m\pi \boldsymbol{s}_x \cdot \boldsymbol{a})}{\sin(\pi \boldsymbol{s}_x \cdot \boldsymbol{a})} \right|^2 \cdot \left| \frac{\sin(n\pi \boldsymbol{s}_y \cdot \boldsymbol{b})}{\sin(\pi \boldsymbol{s}_y \cdot \boldsymbol{b})} \right|^2$$

Exercise 5.15

1. The diffraction intensity I is obtained by multiplying the amplitude of a scattered wave G and its complex conjugate.

$$I = G^* G = \frac{\sin^2 \left\{ \frac{1}{2} m (\boldsymbol{a} \cdot \boldsymbol{q}) \right\}}{\sin^2 \left\{ \frac{1}{2} (\boldsymbol{a} \cdot \boldsymbol{q}) \right\}}$$

2. Setting to $\boldsymbol{a} \cdot \boldsymbol{q} = 2\pi h + \delta (\delta > 0)$, the sine function of the numerator is taken into account. Under the condition of $\delta > 0$, the minimum value is obtained in the case of $\frac{m}{2}\delta = \pi$.

Exercise 5.16

$$I(\boldsymbol{q}) = f_{\mathrm{C}}^2 + 4 f_{\mathrm{Cl}}^2 + 12 f_{\mathrm{Cl}}^2 \frac{\sin(2\pi \boldsymbol{q} \boldsymbol{r}_{\mathrm{Cl-Cl}})}{2\pi \boldsymbol{q} \boldsymbol{r}_{\mathrm{Cl-Cl}}} + 8 f_{\mathrm{C}} f_{\mathrm{Cl}} \frac{\sin(2\pi \boldsymbol{q} \boldsymbol{r}_{\mathrm{C-Cl}})}{2\pi \boldsymbol{q} \boldsymbol{r}_{\mathrm{C-Cl}}}$$

Exercise 5.17

The form of a diffraction peak may be discussed with Laue function, because the scattering intensity is proportional to the square of its amplitude (see Question 5.9).

Let us to set the number of a unit cell to N and the wave vector to Q. For example, a peak is found if Qa is given by integer multiple of 2π or if the relation of $Qa = 2\pi n + 2\pi/N$ (N is an integer) is satisfied, Qa becomes zero for the first time, so that the value of Q, which the Laue function becomes zero, is given by $Q = (2\pi/aN)$. Therefore, with respect to the case of $\boldsymbol{a} = 0.25$ nm, we obtain $Q = 2.513 \times 10^{-3}\,\mathrm{nm}^{-1}$ for $N = 10^4$ and $Q = 2.513\,\mathrm{nm}^{-1}$ for $\boldsymbol{N} = 10$. For discussion, we also include that the length of the reciprocal lattice vector is equal to the reciprocal of the spacing d_{hkl}; $4\,\text{nm}^{-1}$ and $a = 0.25\,\text{nm}$ in the present case.

1. Each reciprocal lattice point is located at the vertices of a cube with the side of $4\pi\ \text{nm}^{-1}$ and each lattice point is very sharp of the order of $2.5 \times 10^{-3}\ \text{nm}^{-1}$.
2. In this case, we may find the streaked reciprocal lattice extended to the direction of the N_3 in a thin film by about 1,000 times in comparison to those of $\boldsymbol{N}_1$ and $\boldsymbol{N}_2$.
3. This case corresponds to a narrow string-like sample. We may find a very sharp reciprocal lattice along the direction of the narrow string-like sample, whereas the reciprocal lattice is rather spread with the order of $2.513\,\text{nm}^{-1}$ in the plane perpendicular to the string-like sample.

Exercise 5.18

1.
$$\begin{aligned} F = &\left[f_{\text{Ca}} + 2 f_{\text{F}} \cos \frac{\pi}{2}(h+k+l) \left\{ 2\cos^2 \frac{\pi}{2}(h+k+l) - 1 \right\} \right] \\ &\times \{1 + \cos\pi(h+k) + \cos\pi(h+l) + \cos\pi(l+h)\} \end{aligned}$$

2.
$$\begin{aligned} F_{111}^2 &= 16 f_{\text{Ca}}^2,\ F_{222}^2 = 16(f_{\text{Ca}} - 2f_{\text{F}})^2 \\ F_{111}^2 &= 16 \times (15.43)^2 = 3809.4 \quad (f_{\text{Ca}} = 15.43) \\ F_{222}^2 &= 16 \times (11.24 - 2 \times 4.76)^2 = 47.3 \quad (f_{\text{Ca}} = 11.24,\ f_{\text{F}} = 4.76) \end{aligned}$$

3. If mixed, $F = 0$ and if unmixed, the structure factor is as follows.

$$F^2 = 16\left[\left\{ f_{0\text{Ca}} + f'_{\text{Ca}} + 2 f_{\text{F}} \cos \frac{\pi}{2}(h+k+l)\left(2\cos^2 \frac{\pi}{2}(h+k+l) - 1 \right) \right\}^2 + f''^2_{\text{Ca}} \right]$$

Exercise 6.1

Set the standard to ***abc***, obtain the variation of Hermann–Morguin symbols when changing ***cab*** → ***bca*** → $\boldsymbol{a\bar{c}b}$ → $\boldsymbol{ba\bar{c}}$ → $\boldsymbol{\bar{c}ba}$.

Exercise 6.3

$$\begin{aligned} F_{hkl} &= f\,\mathrm{e}^{2\pi\mathrm{i}(ky+lz)}(\mathrm{e}^{2\pi\mathrm{i}hx} + \mathrm{e}^{\mathrm{i}\pi l}\mathrm{e}^{-2\pi\mathrm{i}hx}) \\ |F_{hkl}|^2 &= 4f^2\cos^2 2\pi hx (\text{for } l = 2n) \\ |F_{hkl}|^2 &= 4f^2\sin^2 2\pi hx (\text{for } l = 2n+1) \end{aligned}$$

With respect to the $0kl$ peaks, the intensity is detected only when $l = 2n$.

Exercise 6.4

For example, let us consider the case where there is the 2_1-screw axis parallel to b-axis through the origin. In this case, if an atom is found in $(x,\ y,\ z)$, there is necessarily an atom in $\left(-x, y+\dfrac{1}{2}, -z\right)$. In this case the structure factor is given in the following.

$$F(hkl) = \sum_{j=1}^{N/2} f_i \Big[\exp\{2\pi i(hx_j + ky_j + lz_j)\} + \exp\left\{2\pi i\left(-hx_j + ky_j + lz_j + \frac{k}{2}\right)\right\}\Big]$$

When considering that both h and l are zero, one obtains;

$$F(0k0) = \sum_{j=1}^{N/2} f_i \exp(2\pi i k y_j) + \{1 + \exp(\pi i k)\}$$

The extinction conditions are given as follows.

$$\left.\begin{aligned} F(0k0) &= \sum_{j=1}^{N/2} f_i \exp(2\pi i k y_j) \qquad & k = 2n \\ F(0k0) &= 0 & k = 2n+1 \end{aligned}\right\}$$

Thus, with respect to the 2_1-screw axis, we can not detect the diffraction intensity from the $0k0$ peak where k is odd number and it corresponds to the direction of a screw axis. Discussion is possible for other screw axes, along the way similar to the 2_1-screw axis case.

Similarly, the extinction condition appears for a plane of glide reflection. For example, when there is a c-glide plane perpendicular to b-axis, if there is an atom at (x, y, z), we always find an atom at $\left(x, -y, -z+\frac{1}{2}\right)$.

$$\left.\begin{aligned} F(h0l) &= 2\sum_{j=1}^{N/2} f_i \exp\{2\pi i(ky_j + lz_j)\} \qquad & l = 2n \\ F(h0l) &= 0 & l = 2n+1 \end{aligned}\right\}$$

Thus, the diffraction intensity is not detected when the index for giving the direction perpendicular to a glide plane is zero and the index for the direction of a translation axis is odd number.

Exercise 6.5

1. $$F = f_{\mathrm{Ti}}\left[1 + e^{2\pi i\left(\frac{h+k+l}{2}\right)}\right] + f_{\mathrm{O}}\left[e^{2\pi i(uh+uk)} + e^{2\pi i(-uh-uk)} + e^{2\pi i\left(\frac{h+k+l}{2} - uh + uk\right)} + e^{2\pi i\left(\frac{h+k+l}{2} + uh - uk\right)}\right]$$

2.
$$F = f_{\mathrm{Ti}}\left[1 + \cos 2\pi\left(\frac{h+k+l}{2}\right) - \mathrm{i}\sin 2\pi\left(\frac{h+k+l}{2}\right)\right]$$
$$+ f_{\mathrm{O}}\left[2\cos 2\pi u(h+k) + \left\{\cos 2\pi\left(\frac{h+k+l}{2}\right) + \mathrm{i}\sin 2\pi\left(\frac{h+k+l}{2}\right)\right\} \cdot 2\cos 2\pi u(h+k)\right]$$

(i) $h + k + l = 2n$

$$F = f_{\mathrm{Ti}} + f_{\mathrm{O}}[2\cos 2\pi u(h+k) + 2\cos 2\pi u(h-k)] \neq 0$$

(ii) $h + k + l = 2n + 1$

$$F = f_{\mathrm{O}}[2\cos 2\pi u(h+k) - 2\cos 2\pi u(h-k)] \neq 0$$

However, it should be kept in mind that $F = 0$ in the case of $h+k+l = 2n+1$ with h or k is zero.

3. The structure factors can be estimated from the given atomic positions.

(i) $h + k + l = 2n$

$$F = 2\left[f_{\mathrm{Ti}} + f_{\mathrm{O}}\left\{\mathrm{e}^{2\pi\mathrm{i}(uh+uk)} + \mathrm{e}^{2\pi\mathrm{i}(-uh-uk)}\right\}\right]$$

(ii) $h + k + l = 2n + 1$

$$F = 0$$

In conclusion, Bravais lattice is considered to be body-centered tetragonal.

Exercise 6.6

1. For example, Bragg equation of the plane $h_1k_1l_1$ is as follows.

$$\frac{S_1 - S_0}{\lambda} = h_1b_1 + k_1b_2 + l_1b_3$$

2. According to the results obtained in (1),

$$2 = -1 + h_2, 2 = -1 + k_2, 2 = -1 + l_2 \rightarrow (h_2k_2l_2) = (331)$$

In the orthogonal-axes, a_1a_2 and a_3, the formulas of $(\bar{1}\bar{1}1)$, (331) and (222) planes are given as follows. For $(\bar{1}\bar{1}1)$ plane;

$$\frac{a_1}{-a} + \frac{a_2}{-a} + \frac{a_3}{a} = 1 \qquad \therefore a_1 + a_2 - a_3 + a = 0$$

Similarly, for (331) plane; $3a_1 + 3a_2 + a_3 - a = 0$ and for (222) plane; $2a_1 + 2a_2 + a_3 - a = 0$.

The direction cosine of S_0 is set as x, y, z, the formula of S_0 is given as follows.

$$\frac{a_1}{x} + \frac{a_2}{y} + \frac{a_3}{z}, \qquad \therefore x^2 + y^2 + z^2 = 1$$

When the angle formed by S_0 and $(\bar{1}\bar{1}1)$ plane is set as θ_1, the following relation is readily obtained.

$$\sin\theta_1 = \frac{x + y - z}{\sqrt{1^2 + 1^2 + (-1)^2}} \quad \text{and} \quad 2d_{(\bar{1}\bar{1}1)} \sin\theta_1 = \lambda$$

$$x + y - z = \frac{3}{2a}\lambda$$

For S_0 and (222); $\quad 2x + 2y + 2z = \dfrac{6\lambda}{a}$

Then one obtains; $z = \dfrac{3}{4a}\lambda,\ x + y = \dfrac{9}{4a}\lambda$

Using the values of $a = 0.3567$ nm, $\lambda = 0.1542$ nm and $x^2 + y^2 + z^2 = 1$, (x, y, z) can be estimated.

$$(x, y, z) = (0.946, 0.027, 0.324)$$
$$(0.027, 0.946, 0.324)$$

Exercise 6.8

1.
$$F = f_{\text{Fe}}\left[1 + e^{2\pi i\left(\frac{h+k}{2}\right)} + e^{2\pi i\left(\frac{k+l}{2}\right)} + e^{2\pi i\left(\frac{h+l}{2}\right)}\right]$$
$$+ 2f_S\left\{e^{2\pi iu[h+k+l]} + e^{2\pi i\left(\frac{h+k}{2}\right)} \cdot e^{2\pi iu[h-k-l]}\right.$$
$$\left. + e^{2\pi i\left(\frac{k+l}{2}\right)} \cdot e^{2\pi iu[-h+k-l]} + e^{2\pi i\left(\frac{h+l}{2}\right)} \cdot e^{2\pi iu[-h-k+l]}\right\}$$

For the unmixed case,

$$F = 4_{\text{Fe}} + 2f_S\{2\cos 2\pi ul \cdot \cos 2\pi u(h + k) + 2\cos 2\pi ul \cdot \cos 2\pi u(h - k)\}$$

2. The area of a diffraction peak A is proportional to the intensity per unit length P' which is given by the multiplicity factor m and the structure factor F. Then the following relation is obtained.

$$\frac{A_{200}}{A_{111}} = \frac{P'_{200}}{P'_{111}} = \frac{m_{200}\left(\frac{F_{200}}{4f_{\text{Fe}}+8f_S}\right)^2}{m_{111}\left(\frac{F_{111}}{4f_{\text{Fe}}+8f_S}\right)^2}$$

The structure factors are given as follows,

$$F_{200} = 4f_{\text{Fe}} + 8f_{\text{S}} \cos 4\pi u$$
$$F_{111} = 4f_{\text{Fe}} + 4f_{\text{S}} \cos 2\pi u[\cos 4\pi u + 1]$$

For cubic systems $m_{111} = 8$ and $m_{200} = 6$. Estimate the atomic scattering factors of $f(\text{S})$ and $f(\text{Fe})$ for two peaks of 200 and 111 from Appendix A.3 and the lattice parameter.

$$f(\text{Fe})_{111} = 20.6 \qquad f(\text{S})_{111} = 12.0$$

$$f(\text{Fe})_{200} = 19.8 \qquad f(\text{S})_{200} = 11.6$$

If we use the given values of $A_{111} = 69.2$ and $A_{200} = 277.5$ as well as $m_{111}/m_{200} = 1.33$, we obtain $u = 0.387$.

Appendix A

A.1 Fundamental Units and Some Physical Constants

SI: LeSystèmac Internation d'Unitès

Seven SI base units

Length	meter	m
Mass	kilogram	kg
Time	second	s
Electric current	ampere	A
Thermodynamic temperature	kelvin	K
Amounts of substance	mole	mol
Luminous intensity	candera	cd

Derived SI units with a specific name

Quantity	Term・Symbol	Relevance to other units
Frequency	Hertz Hz	s^{-1}
Force	Newton N	$m \cdot kg \cdot s^{-2}$
Pressure, Stress	Pascal Pa	N/m^2 $m^{-1} \cdot kg \cdot s^{-2}$
Energy Work, Calorific value	Joule J	N/m $m^2 \cdot kg \cdot s^{-2}$
Power	Watt W	J/s $m^2 \cdot kg \cdot s^{-3}$
Electric charge	Coulomb C	$A \cdot s$ $s \cdot A$
Voltage, Potential	Volt V	W/A $m^2 \cdot kg \cdot s^{-3} \cdot A^{-1}$
Electric capacity	Farad F	C/V $m^{-2} \cdot kg^{-1} \cdot s^2 \cdot A^2$
Electrical resistivity	Ohm Ω	V/A $m^2 \cdot kg \cdot s^{-3} \cdot A^{-2}$
Radioactivity	Becquerel Bq	s^{-1}
Absorbed dose	Gray Gy	J/kg $m^2 \cdot s^{-2}$
Dose equivalent	Sievert Sv	J/kg $m^2 \cdot s^{-2}$
Plane angle	Radian rad	
Solid angle	Steradian sr	

Time: minute and hour, Plane angle: degree, minute, second, Volume: liter and Mass: metric ton. These units are non-SI units, but they are accepted for use with the SI units.

Physical constants

Quantity	Symbol	Value	Unit	
			SI	CGS
Velocity of light in vacuum	c	2.997925	10^8 m/s	10^{10} cm/s
Planck's constant	h	6.6260	10^{-34} J·s	10^{-27} erg·s
Avogadro's number	N_A	6.02217	10^{23}/mol	10^{23}/mol
Atomic mass unit[a]	amu	1.66053	10^{-27} kg	10^{-24} g
Universal gravitation constant	G	6.67259	10^{-11} m^3/s^2·kg	10^{-8} cm^3/s^2·g
Permeability of vacuum free space	μ_0	4π=12.56637	10^{-7} H/m	—
Permittivity of vacuum free space	ε_0	8.854188	10^{-12} F/m	—
Charge of electron	e	1.60219	10^{-19} C	10^{-20} emu
		4.80320	—	10^{-10} esu
Faraday's constant	F	9.64853	10^4 C/mol	10^3 emu/mol
Electron rest mass	m_e	9.10956	10^{-31} g	10^{-28} g
Electron specific charge	e/m_e	1.7588	10^{11} C/kg	10^7 emu/g
Electron radius	r_e	2.81794	10^{-15} m	10^{-13} cm
Electron Compton wavelength	λ_e	2.42631	10^{-12} m	10^{-10} cm
Proton rest mass	m_p	1.67262	10^{-27} kg	10^{-24} g
	m_p/m_e	1836.15	1	1
Fine structure constant	α	7.29735	10^{-3}	10^{-3}
	α^{-1}	137.036	1	1
Rydberg constant	R_∞	1.09737	10^7/m	10^5/cm
Bohr radius	a_0	5.29177	10^{-11} m	10^{-9} cm
Gas constant	R	8.31451	J/mol·K	10^7 erg/mol·K
Molar volume of ideal gas at N.T.P.[b]	V_m	22.414	10^{-3} m^3/mol	10^3 cm^3/mol
Boltzmann constant	k_B	1.38062	10^{-23} J/K	10^{-16} erg/K
Stefan-Boltzmann constant	σ	5.67051	10^{-8} W/m^2·K^4	10^{-5} erg/s·cm^2·K^4

[a]One twelfth of mass of ^{12}C.

[b]Temperature 273.15 K, Pressure 101325 Pa(1 atm).

Units frequently used with SI units

Quantity	Symbol	Value	Unit	
			SI	CGS
electron volt	eV	1.60219	10^{19} J	10^{-12} erg
angstrom	A[a]		0.1nm=10^{-10} m	10^{-8} cm
barn	b		10^{-28} m^2	

[a]Å is also used in comparison to the electric current A.

A.2 Atomic Weight, Density, Debye Temperature and Mass Absorption Coefficients (cm^2/g) for Elements

Characteristic radiation	Wavelength (Å)	1 Hydrogen	2 Helium	3 Lithium	4 Beryllium	5 Boron	6 Carbon	7 Nitrogen	8 Oxygen
	Atomic weight	1.0079	4.0026	6.941	9.0122	10.811	12.011	14.0067	15.9994
	Density	8.375E-05	1.664E-04	0.533	1.86	2.47	2.27	1.165E-03	1.332E-03
	Θ(K)			344	1440		2230		
Cr Ka	2.2910	4.12E-01	4.98E-01	1.30E+00	3.44E+00	7.59E+00	1.50E+01	2.47E+01	3.78E+01
Cr Kb_1	2.0849	4.05E-01	4.25E-01	1.01E+00	2.59E+00	5.69E+00	1.12E+01	1.86E+01	2.84E+01
Fe Ka	1.9374	4.00E-01	3.81E-01	8.39E-01	2.09E+00	4.55E+00	8.99E+00	1.49E+01	2.28E+01
Fe Kb_1	1.7566	3.96E-01	3.35E-01	6.63E-01	1.58E+00	3.39E+00	6.68E+00	1.10E+01	1.70E+01
Co Ka	1.7903	3.97E-01	3.43E-01	6.93E-01	1.67E+00	3.59E+00	7.07E+00	1.17E+01	1.80E+01
Co Kb_1	1.6208	3.93E-01	3.07E-01	5.55E-01	1.27E+00	2.67E+00	5.24E+00	8.66E+00	1.33E+01
Cu Ka	1.5418	3.91E-01	2.92E-01	5.00E-01	1.11E+00	2.31E+00	4.51E+00	7.44E+00	1.15E+01
Cu Kb_1	1.3922	3.88E-01	2.68E-01	4.12E-01	8.53E-01	1.73E+00	3.33E+00	5.48E+00	8.42E+00
Mo Ka	0.7107	3.73E-01	2.02E-01	1.98E-01	2.56E-01	3.68E-01	5.76E-01	8.45E-01	1.22E+00
Mo Kb_1	0.6323	3.70E-01	1.97E-01	1.87E-01	2.29E-01	3.09E-01	4.58E-01	6.45E-01	9.08E-01

Characteristic radiation	Wavelength (Å)	9 Fluorine	10 Neon	11 Sodium	12 Magnesium	13 Aluminium	14 Silicon	15 Phosphorus	16 Sulfur
	Atomic weight	18.9984	20.1797	22.9898	24.305	26.9815	28.0855	30.9738	32.066
	Density	1.696E-03	8.387E-04	0.966	1.74	2.70	2.33	1.82(yellow)	2.09
	Θ(K)		75	158	400	428	645		
Cr Ka	2.2910	5.15E+01	7.41E+01	9.49E+01	1.26E+02	1.55E+02	1.96E+02	2.30E+02	2.81E+02
Cr Kb_1	2.0849	3.89E+01	5.61E+01	7.21E+01	9.62E+01	1.18E+02	1.51E+02	1.77E+02	2.17E+02
Fe Ka	1.9374	3.13E+01	4.52E+01	5.82E+01	7.78E+01	9.59E+01	1.22E+02	1.44E+02	1.77E+02
Fe Kb_1	1.7566	2.33E+01	3.38E+01	4.37E+01	5.85E+01	7.23E+01	9.27E+01	1.09E+02	1.35E+02
Co Ka	1.7903	2.47E+01	3.58E+01	4.62E+01	6.19E+01	7.64E+01	9.78E+01	1.15E+02	1.42E+02
Co Kb_1	1.6208	1.83E+01	2.66E+01	3.45E+01	4.63E+01	5.73E+01	7.36E+01	8.70E+01	1.07E+02
Cu Ka	1.5418	1.58E+01	2.29E+01	2.97E+01	4.00E+01	4.96E+01	6.37E+01	7.55E+01	9.33E+01
Cu Kb_1	1.3922	1.16E+01	1.69E+01	2.20E+01	2.96E+01	3.68E+01	4.75E+01	5.64E+01	6.98E+01
Mo Ka	0.7107	1.63E+00	2.35E+00	3.03E+00	4.09E+00	5.11E+00	6.64E+00	7.97E+00	9.99E+00
Mo Kb_1	0.6323	1.19E+00	1.69E+00	2.17E+00	2.92E+00	3.64E+00	4.73E+00	5.67E+00	7.11E+00

Characteristic radiation	Wavelength (Å)	17 Chlorine	18 Argon	19 Potassium	20 Calcium	21 Scandium	22 Titanium	23 Vanadium	24 Chromium
	Atomic weight	35.4527	39.948	39.0983	40.078	44.9559	47.867	50.9415	51.9961
	Density	3.214E-03	1.663E-03	0.862	1.53	2.99	4.51	6.09	7.19
	Θ(K)		92	91	230	360	420	380	630
Cr Ka	2.2910	3.16E+02	3.42E+02	4.21E+02	4.90E+02	5.16E+02	5.90E+02	7.47E+01	8.68E+01
Cr Kb_1	2.0849	2.44E+02	2.66E+02	3.28E+02	3.82E+02	4.03E+02	4.44E+02	4.79E+02	6.70E+01
Fe Ka	1.9374	2.00E+02	2.18E+02	2.70E+02	3.14E+02	3.32E+02	3.58E+02	3.99E+02	4.92E+02
Fe Kb_1	1.7566	1.52E+02	1.67E+02	2.07E+02	2.42E+02	2.56E+02	2.77E+02	3.09E+02	3.85E+02
Co Ka	1.7903	1.61E+02	1.76E+02	2.18E+02	2.55E+02	2.69E+02	2.91E+02	3.25E+02	4.08E+02
Co Kb_1	1.6208	1.22E+02	1.34E+02	1.66E+02	1.95E+02	2.06E+02	2.27E+02	2.50E+02	2.93E+02
Cu Ka	1.5418	1.06E+02	1.16E+02	1.45E+02	1.70E+02	1.80E+02	2.00E+02	2.19E+02	2.47E+02
Cu Kb_1	1.3922	7.95E+01	8.75E+01	1.09E+02	1.29E+02	1.37E+02	1.52E+02	1.66E+02	1.85E+02
Mo Ka	0.7107	1.15E+01	1.28E+01	1.62E+01	1.93E+01	2.08E+01	2.34E+01	2.60E+01	2.99E+01
Mo Kb_1	0.6323	8.20E+00	9.14E+00	1.16E+01	1.38E+01	1.49E+01	1.68E+01	1.87E+01	2.15E+01

Θ: Debye temperature, Unit of density: Mg/m^3.

Characteristic radiation	Wavelength (Å)	25 Manganese	26 Iron	27 Cobalt	28 Nickel	29 Copper	30 Zinc	31 Gallium	32 Germanium
	Atomic weight	54.9381	55.845	58.9332	58.6934	63.546	65.39	69.723	72.61
	Density	7.47	7.87	8.8	8.91	8.93	7.13	5.91	5.32
	Θ(K)	410	470	445	450	343	327	320	374
Cr *K*a	2.2910	9.75E+01	1.13E+02	1.24E+02	1.44E+02	1.53E+02	1.71E+02	1.83E+02	1.99E+02
Cr Kb_1	2.0849	7.53E+01	8.69E+01	9.60E+01	1.12E+02	1.18E+02	1.32E+02	1.42E+02	1.55E+02
Fe *K*a	1.9374	6.16E+01	7.10E+01	7.85E+01	9.13E+01	9.68E+01	1.08E+02	1.16E+02	1.27E+02
Fe Kb_1	1.7566	3.75E+02	5.43E+01	6.00E+01	6.98E+01	7.40E+01	8.27E+01	8.86E+01	9.69E+01
Co *K*a	1.7903	3.93E+02	5.72E+01	6.32E+01	7.35E+01	7.80E+01	8.71E+01	9.34E+01	1.02E+02
Co Kb_1	1.6208	3.06E+02	3.42E+02	4.81E+01	5.60E+01	5.94E+01	6.64E+01	7.12E+01	7.78E+01
Cu *K*a	1.5418	2.70E+02	3.02E+02	3.21E+02	4.88E+01	5.18E+01	5.79E+01	6.21E+01	6.79E+01
Cu Kb_1	1.3922	2.07E+02	2.32E+02	2.48E+02	2.79E+02	3.92E+01	4.38E+01	4.70E+01	5.14E+01
Mo *K*a	0.7107	3.31E+01	3.76E+01	4.10E+01	4.69E+01	4.91E+01	5.40E+01	5.70E+01	6.12E+01
Mo Kb_1	0.6323	2.38E+01	2.71E+01	2.96E+01	3.40E+01	3.57E+01	3.93E+01	4.15E+01	4.46E+01

Characteristic radiation	Wavelength (Å)	33 Arsenic	34 Selenium	35 Bromine	36 Krypton	37 Rubidium	38 Strontium	39 Yttrium	40 Zirconium
	Atomic weight	74.9216	78.96	79.904	83.8	85.4678	87.62	88.9059	91.224
	Density	5.78	4.81	3.12 (liq.)	3.488E-03	1.53	2.58	4.48	6.51
	Θ(K)	282	90		72	56	147	280	291
Cr *K*a	2.2910	2.19E+02	2.34E+02	2.60E+02	2.77E+02	3.03E+02	3.28E+02	3.58E+02	3.86E+02
Cr Kb_1	2.0849	1.70E+02	1.82E+02	2.02E+02	2.15E+02	2.36E+02	2.56E+02	2.79E+02	3.00E+02
Fe *K*a	1.9374	1.39E+02	1.49E+02	1.65E+02	1.76E+02	1.93E+02	2.10E+02	2.29E+02	2.47E+02
Fe Kb_1	1.7566	1.06E+02	1.14E+02	1.27E+02	1.35E+02	1.48E+02	1.61E+02	1.76E+02	1.91E+02
Co *K*a	1.7903	1.12E+02	1.20E+02	1.33E+02	1.42E+02	1.56E+02	1.70E+02	1.85E+02	2.00E+02
Co Kb_1	1.6208	8.55E+01	9.16E+01	1.02E+02	1.09E+02	1.19E+02	1.30E+02	1.42E+02	1.54E+02
Cu *K*a	1.5418	7.47E+01	8.00E+01	8.90E+01	9.52E+01	1.04E+02	1.13E+02	1.24E+02	1.39E+02
Cu Kb_1	1.3922	5.65E+01	6.05E+01	6.74E+01	7.21E+01	7.90E+01	8.59E+01	9.40E+01	1.01E+02
Mo *K*a	0.7107	6.61E+01	6.95E+01	7.56E+01	7.93E+01	8.51E+01	9.06E+01	9.70E+01	1.63E+01
Mo Kb_1	0.6323	4.82E+01	5.08E+01	5.55E+01	5.84E+01	6.30E+01	6.72E+01	7.21E+01	7.61E+01

Characteristic radiation	Wavelength (Å)	41 Niobium	42 Molybdenum	43 Technetium	44 Ruthenium	45 Rhodium	46 Palladium	47 Silver	48 Cadmium
	Atomic weight	92.9064	95.94	[99]	101.07	102.9055	106.42	107.8682	112.411
	Density	8.58	10.22	11.50	12.36	12.42	12.00	10.50	8.65
	Θ(K)	275	450		600	480	274	225	209
Cr *K*a	2.2910	4.16E+02	4.42E+02	4.74E+02	5.01E+02	5.36E+02	5.63E+02	6.02E+02	6.26E+02
Cr Kb_1	2.0849	3.25E+02	3.45E+02	3.70E+02	3.92E+02	4.20E+02	4.41E+02	4.72E+02	4.90E+02
Fe *K*a	1.9374	2.67E+02	2.84E+02	3.05E+02	3.23E+02	3.46E+02	3.63E+02	3.89E+02	4.05E+02
Fe Kb_1	1.7566	2.05E+02	2.19E+02	2.35E+02	2.49E+02	2.67E+02	2.81E+02	3.01E+02	3.13E+02
Co *K*a	1.7903	2.16E+02	2.30E+02	2.47E+02	2.62E+02	2.80E+02	2.95E+02	3.16E+02	3.29E+02
Co Kb_1	1.6208	1.66E+02	1.76E+02	1.90E+02	2.01E+02	2.16E+02	2.27E+02	2.43E+02	2.53E+02
Cu *K*a	1.5418	1.45E+02	1.54E+02	1.66E+02	1.76E+02	1.89E+02	1.99E+02	2.13E+02	2.22E+02
Cu Kb_1	1.3922	1.10E+02	1.17E+02	1.26E+02	1.34E+02	1.44E+02	1.51E+02	1.63E+02	1.69E+02
Mo *K*a	0.7107	1.77E+01	1.88E+01	2.04E+01	2.17E+01	2.33E+01	2.47E+01	2.65E+01	2.78E+01
Mo Kb_1	0.6323	8.10E+01	1.38E+01	1.49E+01	1.58E+01	1.70E+01	1.80E+01	1.94E+01	2.02E+01

Θ: Debye temperature, Unit of density: Mg/m^3.

Characteristic radiation	Wavelength (Å)	49 Indium	50 Tin	51 Antimony	52 Tellurium	53 Iodine	54 Xenon	55 Caesium	56 Barium
	Atomic weight	114.818	118.71	121.76	127.6	126.9045	131.29	132.9054	137.327
	Density	7.29	7.29	6.69	6.25	4.95	5.495E-03	1.91(263K)	3.59
	Θ(K)	108	200	211	153		64	38	110
Cr *Ka*	2.2910	6.63E+02	6.91E+02	7.23E+02	7.40E+02	7.96E+02	7.21E+02	7.60E+02	5.70E+02
Cr Kb_1	2.0849	5.19E+02	5.42E+02	5.70E+02	5.85E+02	6.31E+02	6.52E+02	6.86E+02	6.45E+02
Fe *Ka*	1.9374	4.28E+02	4.47E+02	4.71E+02	4.83E+02	5.22E+02	5.40E+02	5.69E+02	5.86E+02
Fe Kb_1	1.7566	3.32E+02	3.47E+02	3.65E+02	3.74E+02	4.08E+02	4.22E+02	4.46E+02	4.61E+02
Co *Ka*	1.7903	3.49E+02	3.64E+02	3.83E+02	3.94E+02	4.25E+02	4.40E+02	4.65E+02	4.80E+02
Co Kb_1	1.6208	2.69E+02	2.81E+02	2.96E+02	3.04E+02	3.30E+02	3.43E+02	3.63E+02	3.76E+02
Cu *Ka*	1.5418	2.36E+02	2.47E+02	2.59E+02	2.67E+02	2.88E+02	2.99E+02	3.17E+02	3.25E+02
Cu Kb_1	1.3922	1.80E+02	1.88E+02	1.98E+02	2.04E+02	2.20E+02	2.29E+02	2.43E+02	2.52E+02
Mo *Ka*	0.7107	2.95E+01	3.10E+01	3.27E+01	3.38E+01	3.67E+01	3.82E+01	4.07E+01	4.23E+01
Mo Kb_1	0.6323	2.16E+01	2.26E+01	2.39E+01	2.47E+01	2.68E+01	2.80E+01	2.98E+01	3.10E+01

Characteristic radiation	Wavelength (Å)	57 Lanthanum	58 Cerium	59 Praseodymium	60 Neodymium	61 Promethium	62 Samarium	63 Europium	64 Gadolinium
	Atomic weight	138.9055	140.115	140.9077	144.24	[145]	150.36	151.965	157.25
	Density	6.17	6.77	6.78	7.00		7.54	5.25	7.87
	Θ(K)	142							200
Cr *Ka*	2.2910	2.25E+02	2.38E+02	2.38E+02	2.51E+02	2.94E+02	2.79E+02	3.09E+02	2.98E+02
Cr Kb_1	2.0849	7.44E+02	4.94E+02	1.88E+02	1.98E+02	2.32E+02	2.21E+02	2.44E+02	2.35E+02
Fe *Ka*	1.9374	6.18E+02	5.61E+02	4.48E+02	4.55E+02	1.94E+02	2.04E+02	2.03E+02	1.95E+02
Fe Kb_1	1.7566	4.83E+02	5.10E+02	5.39E+02	4.92E+02	5.88E+02	1.63E+02	4.08E+02	1.53E+02
Co *Ka*	1.7903	5.07E+02	5.35E+02	5.65E+02	5.05E+02	4.00E+02	1.76E+02	4.19E+02	1.61E+02
Co Kb_1	1.6208	3.95E+02	4.17E+02	4.41E+02	4.57E+02	4.82E+02	3.54E+02	4.80E+02	3.35E+02
Cu *Ka*	1.5418	3.48E+02	3.68E+02	3.90E+02	4.04E+02	4.26E+02	4.34E+02	4.34E+02	4.03E+02
Cu Kb_1	1.3922	2.66E+02	2.82E+02	2.99E+02	3.10E+02	3.28E+02	3.35E+02	3.52E+02	3.60E+02
Mo *Ka*	0.7107	4.49E+01	4.77E+01	5.07E+01	5.30E+01	5.63E+01	5.78E+01	6.09E+01	6.26E+01
Mo Kb_1	0.6323	3.29E+01	3.49E+01	3.72E+01	3.88E+01	4.13E+01	4.24E+01	4.47E+01	4.60E+01

Characteristic radiation	Wavelength (Å)	65 Terbium	66 Dysprosium	67 Holmium	68 Erbium	69 Thulium	70 Ytterbium	71 Lutetium	72 Hafnium
	Atomic weight	158.9253	162.5	164.9303	167.26	168.9342	173.04	174.967	178.49
	Density	8.27	8.53	8.80	9.04	9.33	6.97	9.84	13.28
	Θ(K)		210				120	210	252
Cr *Ka*	2.2910	3.32E+02	3.25E+02	3.47E+02	3.52E+02	3.86E+02	3.87E+02	4.31E+02	4.25E+02
Cr Kb_1	2.0849	2.63E+02	2.57E+02	2.72E+02	2.78E+02	3.05E+02	3.04E+02	3.39E+02	3.34E+02
Fe *Ka*	1.9374	2.19E+02	2.14E+02	2.28E+02	2.32E+02	2.53E+02	2.51E+02	2.80E+02	2.77E+02
Fe Kb_1	1.7566	1.71E+02	1.68E+02	1.78E+02	1.82E+02	1.96E+02	1.96E+02	2.18E+02	2.16E+02
Co *Ka*	1.7903	1.80E+02	1.76E+02	1.87E+02	1.91E+02	2.06E+02	2.06E+02	2.29E+02	2.27E+02
Co Kb_1	1.6208	3.60E+02	1.38E+02	1.46E+02	1.49E+02	1.59E+02	1.59E+02	1.78E+02	1.76E+02
Cu *Ka*	1.5418	3.21E+02	3.62E+02	1.29E+02	1.32E+02	1.40E+02	1.42E+02	1.56E+02	1.55E+02
Cu Kb_1	1.3922	3.76E+02	3.87E+02	4.02E+02	4.17E+02	1.08E+02	1.08E+02	1.21E+02	1.20E+02
Mo *Ka*	0.7107	6.58E+01	6.83E+01	7.13E+01	7.44E+01	7.79E+01	8.04E+01	8.40E+01	8.69E+01
Mo Kb_1	0.6323	4.83E+01	5.02E+01	5.24E+01	5.48E+01	5.74E+01	5.93E+01	6.19E+01	6.41E+01

Θ: Debye temperature, Unit of density: Mg/m^3.

Characteristic radiation	Wavelength (Å)	73 Tantalum	74 Tungsten	75 Rhenium	76 Osmium	77 Iridium	78 Platinum	79 Gold	80 Mercury
	Atomic weight	180.9479	183.84	186.207	190.23	192.217	195.08	196.9665	200.59
	Density	16.67	19.25	21.02	22.58	22.55	21.44	19.28	13.55
	Θ(K)	240	400	430	500	420	240	165	71.9
Cr Ka	2.2910	4.32E+02	4.57E+02	5.01E+02	4.99E+02	5.20E+02	5.41E+02	5.51E+02	5.41E+02
Cr Kb_1	2.0849	3.39E+02	3.61E+02	3.94E+02	3.92E+02	4.11E+02	4.23E+02	4.34E+02	4.16E+02
Fe Ka	1.9374	2.83E+02	3.01E+02	3.27E+02	3.27E+02	3.40E+02	3.57E+02	3.61E+02	3.39E+02
Fe Kb_1	1.7566	2.20E+02	2.34E+02	2.57E+02	2.55E+02	2.65E+02	2.61E+02	2.79E+02	2.60E+02
Co Ka	1.7903	2.31E+02	2.46E+02	2.68E+02	2.68E+02	2.78E+02	2.76E+02	2.95E+02	2.73E+02
Co Kb_1	1.6208	1.79E+02	1.91E+02	2.09E+02	2.09E+02	2.16E+02	2.14E+02	2.29E+02	2.16E+02
Cu Ka	1.5418	1.58E+02	1.68E+02	1.87E+02	1.84E+02	1.91E+02	1.88E+02	2.01E+02	1.88E+02
Cu Kb_1	1.3922	1.22E+02	1.30E+02	1.43E+02	1.42E+02	1.48E+02	1.45E+02	1.55E+02	1.41E+02
Mo Ka	0.7107	9.04E+01	9.38E+01	9.74E+01	1.00E+02	1.04E+02	1.07E+02	1.12E+02	1.15E+02
Mo Kb_1	0.6323	6.67E+01	6.92E+01	7.19E+01	7.41E+01	7.70E+01	7.97E+01	8.29E+01	8.54E+01

Characteristic radiation	Wavelength (Å)	81 Thallium	82 Lead	83 Bismuth	84 Polonium	85 Astatine	86 Radon	87 Francium	88 Radium
	Atomic weight	204.3833	207.2	208.9804	[210]	[210]	[222]	[223]	[226]
	Density	11.87	11.34	9.80			4.40 (liq. , 211K)		
	Θ(K)	78.5	105	119					
Cr Ka	2.2910	5.97E+02	6.43E+02	6.66E+02	6.91E+02	6.80E+02	7.34E+02	7.58E+02	7.43E+02
Cr Kb_1	2.0849	4.87E+02	5.07E+02	5.24E+02	5.44E+02	5.33E+02	5.76E+02	5.97E+02	5.85E+02
Fe Ka	1.9374	4.03E+02	4.20E+02	4.34E+02	4.52E+02	4.44E+02	4.77E+02	4.93E+02	4.87E+02
Fe Kb_1	1.7566	3.14E+02	3.27E+02	3.39E+02	3.54E+02	3.45E+02	3.73E+02	3.84E+02	3.80E+02
Co Ka	1.7903	3.31E+02	3.43E+02	3.55E+02	3.70E+02	3.63E+02	3.92E+02	4.03E+02	3.98E+02
Co Kb_1	1.6208	2.57E+02	2.67E+02	2.76E+02	2.88E+02	2.82E+02	3.04E+02	3.12E+02	3.10E+02
Cu Ka	1.5418	2.26E+02	2.35E+02	2.44E+02	2.54E+02	2.48E+02	2.67E+02	2.77E+02	2.73E+02
Cu Kb_1	1.3922	1.75E+02	1.81E+02	1.88E+02	1.96E+02	1.86E+02	2.05E+02	2.13E+02	2.10E+02
Mo Ka	0.7107	1.18E+02	1.22E+02	1.26E+02	1.32E+02	1.17E+02	1.08E+02	8.70E+01	8.80E+01
Mo Kb_1	0.6323	8.79E+01	9.08E+01	9.41E+01	9.83E+01	1.02E+02	1.01E+02	1.04E+02	1.08E+01

Characteristic radiation	Wavelength (Å)	89 Actinium	90 Thorium	91 Protactinium	92 Uranium	93 Neptunium	94 Plutonium	95 Americium	96 Curium
	Atomic weight	[227]	232.0381	231.0359	238.0289	[237]	[239]	[243]	[247]
	Density		11.72		19.05		19.81		
	Θ(K)		163		207				
Cr Ka	2.2910	7.39E+02	7.68E+02	7.38E+02	7.66E+02	8.00E+02	7.60E+02	7.95E+02	8.12E+02
Cr Kb_1	2.0849	6.18E+02	5.09E+02	5.82E+02	6.17E+02	6.30E+02	6.00E+02	6.27E+02	6.40E+02
Fe Ka	1.9374	5.30E+02	4.85E+02	4.82E+02	5.28E+02	5.52E+02	4.98E+02	5.81E+02	5.90E+02
Fe Kb_1	1.7566	4.44E+02	3.89E+02	3.75E+02	4.00E+02	4.10E+02	3.89E+02	4.07E+02	4.21E+02
Co Ka	1.7903	4.61E+02	4.06E+02	3.94E+02	4.20E+02	4.30E+02	4.08E+02	4.26E+02	4.37E+02
Co Kb_1	1.6208	3.81E+02	3.48E+02	3.06E+02	3.26E+02	3.35E+02	3.17E+02	3.33E+02	3.43E+02
Cu Ka	1.5418	3.17E+02	3.06E+02	2.71E+02	2.88E+02	3.14E+02	2.80E+02	3.22E+02	3.38E+02
Cu Kb_1	1.3922	2.85E+02	2.19E+02	2.08E+02	2.22E+02	2.27E+02	2.16E+02	2.27E+02	2.32E+02
Mo Ka	0.7107	9.08E+01	9.65E+01	1.01E+02	1.02E+02	4.22E+01	3.99E+01	4.81E+01	4.90E+01
Mo Kb_1	0.6323	1.10E+02	9.87E+01	1.19E+02	7.49E+01	1.25E+02	1.29E+02	1.31E+02	1.34E+02

Θ: Debye temperature, Unit of density: Mg/m^3.

A.3 Atomic Scattering Factors as a Function of sin θ/λ

$\frac{\sin\theta}{\lambda}$ (Å^{-1})	0.0	0.1	0.2	0.3	0.4	0.5	0.6	0.7	0.8	0.9
H	1	0.81	0.48	0.25	0.13	0.07	0.04	0.03	0.02	0.01
He	2	1.88	1.46	1.05	0.75	0.52	0.35	0.24	0.18	0.14
Li	3	2.2	1.8	1.5	1.3	1.0	0.8	0.6	0.5	0.4
Be	4	2.9	1.9	1.7	1.6	1.4	1.2	1.0	0.9	0.7
B	5	3.5	2.4	1.9	1.7	1.5	1.4	1.2	1.2	1.0
C	6	4.6	3.0	2.2	1.9	1.7	1.6	1.4	1.3	1.16
N	7	5.8	4.2	3.0	2.3	1.9	1.65	1.54	1.49	1.39
O	8	7.1	5.3	3.9	2.9	2.2	1.8	1.6	1.5	1.4
F	9	7.8	6.2	4.45	3.35	2.65	2.15	1.9	1.7	1.6
Ne	10	9.3	7.5	5.8	4.4	3.4	2.65	2.2	1.9	1.65
Na	11	9.65	8.2	6.7	5.25	4.05	3.2	2.65	2.25	1.95
Mg	12	10.5	8.6	7.25	5.95	4.8	3.85	3.15	2.55	2.2
Al	13	11.0	8.95	7.75	6.6	5.5	4.5	3.7	3.1	2.65
Si	14	11.35	9.4	8.2	7.15	6.1	5.1	4.2	3.4	2.95
P	15	12.4	10.0	8.45	7.45	6.5	5.65	4.8	4.05	3.4
S	16	13.6	10.7	8.95	7.85	6.85	6.0	5.25	4.5	3.9
Cl	17	14.6	11.3	9.25	8.05	7.25	6.5	5.75	5.05	4.4
A	18	15.9	12.6	10.4	8.7	7.8	7.0	6.2	5.4	4.7
K	19	16.5	13.3	10.8	9.2	7.9	6.7	5.9	5.2	4.6
Ca	20	17.5	14.1	11.4	9.7	8.4	7.3	6.3	5.6	4.9
Sc	21	18.4	14.9	12.1	10.3	8.9	7.7	6.7	5.9	5.3
Ti	22	19.3	15.7	12.8	10.9	9.5	8.2	7.2	6.3	5.6
V	23	20.2	16.6	13.5	11.5	10.1	8.7	7.6	6.7	5.9
Cr	24	21.1	17.4	14.2	12.1	10.6	9.2	8.0	7.1	6.3
Mn	25	22.1	18.2	14.9	12.7	11.1	9.7	8.4	7.5	6.6
Fe	26	23.1	18.9	15.6	13.3	11.6	10.2	8.9	7.9	7.0
Co	27	24.1	19.8	16.4	14.0	12.4	10.7	9.3	8.3	7.3
Ni	28	25.0	20.7	17.2	14.6	12.7	11.2	9.8	8.7	7.7
Cu	29	25.9	21.6	17.9	15.2	13.3	11.7	10.2	9.1	8.1
Zn	30	26.8	22.4	18.6	15.8	13.9	12.2	10.7	9.6	8.5

(Continued)

$\frac{\sin\theta}{\lambda}$ (Å^{-1})	0.0	0.1	0.2	0.3	0.4	0.5	0.6	0.7	0.8	0.9
Ga	31	27.8	23.3	19.3	16.5	14.5	12.7	11.2	10.0	8.9
Ge	32	28.8	24.1	20.0	17.1	15.0	13.2	11.6	10.4	9.3
As	33	29.7	25.0	20.8	17.7	15.6	13.8	12.1	10.8	9.7
Se	34	30.6	25.8	21.5	18.3	16.1	14.3	12.6	11.2	10.0
Br	35	31.6	26.6	22.3	18.9	16.7	14.8	13.1	11.7	10.4
Kr	36	32.5	27.4	23.0	19.5	17.3	15.3	13.6	12.1	10.8
Rb	37	33.5	28.2	23.8	20.2	17.9	15.9	14.1	12.5	11.2
Sr	38	34.4	29.0	24.5	20.8	18.4	16.4	14.6	12.9	11.6
Y	39	35.4	29.9	25.3	21.5	19.0	17.0	15.1	13.4	12.0
Zr	40	36.3	30.8	26.0	22.1	19.7	17.5	15.6	13.8	12.4
Nb	41	37.3	31.7	26.8	22.8	20.2	18.1	16.0	14.3	12.8
Mo	42	38.2	32.6	27.6	23.5	20.8	18.6	16.5	14.8	13.2
Tc	43	39.1	33.4	28.3	24.1	21.3	19.1	17.0	15.2	13.6
Ru	44	40.0	34.3	29.1	24.7	21.9	19.6	17.5	15.6	14.1
Rh	45	41.0	35.1	29.9	25.4	22.5	20.2	18.0	16.1	14.5
Pd	46	41.9	36.0	30.7	26.2	23.1	20.8	18.5	16.6	14.9
Ag	47	42.8	36.9	31.5	26.9	23.8	21.3	19.0	17.1	15.3
Cd	48	43.7	37.7	32.2	27.5	24.4	21.8	19.6	17.6	15.7
In	49	44.7	38.6	33.0	28.1	25.0	22.4	20.1	18.0	16.2
Sn	50	45.7	39.5	33.8	28.7	25.6	22.9	20.6	18.5	16.6
Sb	51	46.7	40.4	34.6	29.5	26.3	23.5	21.1	19.0	17.0
Te	52	47.7	41.3	35.4	30.3	26.9	24.0	21.7	19.5	17.5
I	53	48.6	42.1	36.1	31.0	27.5	24.6	22.2	20.0	17.9
Xe	54	49.6	43.0	36.8	31.6	28.0	25.2	22.7	20.4	18.4
Cs	55	50.7	43.8	37.6	32.4	28.7	25.8	23.2	20.8	18.8
Ba	56	51.7	44.7	38.4	33.1	29.3	26.4	23.7	21.3	19.2
La	57	52.6	45.6	39.3	33.8	29.8	26.9	24.3	21.9	19.7
Ce	58	53.6	46.5	40.1	34.5	30.4	27.4	24.8	22.4	20.2
Pr	59	54.5	47.4	40.9	35.2	31.1	28.0	25.4	22.9	20.6
Nd	60	55.4	48.3	41.6	35.9	31.8	28.6	25.9	23.4	21.1
Pm	61	56.4	49.1	42.4	36.6	32.4	29.2	26.4	23.9	21.5
Sm	62	57.3	50.0	43.2	37.3	32.9	29.8	26.9	24.4	22.0
Eu	63	58.3	50.9	44.0	38.1	33.5	30.4	27.5	24.9	22.4
Gd	64	59.3	51.7	44.8	38.8	34.1	31.0	28.1	25.4	22.9
Tb	65	60.2	52.6	45.7	39.6	34.7	31.6	28.6	25.9	23.4

(Continued)

$\frac{\sin\theta}{\lambda}$ (Å^{-1})	0.0	0.1	0.2	0.3	0.4	0.5	0.6	0.7	0.8	0.9
Dy	66	61.1	53.6	46.5	40.4	35.4	32.2	29.2	26.3	23.9
Ho	67	62.1	54.5	47.3	41.1	36.1	32.7	29.7	26.8	24.3
Er	68	63.0	55.3	48.1	41.7	36.7	33.3	30.2	27.3	24.7
Tm	69	64.0	56.2	48.9	42.4	37.4	33.9	30.8	27.9	25.2
Yb	70	64.9	57.0	49.7	43.2	38.0	34.4	31.3	28.4	25.7
Lu	71	65.9	57.8	50.4	43.9	38.7	35.0	31.8	28.9	26.2
Hf	72	66.8	58.6	51.2	44.5	39.3	35.6	32.3	29.3	26.7
Ta	73	67.8	59.5	52.0	45.3	39.9	36.2	32.9	29.8	27.1
W	74	68.8	60.4	52.8	46.1	40.5	36.8	33.5	30.4	27.6
Re	75	69.8	61.3	53.6	46.8	41.1	37.4	34.0	30.9	28.1
Os	76	70.8	62.2	54.4	47.5	41.7	38.0	34.6	31.4	28.6
Ir	77	71.7	63.1	55.3	48.2	42.4	38.6	35.1	32.0	29.0
Pt	78	72.6	64.0	56.2	48.9	43.1	39.2	35.6	32.5	29.5
Au	79	73.6	65.0	57.0	49.7	43.8	39.8	36.2	33.1	30.0
Hg	80	74.6	65.9	57.9	50.5	44.4	40.5	36.8	33.6	30.6
Tl	81	75.5	66.7	58.7	51.2	45.0	41.1	37.4	34.1	31.1
Pb	82	76.5	67.5	59.5	51.9	45.7	41.6	37.9	34.6	31.5
Bi	83	77.5	68.4	60.4	52.7	46.4	42.2	38.5	35.1	32.0
Po	84	78.4	69.4	61.3	53.5	47.1	42.8	39.1	35.6	32.6
At	85	79.4	70.3	62.1	54.2	47.7	43.4	39.6	36.2	33.1
Rn	86	80.3	71.3	63.0	55.1	48.4	44.0	40.2	36.8	33.5
Fr	87	81.3	72.2	63.8	55.8	49.1	44.5	40.7	37.3	34.0
Ra	88	82.2	73.2	64.6	56.5	49.8	45.1	41.3	37.8	34.6
Ac	89	83.2	74.1	65.5	57.3	50.4	45.8	41.8	38.3	35.1
Th	90	84.1	75.1	66.3	58.1	51.1	46.5	42.4	38.8	35.5
Pa	91	85.1	76.0	67.1	58.8	51.7	47.1	43.0	39.3	36.0
U	92	86.0	76.9	67.9	59.6	52.4	47.7	43.5	39.8	36.5
Np	93	87	78	69	60	53	48	44	40	37
Pu	94	88	79	69	61	54	49	44	41	38
Am	95	89	79	70	62	55	50	45	42	38
Cm	96	90	80	71	62	55	50	46	42	39
Bk	97	91	81	72	63	56	51	46	43	39
Cf	98	92	82	73	64	57	52	47	43	40

A.4 Quadratic Forms of Miller Indices for Cubic and Hexagonal Systems

Cubic					Hexagonal	
$h^2+k^2+l^2$	$h\ k\ l$				h^2+hk+k^2	$h\ k$
	Simple	Face centerd	Body centerd	Diamond		
1	100				1	10
2	110		110		2	
3	111	111		111	3	11
4	200	200	200		4	20
5	210				5	
6	211		211		6	
7					7	21
8	220	220	220	220	8	
9	300,221				9	30
10	310		310		10	
11	311	311		311	11	
12	222	222	222		12	22
13	320				13	31
14	321		321		14	
15					15	
16	400	400	400	400	16	40
17	410,322				17	
18	411,330		411,330		18	
19	331	331		331	19	32
20	420	420	420		20	
21	421				21	41
22	332		332		22	
23					23	
24	422	422	422	422	24	
25	500,430				25	50
26	510,431		510,431		26	
27	511,333	511,333		511,333	27	33
28					28	42
29	520,432				29	
30	521		521		30	
31					31	51
32	440	440	440	440	32	
33	522,441				33	
34	530,433		530,433		34	
35	531	531		531	35	
36	600,442	600,442	600,442		36	60
37	610				37	43
38	611,532		611,532		38	
39					39	52
40	620	620	620	620	40	
41	621,540,443				41	
42	541		541		42	
43	533	533		533	43	61
44	622	622	622		44	
45	630,542				45	
46	631		631		46	
47					47	
48	444	444	444	444	48	44
49	700,632				49	70,53
50	710,550,543		710,550,543		50	
51	711,551	711,551		711,551	51	
52	640	640	640		52	62
53	720,641				53	
54	721,633,552		721,633,552		54	
55					55	
56	642	642	642	642	56	
57	722,544				57	71
58	730		730		58	
59	731,553	731,553		731,553	59	

A.5 Volume and Interplanar Angles of a Unit Cell

Cellvolumes

Cubic : $V = a^3$

Tetragonal : $V = a^2 c$

Hexagonal : $V = \frac{\sqrt{3}a^2 c}{2} = 0.866a^2 c$

Trigonal : $V = a^3\sqrt{1 - 3\cos^2\alpha + 2\cos^3\alpha}$

Orthorhombic : $V = abc$

Monoclinic : $V = abc\sin\beta$

Triclinic : $V = abc\sqrt{1 - \cos^2\alpha - \cos^2\beta - \cos^2\gamma + 2\cos\alpha\cos\beta\cos\gamma}$

Interplanar angles

The angle ϕ between the plane $(h_1 k_1 l_1)$ of spacing d_1 and the plane $(h_2 k_2 l_2)$ of d_2 is estimated from the following equation, where V is the volume of a unit cell:

$$\text{Cubic : } \cos\phi = \frac{h_1h_2 + k_1k_2 + l_1l_2}{\sqrt{({h_1}^2 + {k_1}^2 + {l_1}^2)({h_2}^2 + {k_2}^2 + {l_2}^2)}}$$

$$\text{Tetragonal : } \cos\phi = \frac{\frac{h_1h_2 + k_1k_2}{a^2} + \frac{l_1l_2}{c^2}}{\sqrt{\left(\frac{{h_1}^2 + {k_1}^2}{a^2} + \frac{{l_1}^2}{c^2}\right)\left(\frac{{h_2}^2 + {k_2}^2}{a^2} + \frac{{l_2}^2}{c^2}\right)}}$$

$$\text{Hexagonal : } \cos\phi = \frac{h_1h_2 + k_1k_2 + \frac{1}{2}(h_1k_2 + h_2k_1) + \frac{3a^2}{4c^2}l_1l_2}{\sqrt{({h_1}^2 + {k_1}^2 + h_1k_1 + \frac{3a^2}{4c^2}{l_1}^2)({h_2}^2 + {k_2}^2 + h_2k_2 + \frac{3a^2}{4c^2}{l_2}^2)}}$$

$$\text{Trigonal : } \cos\phi = \frac{a^4 d_1 d_2}{V^2}[\sin^2\alpha(h_1h_2 + k_1k_2 + l_1l_2) + (\cos^2\alpha - \cos\alpha)(k_1l_2 + k_2l_1 + l_1h_2 + l_2h_1 + h_1k_2 + h_2k_1)]$$

$$\text{Orthorhombic : } \cos\phi = \frac{\frac{h_1h_2}{a^2} + \frac{k_1k_2}{b^2} + \frac{l_1l_2}{c^2}}{\sqrt{\left(\frac{{h_1}^2}{a^2} + \frac{{k_1}^2}{b^2} + \frac{{l_1}^2}{c^2}\right)\left(\frac{{h_2}^2}{a^2} + \frac{{k_2}^2}{b^2} + \frac{{l_2}^2}{c^2}\right)}}$$

$$\text{Monoclinic : } \cos\phi = \frac{d_1 d_2}{\sin^2\beta}\left[\frac{h_1h_2}{a^2} + \frac{k_1k_2\sin^2\beta}{b^2} + \frac{l_1l_2}{c^2} - \frac{(l_1h_2 + l_2h_1)\cos\beta}{ac}\right]$$

$$\text{Triclinic : } \cos\phi = \frac{d_1 d_2}{V^2}[S_{11}h_1h_2 + S_{22}k_1k_2 + S_{33}l_1l_2 + S_{23}(k_1l_2 + k_2l_1) + S_{13}(l_1h_2 + l_2h_1) + S_{12}(h_1k_2 + h_2k_1)]$$

$$S_{11} = b^2c^2\sin^2\alpha \qquad S_{12} = abc^2(\cos\alpha\cos\beta - \cos\gamma)$$
$$S_{22} = a^2c^2\sin^2\beta \qquad S_{23} = a^2bc(\cos\beta\cos\gamma - \cos\alpha)$$
$$S_{33} = a^2b^2\sin^2\gamma \qquad S_{13} = ab^2c(\cos\gamma\cos\alpha - \cos\beta)$$

A.6 Numerical Values for Calculation of the Temperature Factor

Values of $\phi(x) = \dfrac{1}{x}\displaystyle\int_0^x \dfrac{\xi}{\mathrm{e}^{\xi}-1}\mathrm{d}\xi \quad x = \dfrac{\Theta}{T}$, Θ: Debye temperature

x	.0	.1	.2	.3	.4	.5	.6	.7	.8	.9
0	1.000	0.975	0.951	0.928	0.904	0.882	0.860	0.839	0.818	0.797
1	0.778	0.758	0.739	0.721	0.703	0.686	0.669	0.653	0.637	0.622
2	0.607	0.592	0.578	0.565	0.552	0.539	0.526	0.514	0.503	0.491
3	0.480	0.470	0.460	0.450	0.440	0.431	0.422	0.413	0.404	0.396
4	0.388	0.380	0.373	0.366	0.359	0.352	0.345	0.339	0.333	0.327
5	0.321	0.315	0.310	0.304	0.299	0.294	0.289	0.285	0.280	0.276
6	0.271	0.267	0.263	0.259	0.255	0.251	0.248	0.244	0.241	0.237

For x lager than 7, $\phi(x)$ values are approximated by $(1.642/x)$.
Debye temperatures are compiled in Appendix A.2 using the following reference: (C.Kittel: *Introduction to Solid State Physics*, 6th Edition, John Wiley & Sons, New York (1986), p.110.)

A.7 Fundamentals of Least-Squares Analysis

Let us consider that the number of n-points have coordinates(x_1, y_1), $(x_2, y_2) \cdots (x_n, y_n)$, and the x and y are related by a straight line with the form of $y = a + bx$. Our problem is to find the best value of a and b which makes the sum of the squared errors a minimum by using the least-squares method. In this case, we use the following two normal equations:

$$\sum y = \sum a + b \sum x, \tag{1}$$

$$\sum xy = a \sum x + b \sum x^2. \tag{2}$$

For given n-points, the following four steps are suggested:

(i) Substitute the experimental data into $y = a + bx$ for obtaining n-equations.

$$\left.\begin{aligned} y_1 &= a + bx_1 \\ y_2 &= a + bx_2 \\ &\vdots \\ y_n &= a + bx_n \end{aligned}\right\}. \tag{3}$$

(ii) Multiply each equation by the coefficient of a (1 in the present case) and add for obtaining the first normal equation.

$$\begin{aligned} y_1 &= a + bx_1, \\ y_2 &= a + bx_2, \\ &\vdots \\ y_n &= a + bx_n, \end{aligned}$$

$$\overline{}$$

$$\sum^{n} y = \sum a + b \sum x. \tag{4}$$

(iii) Multiply each equation by the coefficient b and add for obtaining the second normal equation.

$$\begin{aligned} x_1 y_1 &= x_1 a + b{x_1}^2, \\ x_2 y_2 &= x_2 a + b{x_2}^2, \\ &\vdots \\ x_n y_n &= x_n a + b{x_n}^2 \end{aligned}$$

$$\overline{}$$

$$\sum^{n} xy = a \sum x + b \sum x^2 \tag{5}$$

(iv) Simultaneous solution of the two equations of (4) and (5) yields the value of a and b.

A.8 Prefixes to Unit and Greek Alphabet

Notation	Symbol	Factor	Notation	Symbol	Factor
exa	E	10^{18}	deci	d	10^{-1}
peta	P	10^{15}	centi	c	10^{-2}
tera	T	10^{12}	milli	m	10^{-3}
giga	G	10^{9}	micro	μ	10^{-6}
mega	M	10^{6}	nano	n	10^{-9}
kilo	k	10^{3}	pico	p	10^{-12}
hecto	h	10^{2}	femto	f	10^{-15}
deca	da	10	atto	a	10^{-18}

Greek alphabet

A, α ··········	Alpha	N, ν ··········	Nu
B, β ··········	Beta	Ξ, ζ ··········	Xi
Γ, γ ··········	Gamma	O, o ··········	Omicron
Δ, δ ··········	Delta	Π, π ··········	Pi
E, ε ··········	Epsilon	P, ρ ··········	Rho
Z, ζ ··········	Zeta	Σ, σ ··········	Sigma
H, η ··········	Eta	T, τ ··········	Tau
$\Theta, \vartheta, \theta$ ········	Theta	Y, υ ··········	Upsilon
I, ι ··········	Iota	Φ, φ, ϕ ·······	Phi
K, κ ··········	Kappa	X, χ ··········	Chi
Λ, λ ··········	Lambda	Ψ, ψ ··········	Psi
M, μ ··········	Mu	Ω, ω ··········	Omega

A.9 Crystal Structures of Some Elements and Compounds

Elements or compounds	Crystal systems	Lattice parameter	
Al	*fcc*	$a = 0.40497$ nm	
α-Al_2O_3	hexagonal (corundum structure)	$a = 0.4763$ nm	$c = 0.13003$ nm
Au	*fcc*	$a = 0.40786$ nm	
CaO	NaCl	$a = 0.48105$ nm	
CaF_2	ZnS	$a = 0.5463$ nm	
Cr	*bcc*	$a = 0.28847$ nm	
CsCl	CsCl	$a = 0.4123$ nm	
Cu	*fcc*	$a = 0.36148$ nm	
CuCl	ZnS	$a = 0.54057$ nm	
Cu_2O	CsCl(cuprite structure)	$a = 0.42696$ nm	
Fe α	*bcc*	$a = 0.28665$ nm	
γ	*fcc*	$a = 0.36469$ nm	
δ	*bcc*	$a = 0.29323$ nm	
FeS	NaCl(pyrite structure)	$a = 0.5408$ nm	
K	*bcc*	$a = 0.5247$ nm	
Mg	*hcp*	$a = 0.32095$ nm	$c = 0.52107$ nm
MgO	NaCl	$a = 0.42112$ nm	
Mo	*bcc*	$a = 0.31469$ nm	
NaCl	NaCl	$a = 0.56406$ nm	
NaF	NaCl	$a = 0.4620$ nm	
Ni	*fcc*	$a = 0.35239$ nm	
NiO	NaCl	$a = 0.41769$ nm	
Pt	*fcc*	$a = 0.39240$ nm	
Si	diamond	$a = 0.54309$ nm	
SiO_2 α-quartz	(Rombohedral)	$a = 0.4913$ nm	$c = 0.5405$nm
β-quartz	(hexagonal)	$a = 0.501$ nm	$c = 0.547$ nm
Tridymite	(hexagonal)	$a = 0.503$ nm	$c = 0.822$ nm
α–crystobalite	(hexagonal)	$a = 0.4973$ nm	$c = 0.6926$ nm
Ti	*hcp*	$a = 0.29512$ nm	$c = 0.46845$nm
TiC	NaCl	$a = 0.43186$ nm	
TiO_2	CsCl(rutile structure)	$a = 0.45929$ nm	$c = 0.29591$nm
W	*bcc*	$a = 0.31653$ nm	
Zn	*hcp*	$a = 0.26650$ nm	$c = 0.49470$ nm
β-ZnS	ZnS	$a = 0.54109$ nm	

These data are taken from the following references.

B.D. Cullity: Elements of X-ray Diffraction (2nd Edition), Addison-Wesley (1978).

F.S. Galasso: Structure and Properties of Inorganic Solids, PergamonPress (1970).

Index

Printed by Publishers' Graphics LLC